The PHYSICAL WORLδ

CLASSICAL PHYSICS OF MATTER

Edited by John Bolton

IoP

Institute of Physics Publishing
Bristol and Philadelphia
in association with

The Physical World Course Team

Course Team Chair	Robert Lambourne
Academic Editors	John Bolton, Alan Durrant, Robert Lambourne, Joy Manners, Andrew Norton
Authors	David Broadhurst, Derek Capper, Andrew Conway, Dan Dubin, Tony Evans, Ian Halliday, Carole Haswell, Keith Higgins, Keith Hodgkinson, Mark Jones, Sally Jordan, Ray Mackintosh, David Martin, John Perring, Michael de Podesta, Sean Ryan, Ian Saunders, Richard Skelding, Tony Sudbery, Stan Zochowski
Consultants	Alan Cayless, Melvin Davies, Graham Farmelo, Stuart Freake, Gloria Medina, Kerry Parker, Alice Peasgood, Graham Read, Russell Stannard, Chris Wigglesworth
Course Managers	Gillian Knight, Michael Watkins
Course Secretaries	Tracey Moore, Tracey Woodcraft
BBC	Deborah Cohen, Tessa Coombs, Steve Evanson, Lisa Hinton, Michael Peet, Jane Roberts
Editors	Gerry Bearman, Rebecca Graham, Ian Nuttall, Peter Twomey
Graphic Designers	Javid Ahmad, Mandy Anton, Steve Best, Sue Dobson, Sarah Hofton, Jenny Nockles, Pam Owen, Andy Whitehead
Centre for Educational Software staff	Geoff Austin, Andrew Bertie, Canan Blake, Jane Bromley, Philip Butcher, Chris Denham, Nicky Heath, Will Rawes, Jon Rosewell, Andy Sutton, Fiona Thomson, Rufus Wondre
Course Assessor	Roger Blin-Stoyle
Picture Researcher	Lydia K. Eaton

The Course Team wishes to thank Richard Skelding, Ian Saunders, John Perring, Robert Lambourne, Mark Jones, Alan Cayless, Gloria Medina and Chris Denham for their contributions to this book.

The Open University, Walton Hall, Milton Keynes MK7 6AA

First published 2000

Written, edited, designed and typeset by The Open University.

Published by Institute of Physics Publishing, wholly owned by The Institute of Physics, London.
IoP Publishing, Dirac House, Temple Back, Bristol BS1 6BE, UK.

US Office: Institute of Physics Publishing, The Public Ledger Building, Suite 1035, 150 South Independence Mall West, Philadelphia, PA 19106, USA.

Printed and bound in the United Kingdom by The Alden Group, Oxford.

ISBN 0 7503 0717 X

Library of Congress Cataloging-in-Publication Data are available.

This text forms part of an Open University course, S207 *The Physical World*. The complete list of texts that make up this course can be found on the back cover. Details of this and other Open University courses can be obtained from the Course Reservations Centre, PO Box 724, The Open University, Milton Keynes MK7 6ZS, United Kingdom: tel. +44 (0) 1908 653231; e-mail ces-gen@open.ac.uk

Alternatively, you may visit the Open University website at http://www.open.ac.uk where you can learn more about the wide range of courses and packs offered at all levels by The Open University.

To purchase other books in the series *The Physical World*, contact IoP Publishing, Dirac House, Temple Back, Bristol BS1 6BE, UK: tel. +44 (0) 117 925 1942, fax +44 (0) 117 930 1186; website http://www.iop.org

1.1

s207book4i1.1

CLASSICAL PHYSICS OF MATTER

Introduction

Understanding and using the properties of matter have gone hand in hand with technological progress and the rise and fall of civilizations. We speak of the Stone Age, the Bronze Age and the Iron Age, because the properties of these materials helped to shape the nature of human existence. More recently, we have emerged from the Age of Steam, flirted with the Age of Uranium and, with the rise of computers, are surely living in the Age of Silicon.

Materials do have a remarkable range of properties. Take a paper clip, a pane of glass and a rubber ball. The paper clip is shiny, can be bent, is attracted to magnets and conducts heat and electricity well. The pane of glass is transparent, brittle, non-magnetic and conducts heat and electricity poorly. The rubber ball is soft and bounces when dropped. It is easy to take these properties for granted as part of the environment we are born into, and most people are content to leave things that way. They simply accept that glass has glassy properties because that is the way glass is.

Science, however, does not thrive on easy acceptance; it is based on a much more adventurous curiosity. Taking nothing for granted, and looking with fresh eyes at the substances around us, we are led to ask a profound question: *why* do these substances behave as they do?

This question becomes more sharply focused once we recognize that all normal matter, whether solid, liquid or gas, is composed of atoms. You can think of an atom as being a tiny sphere, about 10^{-10} m in diameter. There are over 100 different types of atom, corresponding to the different types of chemical element. Once we know that different substances contain a limited number of different types of atom, we can try to explain the transparency of glass, the magnetism of iron or the stretchiness of rubber *on an atomic scale*.

The central importance of atoms has been emphasized by Richard Feynman, Nobel laureate and one of the greatest physicists of recent times. Feynman once asked a group of students:

> 'If in some cataclysm, all of scientific knowledge were to be destroyed and only one sentence passed on to the next generation of creatures, what statement would contain the most information in the fewest words?'

There is no record of the replies given by the students, but we do know Feynman's own answer:

> 'I believe that it is the atomic hypothesis … that all things are made of atoms — little particles that move around in perpetual motion, attracting each other when they are a little distance apart, but repelling one another upon being squeezed together. In that one sentence, you will see, there is an enormous amount of information about the world, if just a little imagination and thinking are applied.'

The relationship between the properties of materials and the properties of atoms can be quite straightforward: the mass of a raindrop, for instance, is just the sum of the masses of its atoms. But many of the properties of materials emerge in a far more subtle way — steel is not strong because it is made from strong atoms; and rubber is not stretchy because it is made from stretchy atoms! Establishing the link between the scale of everyday life and the scale of atoms has proved to be one of the major challenges of physics.

At first, the problem was that very little was known about atoms, and their very existence was seriously in doubt. This is not surprising. Even the tiniest scrap of dust, just visible to the naked eye, contains 10^{15} atoms or more, so individual atoms lie well beyond the limits of our senses. Throughout the nineteenth century, many scientists *suspected* that atoms existed, but had very little idea of their properties. Amid this uncertainty, two different traditions emerged.

One group of physicists simply assumed that atoms exist, picturing them as tiny particles bouncing around in accordance with Newton's laws. The enormous number of particles involved meant that it was impossible to keep detailed track of them as they collided with one another. Nevertheless, a variety of statistical methods allowed predictions to be made about the behaviour of matter on a large scale. The other group of physicists chose to ignore the possible existence of atoms and to deduce what they could by treating matter as a continuous medium; this gave rise to two subjects — thermodynamics and fluid mechanics — which remain important today.

This book will look at both traditions. Although I have stressed the ultimate aim of looking for explanations at an atomic level, I do not wish to downplay the value of thermodynamics and fluid mechanics. The truth is that many questions of practical and technological importance are best answered within these two subjects. This is partly because the goal of understanding matter in terms of the properties of atoms is yet to be completely fulfilled. Where it has been attained, we can be very satisfied; elsewhere, thermodynamics and fluid mechanics remain the best tools available.

Ironically, just as the existence of atoms was being demonstrated beyond any reasonable doubt at the beginning of the twentieth century, the foundations of Newtonian mechanics were being shaken by the first glimmerings of quantum theory. The arrival of quantum mechanics gave new insights into the behaviour of matter, and it is now apparent that many important and useful properties of materials are due to quantum mechanical effects. In this course, our discussion of the physics of matter is split into two parts. The present book is based on classical physics. Many aspects of the properties of matter can be explained in this way, without any need for quantum mechanics, and we shall explore these aspects now. Following an introduction to quantum mechanics, the course will return to the properties of matter in *Quantum physics of matter* and discuss those properties that can only be understood in quantum mechanical terms.

Open University students should note that the video associated with this book is Video 4 *Maiden Flights*; this is best viewed towards the end of Chapter 4.

Chapter 1 From atoms to the phases of matter

1 A puzzle

Suppose you go into a cold room and turn on an electric fire. After a while, you begin to feel more comfortable — the chill has been taken out of the room, and you settle down to take a well-deserved snooze. Would it surprise you to learn that the total energy of the air in the warm room is the same as it was in the cold room? Turning on the fire did not increase the energy of the air in the room at all!

This surprising fact illustrates some of the subtleties that arise in discussing the properties of matter in general, and the properties of gases in particular. We leave it as a puzzle, with the promise that we will come back to it before this chapter closes.

2 Atoms: the building blocks of matter

2.1 The concept of an atom

> 'All these things being consider'd, it seems probably to me, that God form'd matter in solid, massy, hard, impenetrable particles...'
>
> Isaac Newton, Query 31, *Opticks*, 1706.

The concept of an atom is an extremely ancient one, dating back at least to the fifth century BC when the Greek philosophers Leucippus and Democritus proposed that all matter is made up of atoms in ceaseless motion. Different substances were supposed to be produced by different combinations of the fundamental atoms. This opposed the view that matter was a continuum which could be divided into smaller and smaller parts, *ad infinitum*. The atomic idea was never part of the mainstream of Greek philosophy, but it survived as a minority view, and Newton clearly favoured it, though he lacked convincing experimental evidence.

John Dalton (*c*. 1766–1844)

The first persuasive arguments for atoms came at the beginning of the nineteenth century, from chemists who had found that certain substances could not be broken down into simpler components. These substances were called **elements**, and each pure substance was regarded as being either an element or a combination of different elements (a **compound**). It gradually became clear that each compound contained a fixed proportion (by mass) of its constituent elements. And, when one element combined with another to produce more than one compound, the ratios of the masses of the elements could be expressed as simple integers.

John Dalton (*c*. 1766–1844) realized that these facts could be explained quite naturally if matter were made up of atoms, and if different elements had different atoms. (Newton did not have this understanding because the chemical notion of an element was not fully developed in his day.) Dalton gave lectures on this idea and, in 1808, published a revolutionary book called *A New System of Chemical Philosophy* which asserted that:

Each element has a distinctive kind of **atom** with its own characteristic properties. A macroscopic sample of an element consists of a large number of these atoms. Atoms of one type cannot be converted into atoms of another type, but atoms can bind together in specific ways.

The images shown in Figure 1.1 are taken from Dalton's book. Although they should not be taken too literally, they clearly show the essential idea that elements have atoms of distinct types and these can combine together to form new substances.

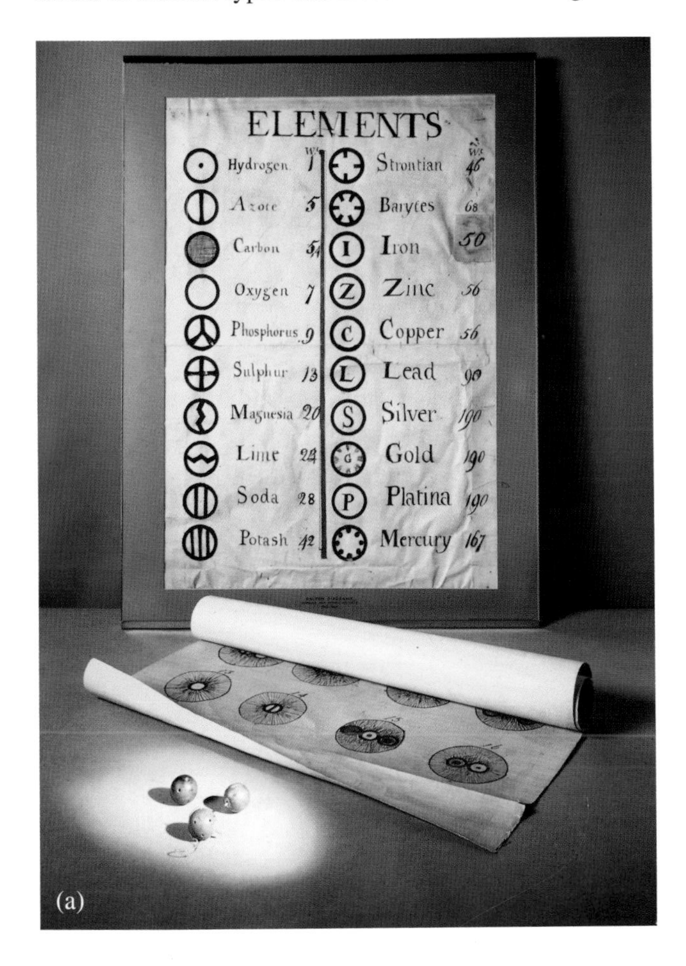

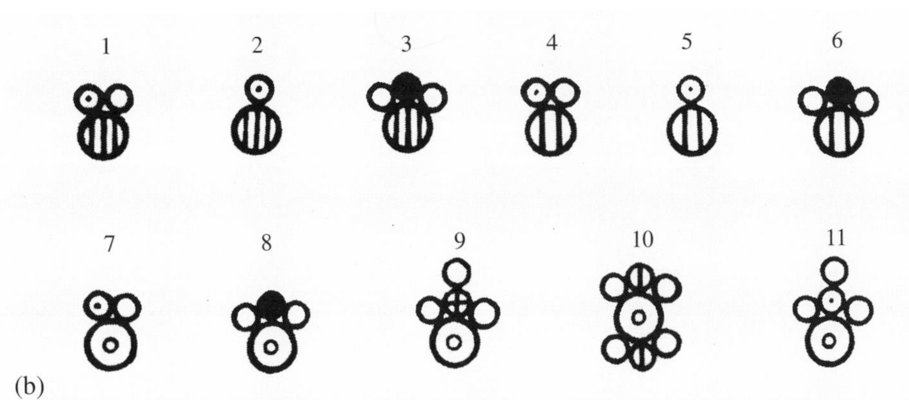

Figure 1.1 (a) Dalton's symbols for substances that he thought were elements; several of them subsequently were found to be compounds. (b) Part of Dalton's original picture of atoms combining together to form molecules.

Each arrangement of atoms is now called a **molecule**. There are several cases to consider (Figure 1.2):

1. Atoms of different types can combine together to form a molecule of a compound. Thus, a water molecule is formed when two hydrogen atoms bind with one oxygen atom. Much larger and more complicated molecules can form, such as DNA, the famous carrier of the genetic code.
2. Atoms of the same type can also join together to form a molecule. Thus, two oxygen atoms join together to form an oxygen molecule. It is also possible for three oxygen atoms to join together to form a different type of molecule, known as an ozone molecule.
3. Finally, some elements, such as argon, have atoms that do not easily combine with others — argon gas consists of individual argon atoms. It is convenient to continue to use the terminology of molecules in this case, so an argon molecule is just a single argon atom.

The words *monatomic*, *diatomic* and *triatomic* are often used to describe molecules containing one, two and three atoms respectively.

We are perhaps fortunate that Dalton lived in Cumbria, a rainy part of Britain. It is said that his concerns about the origin of rain led him to a detailed study of gases which culminated in the atomic theory. By the time of his death, in 1844, the idea of an atom had spread far beyond the small community of scientists; Dalton's lying-in-state in Manchester Town Hall was attended by no fewer than 40 000 people!

The popular appeal of the atomic model was justified, as simplicity is generally an excellent guide to truth. But it was not to everyone's taste. It neatly fitted the chemical facts but there was no direct evidence or proof. That is why it was referred

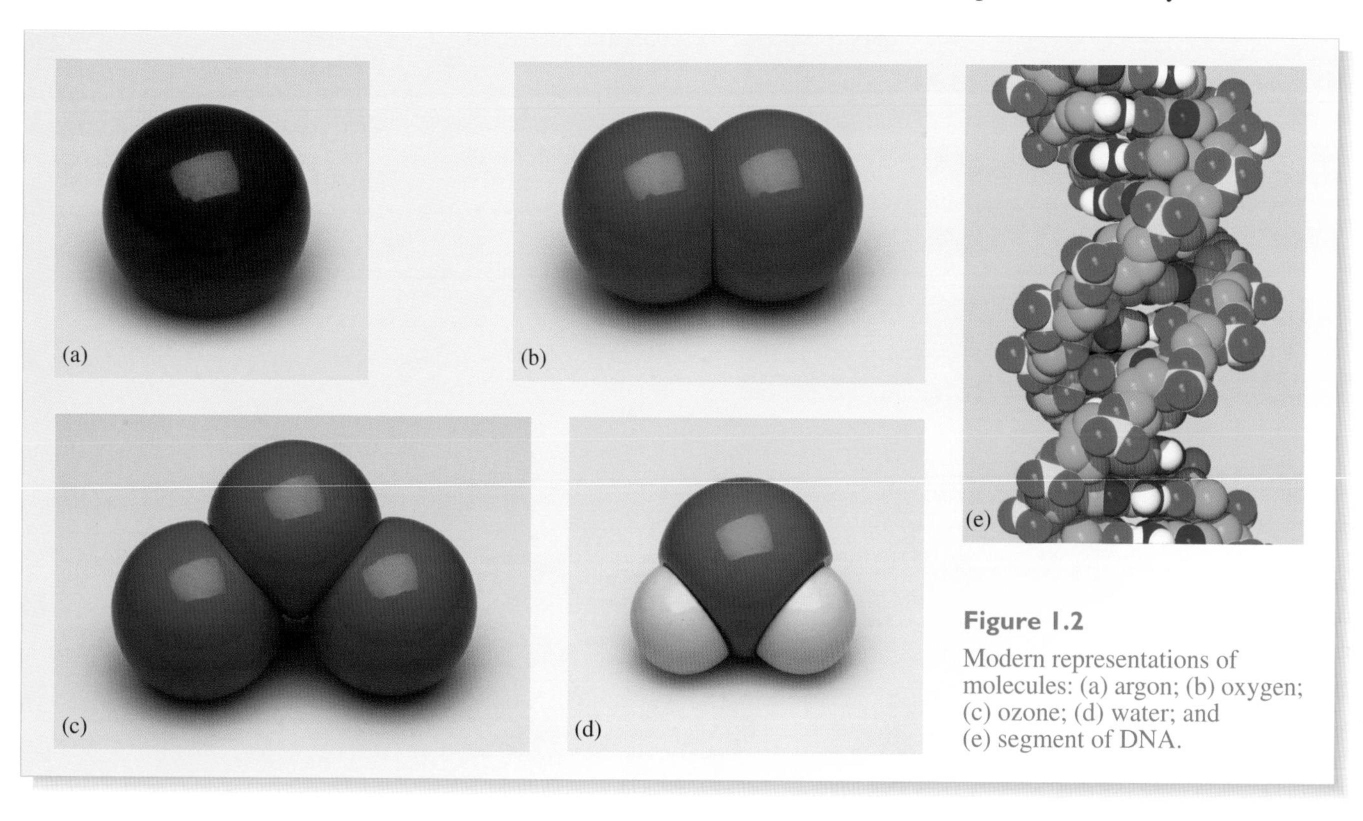

Figure 1.2

Modern representations of molecules: (a) argon; (b) oxygen; (c) ozone; (d) water; and (e) segment of DNA.

to as the atomic *hypothesis*, and why some scientists regarded it as a fiction, useful for accounting purposes but not reflecting deep reality. The debate about whether atoms exist or not was destined to run for the whole of the nineteenth century, only to be finally settled around 1908, the centenary of Dalton's original insight.

2.2 Atomic sizes

Once the ideas of atoms and molecules had been proposed, physicists worked hard to see if this structure of matter would have any large-scale consequences. At the same time, knowledge of atomic properties became of great intrinsic interest. Many questions had been left unanswered by Dalton's ideas: what size and mass do atoms have and what forces act between them?

We will begin with atomic sizes. Clearly, atoms and molecules must be small — below the limits of sight — but how small, exactly? Is their diameter 10^{-10} m, or is it more like 10^{-20} m?

A simple observation, first performed by Lord Rayleigh, provides a clue. If a drop of oil is lowered carefully onto the surface of a lake, it spreads out, forming a circular patch on the surface. The patch soon becomes too thin to be seen directly, but makes its presence felt because the water is particularly calm and unruffled over the area of the patch. The patch is surprisingly large: a tiny drop of oil of radius 0.5 mm spreads to a circular patch with a radius of around 0.4 m.

- Use the information given above to estimate the thickness of the oil film.

○ We can take the oil drop to be a sphere of radius r, and treat the oil patch as a very thin circular cylinder of radius R and thickness t (Figure 1.3). Then, setting the volume of the initial oil drop equal to the volume of the oil patch gives

$$\tfrac{4}{3}\pi r^3 = \pi R^2 t$$

so $$t = \frac{4}{3} \times \frac{r^3}{R^2} = \frac{4}{3} \times \frac{(0.5 \times 10^{-3}\,\text{m})^3}{(0.4\,\text{m})^2} \approx 10^{-9}\,\text{m}. \quad \blacksquare$$

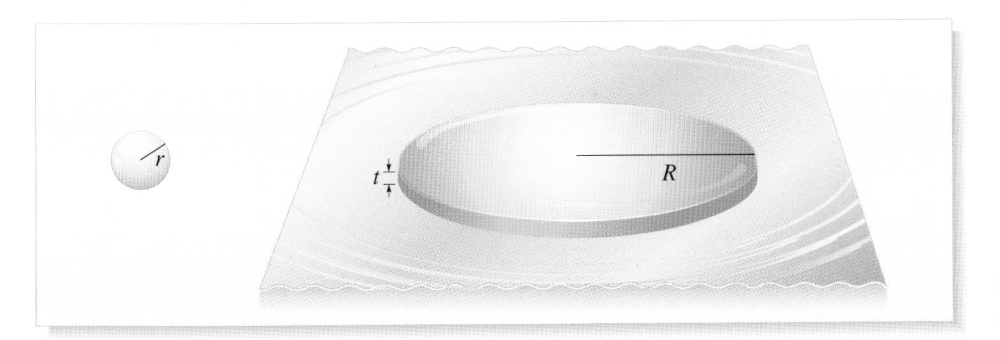

Figure 1.3 A drop of oil spreads out on the surface of a lake.

The oil film cannot be less than one molecule thick, so we have an upper limit for the linear dimensions of an oil molecule — around 10^{-9} m. Clearly, atoms must be smaller than this.

Throughout the nineteenth century, physicists devised several other ways of estimating the sizes of atoms and molecules. Many of these involved the properties of gases — especially the rate at which molecules, energy and momentum spread from one region in a gas to another. When a gas molecule moves from one side of a room to another, it does so at a rate that depends on the number of collisions it makes with other molecules. If it makes no collisions, it will make very rapid progress, but if it collides frequently, it will follow a highly irregular zigzag path and its progress will be much slower. The rate of collisions, of course, depends on the size of the molecule. This fact, and a great deal of careful measurement, quantitative analysis and plausible assumptions, gave physicists a way of estimating the size of molecules. By the end of the nineteenth century the consensus had emerged that atoms would have to be about $1–2 \times 10^{-10}$ m across and small molecules would therefore be slightly larger (depending on the number of atoms they contain).

Although the victory of the atomists has long been complete, it is only recently that we have been able to obtain images that give us the impression of *seeing* atoms. To explain how this is possible, I need to say something about the nature of light. The most important point is that light is a wave-like phenomenon. Rather like water-waves travelling towards a shore, light propagates as a series of peaks and troughs. One of the key parameters that describes a light wave is its wavelength — the distance between one crest in the wave and the next. In fact, the wavelength of a light wave is directly related to its colour: red light has a wavelength of about 7×10^{-7} m while violet light has a wavelength of about 4×10^{-7} m.

Light is discussed more fully in the book *Dynamic fields and waves*.

Wavelength becomes very important when we consider how waves flow around an obstacle. The example of water waves can illustrate this point (Figure 1.4). In Figure 1.4a, the object is larger than the wavelength of the waves. In this case the object has a significant effect on the way the waves propagate, so it should be possible for observers downstream to infer the presence of the object by measuring the waves that reach them. In Figure 1.4b, the object is smaller than the wavelength of the waves. In this case, the waves are scarcely affected by the object and observers would be very hard-pressed to detect it, simply by observing the waves.

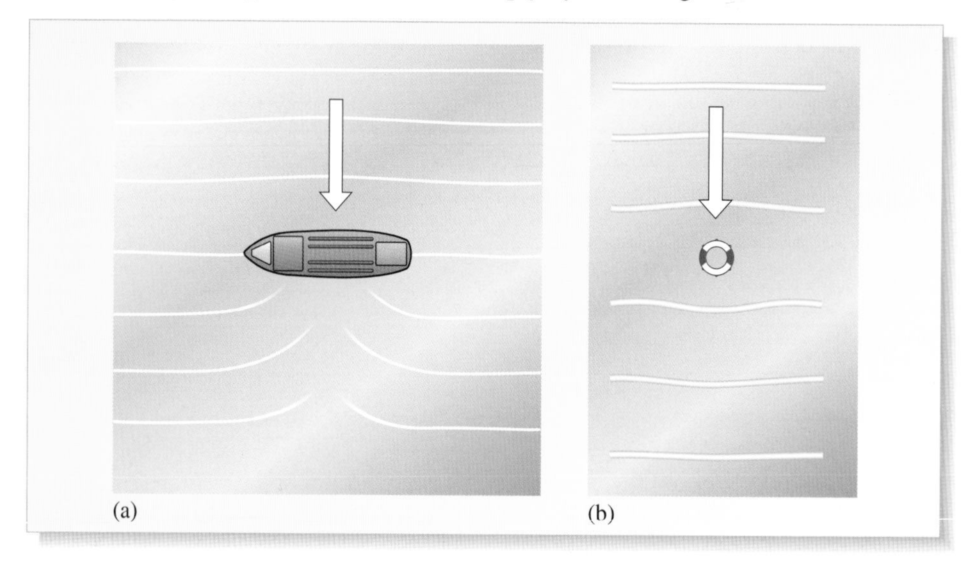

Figure 1.4 Water waves flowing around an object. (a) If the object is larger than the wavelength, the wave propagation is significantly affected. (b) If the object is smaller than the wavelength, the wave propagation is almost unchanged.

Very similar considerations apply to light: it is practically impossible to see any object whose linear dimensions are smaller than the wavelength of the light that is used to illuminate it. With ordinary light, the smallest object that can be seen, even with a microscope, is roughly 10^{-6} m across. This is about 10 000 times larger than the size of an atom.

In order to 'see' atoms we must use waves with much smaller wavelengths than visible light. It turns out that electrons can behave as waves. High energy electrons propagate as waves of very short wavelength, and this fact is used to form very detailed images in **electron microscopes**. The most powerful electron microscopes are able to resolve individual atoms (Figure 1.5).

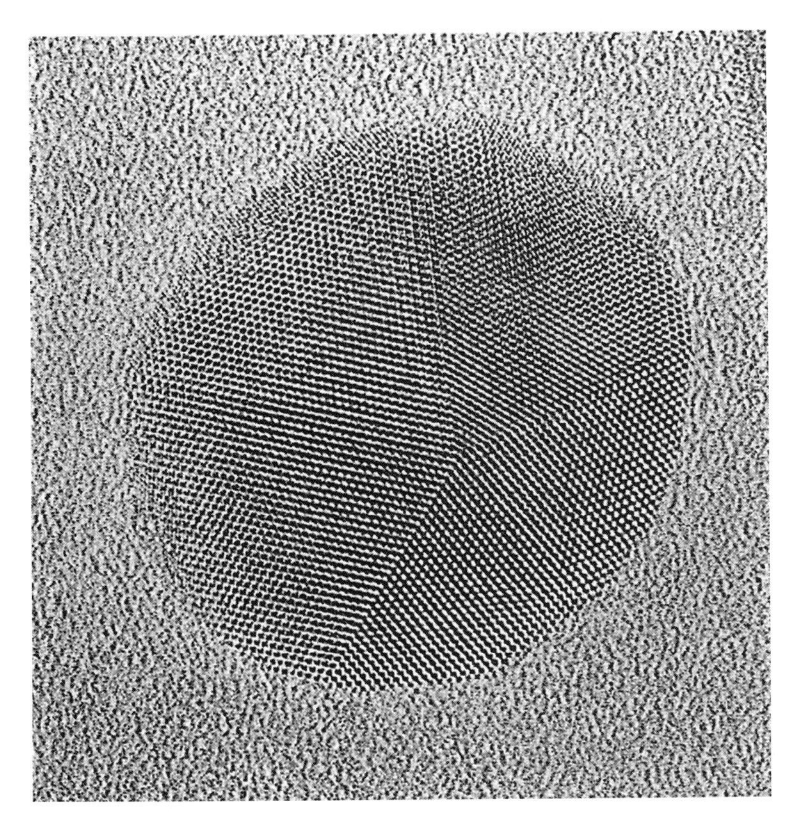

Figure 1.5 An image of gold atoms formed by a very powerful electron microscope. The atoms are arranged in a very regular array so each tiny dot in the central region represents a column of atoms.

In 1982, the **scanning tunnelling microscope** (STM) was invented, which gave an alternative way of viewing atoms. The STM relies on the fact that atoms contain negatively charged electrons which are attracted to positive charges. In order to deduce the arrangement of atoms on the surface of a solid, a very sharply pointed, positively charged metal tip is allowed to approach the surface, coming almost into contact with it. When the metal tip gets close enough to the surface, it starts to suck electrons out of the surface, like a miniature vacuum cleaner. The flow of electrons into the metal tip depends very sensitively on the distance between the tip and the surface.

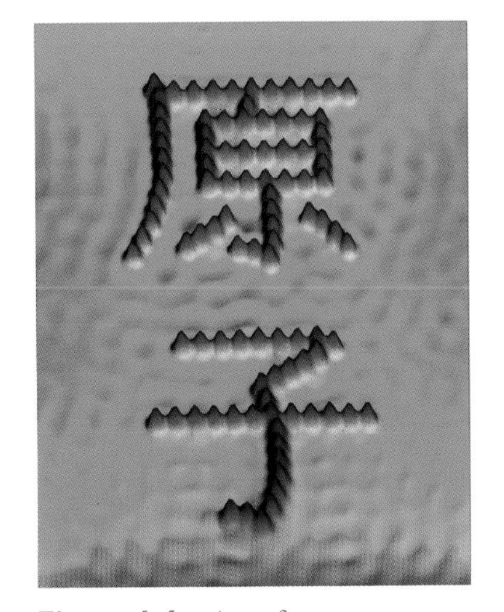

Figure 1.6 A surface map obtained with a STM. Each pinkish-red blob is an atom, though these colours have been assigned by the computer purely to increase the visual impact. The atoms have been carefully positioned (also using the STM) so that they form the Japanese characters for the word 'atom'.

The operator of the STM scans the tip over the surface, whilst making tiny adjustments to its distance from the surface in order to ensure that the flow of electrons remains steady. Since the surface under investigation contains bumps (where the atoms are) and hollows (in between the atoms), the metal tip has to move in and out to follow these tiny surface features. By keeping note of the precise movements of the metal tip, the bumps and hollows can be mapped out, and a picture of the atomic positions formed.

If someone came to you with a plan for a new device, operating along the lines just described, would you be inclined to invest money in it? Perhaps not. The plan requires the position of a metal tip to be controlled to within less than the size of an atom (10^{-10} m), which sounds rather fanciful. But, amazingly enough, this proves to be no problem. Precise positioning of this kind can be achieved with quartz rods, which have the property of changing their lengths in response to applied electrical voltages. Figure 1.6 is the proof of the pudding: an STM image showing exactly where all the surface atoms are located.

2.3 Isotopes and ions

Atoms, as originally conceived by Dalton, were tiny, impenetrable spheres. In order to explore the properties of atoms in more detail, including their masses, we must adopt a more modern viewpoint, and recognize that atoms have internal structure.

Any atom consists of three types of particle: **electrons**, **protons** and **neutrons**, which have the physical properties shown in Table 1.1. An atom is electrically neutral, so it contains equal numbers of negatively charged electrons and positively charged protons. The protons and neutrons are tightly bound together in the **nucleus** — a tiny, dense core at the centre of the atom. Since protons and neutrons are much more massive than electrons, nearly all the atom's mass is locked up in its nucleus. The electrons occupy the region outside the nucleus, so they roam over nearly all of the atom's volume.

Table 1.1 Properties of electrons, protons and neutrons. The coulomb (C) is the SI unit of electric charge; it is discussed fully in *Static fields and potentials*.

	Mass/10^{-30} kg	Charge/1.602×10^{-19} C
electron	0.9109	−1
proton	1673	+1
neutron	1675	0

Each element has atoms with a definite number of electrons and an equal number of protons. For example, iron has atoms with 26 electrons and 26 protons; silver has atoms with 47 electrons and 47 protons. However, atoms of a given element need not all have the same number of neutrons. Most iron atoms contain 30 neutrons, but about 6% contain 28 neutrons and 2% contain 31 neutrons. Each species of atom, with a fixed number of electrons, protons *and* neutrons, is called an **isotope**:

Isotopes

A given element may have a number of different isotopes, each with the same number of electrons, an equal number of protons, but *different* numbers of neutrons. Different isotopes of the same element have slightly different masses, but essentially the same chemical properties.

Isotopes are conventionally labelled by giving two numbers. The **atomic number** Z is the number of protons (or electrons) in the atom. This is the same for all isotopes of a given element. The **mass number** A is the sum of the numbers of protons and neutrons. This number varies from isotope to isotope. For example, the most common isotope of iron, with 26 electrons, 26 protons and 30 neutrons, has $Z = 26$ and $A = 56$. The less common isotopes mentioned above have $A = 54$ and $A = 57$. Very often, the mass number is just tagged on to the name of the element, so that we talk of iron-56, iron-54 or iron-57, for example.

● How would you describe an isotope of silver with 47 protons and 60 neutrons?

❍ The isotope is described as silver-107. ■

One consequence of the internal structure of atoms is the fact that they can gain or lose electrons. In the simplest case, a neutral atom loses a single electron and so becomes positively charged, since there is now one more positively charged proton than there are negatively charged electrons. The positively charged object that remains after the electron has been removed is called an **ion**. Electrons are very light, so the ion has almost the same mass as the atom from which it was prepared, but it behaves quite differently: because it is electrically charged, an ion responds to electrical forces. This means that ions can be controlled more easily than uncharged atoms.

Molecules can also lose electrons to form ions.

2.4 Atomic masses

Dalton's atomic theory developed from the observation that the masses of elements that combine to form compounds can be expressed as simple ratios. If the elements are gases, it turns out that their volumes, at a given temperature and pressure, are also in simple ratios. How can both these facts be true? Shortly after Dalton had published his theory, Amedeo Avogadro (1776–1856) suggested a simple explanation:

Avogadro's hypothesis

Equal volumes of different gases, at the same temperature and pressure, contain the same number of molecules.

This suggestion lay dormant for 50 years before it was realized that it could be put to use — by weighing equal volumes of different gases under specified conditions of temperature and pressure, the mass of one type of molecule could be related to the mass of another. Of course, no absolute measurements could be taken in this way, but a scale of *relative* masses could be established. For example, an oxygen molecule was found to be 16 times more massive than a hydrogen molecule.

In order to establish a definite scale, a particular isotope is chosen as a reference. A natural choice would be hydrogen ($Z = 1$, $A = 1$). This is the lightest and simplest of all atoms — a single electron and a single proton bound together. For technical reasons, a slightly different (but almost equivalent) choice is made:

The atomic mass unit

Masses of atoms and molecules are commonly measured in terms of the **atomic mass unit (amu)**. This is defined to be 1/12 the mass of a carbon-12 atom — an atom with 6 protons, 6 neutrons and 6 electrons. By definition, a carbon-12 atom has a mass of exactly 12 amu. Hydrogen has two stable isotopes with masses close to 1 amu and 2 amu.

Note In some books, the symbol 'u' is used instead of amu.

Relative atomic and molecular masses

The mass of an atom, divided by 1 amu, is known as the **relative atomic mass** of the atom. So, if the relative atomic mass is M_r, the mass of one atom is M_r amu.

Similarly, the mass of a molecule, divided by 1amu, is called the **relative molecular mass** of the molecule. If the relative molecular mass is M_r, the mass of one molecule is M_r amu.

Relative atomic masses and relative molecular masses were formerly known as *atomic weights* and *molecular weights*, and this terminology is still sometimes used today, though it is generally discouraged. Both are ratios of masses, so they have no units.

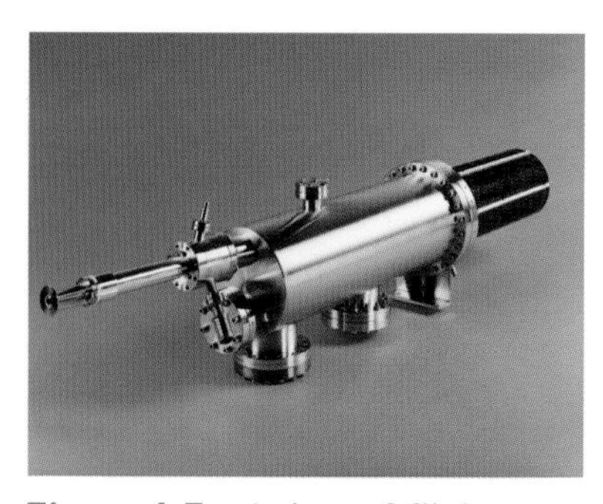

Figure 1.7 A time-of-flight mass spectrometer. Such an instrument can be used to measure the masses of different ions, and their abundance in a given sample. It is very sensitive and quite compact, and so can be loaded on board space probes to analyse the atmospheres of our neighbouring planets.

Nowadays, it is possible to obtain direct measurements of the masses of atoms (more strictly, ions), using a **time-of-flight mass spectrometer** (Figure 1.7). The idea is to accelerate an ion from rest, giving it a known kinetic energy. The speed of the ion can be found by measuring the time taken for it to travel a fixed distance and its mass can then be deduced by rearranging the formula for kinetic energy:

$$E = \tfrac{1}{2}mv^2 .$$

When absolute measurements of this kind are made, a carbon-12 atom is found to have a mass of 1.9926×10^{-26} kg, allowing us to establish that

$$1\,\text{amu} = \tfrac{1}{12} \times 1.9926 \times 10^{-26}\,\text{kg} = 1.6605 \times 10^{-27}\,\text{kg}.$$

Atoms have masses that vary by a factor of about 300, from the lightest, hydrogen, to the heaviest man-made atoms. You might therefore wonder why we quoted the size of an atom as being around 10^{-10} m. Shouldn't the heaviest atoms be much larger than the lightest? In fact, the diameter of atoms only varies by a factor of two or three. Although the heaviest atoms contain more than 100 electrons, these electrons are strongly attracted to the nucleus and are closely packed in around it, so the diameters of all atoms are of the same order of magnitude.

Question 1.1 A molecular ion with kinetic energy 2.0×10^{-19} J takes 3.0×10^{-3} s to travel 0.6 m. What are the mass and the magnitude of the momentum of the ion? ■

2.5 Interactions between atoms

A moment's thought shows that atoms and molecules must interact with one another. The fact that atoms combine together to form molecules shows that attractive forces exist, persuading the atoms to stick together. On the other hand, when a solid metal bar is squeezed, it resists being crushed. This suggests that atoms exert repulsive forces on each other when they get too close. This combination of attractive and repulsive forces makes good sense. If there were only attractive forces, matter would implode. If there were only repulsive forces, it would fly apart. Matter is mechanically stable because atoms attract one another when they are pulled apart, and repel one another when they are crushed together.

The force–separation curve

The simplest case to consider is that of a diatomic molecule which consists of two atoms, A and B, separated by a variable distance r (Figure 1.8). The two atoms exert forces on one another, equal in magnitude and opposite in direction. We concentrate on the force experienced by atom B, and represent this force by its radial component F_r, which is positive when B is repelled by A (Figure 1.8a) and negative when B is attracted to A (Figure 1.8b).

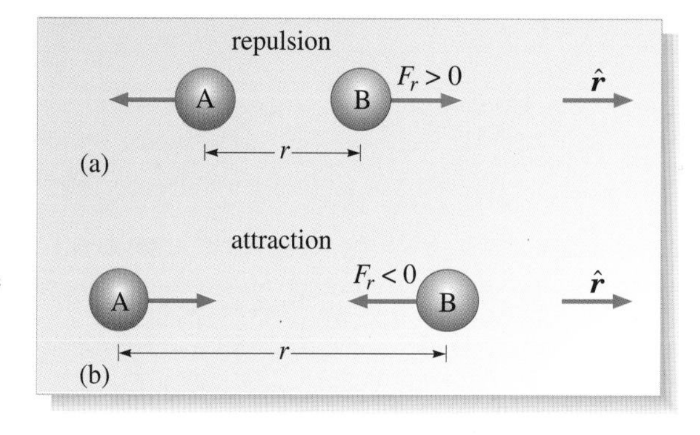

Figure 1.8 Forces exerted by atoms on one another. (a) When the atoms are very close, they repel one another. (b) When the atoms are further apart, they attract one another.

Figure 1.9 shows a typical graph of F_r against r. This type of graph is called a force–separation curve. There are a number of points to notice:

- There is one separation, r_0, at which the force is zero. According to Newtonian mechanics, the atoms can remain stationary when separated by this distance, so it is called the **equilibrium separation**. A typical value for r_0 is around 3×10^{-10} m.
- For separations smaller than the equilibrium separation, F_r is positive and rises to very high values, indicating a strong repulsion. This shows that Newton's idea that atoms are 'hard impenetrable particles' is not so far from the truth.
- For separations greater than the equilibrium separation, F_r is negative, indicating an attraction. The attraction is significant at separations between r_0 and $2r_0$ but becomes negligible for very large separations. Thus each atom has a relatively small region of influence within which its attraction is strongly felt.

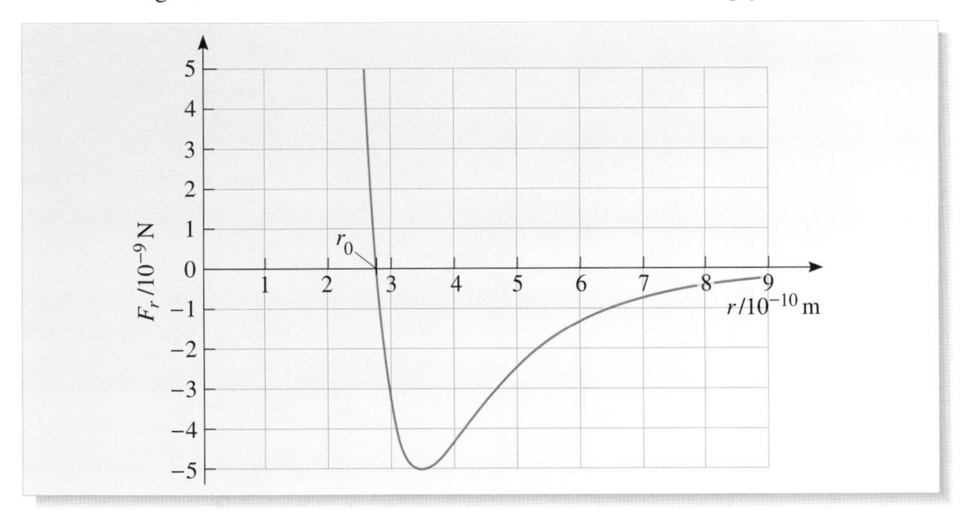

Figure 1.9 A typical force–separation curve.

It is possible for the atoms to remain stationary at their equilibrium separation, r_0, where the force is zero. More commonly though, the atoms oscillate as if joined by a spring. During one half of the oscillation the separation is greater than r_0, and the atoms attract one another; during the other half of the oscillation the separation is less than r_0, and the atoms repel one another. The frequency of oscillation is of interest because a molecule vibrating at a given frequency will strongly absorb and emit light that has exactly this frequency.

The frequency of the oscillation is related to the shape of the force–separation curve and the masses of the atoms. As shown in Figure 1.9, the force–separation graph is practically a straight line in the immediate vicinity of the equilibrium separation so, provided the amplitude of the oscillation is small, the force can be written as

$$F_r = -C(r - r_0) \qquad (1.1)$$

where C (the *magnitude* of the slope of the force–separation curve) is a constant. Because the force is proportional to $(r - r_0)$, and acts in a direction that restores the atoms towards their equilibrium separation, the oscillation will be simple harmonic.

A little care is needed to predict the frequency of oscillation because the equations of simple harmonic motion refer to the oscillation of a particle about a *fixed* point, and neither of the atoms in the molecule is fixed. To take a definite case, consider an oxygen molecule, composed of two identical oxygen atoms. The centre of mass of the molecule (mid-way between the atoms) can be taken to be stationary, with the atoms vibrating to and fro about it, as shown in Figure 1.10a. Relative to this fixed centre of mass, atom B has position $x = r/2$ and equilibrium position $x_0 = r_0/2$, so Equation 1.1 can be written as

$$F_r = -2C(x - x_0).$$

This shows that the effective force constant for the oscillation is $2C$. Using the standard equation for the frequency of a simple harmonic oscillator, it follows that the frequency of oscillation of an oxygen molecule is

$$f = \frac{1}{2\pi}\sqrt{\frac{2C}{m}}.$$

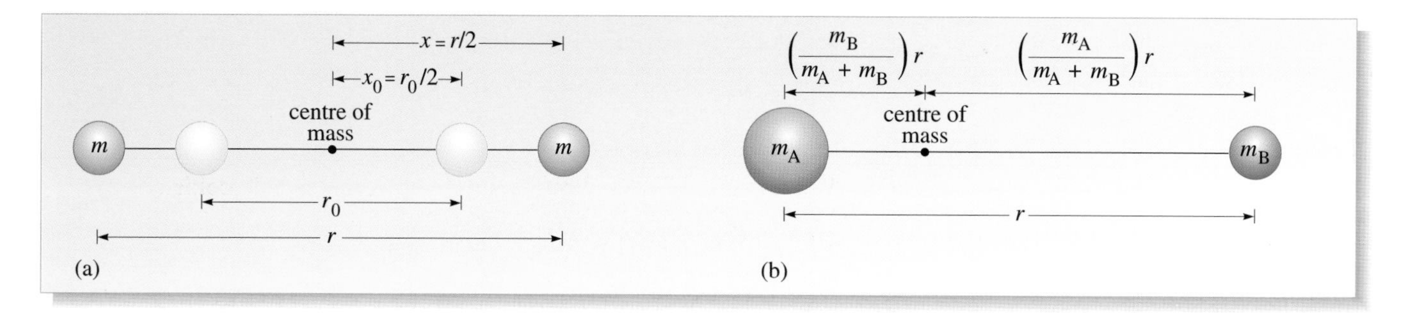

Figure 1.10 (a) Positions of atoms relative to the centre of mass in an oxygen molecule. (b) Positions of atoms relative to the centre of mass in a general diatomic molecule.

Question 1.2 Figure 1.10b shows the position of the centre of mass for a molecule composed of two different atoms, A and B, of masses m_A and m_B. Use an argument similar to that given above to show that the frequency of vibration of this molecule is

$$f = \frac{1}{2\pi}\sqrt{\frac{C(m_A + m_B)}{m_A m_B}}$$

where C is the magnitude of the gradient of the force–separation graph at the equilibrium separation.

Question 1.3 Estimate the frequency of oscillation of a molecule composed of two atoms whose force–separation curve is given in Figure 1.9 and whose masses are 2.2×10^{-25} kg and 8.8×10^{-26} kg. ■

The potential energy–separation curve

The interaction between atoms can also be described in terms of potential energy. Figure 1.11 shows a graph of potential energy against separation, r, with the zero of potential energy corresponding to atoms that are infinitely far apart. Although this graph is superficially similar to Figure 1.9, it is actually very different, and the relationship between the two graphs is quite subtle.

You may remember from Chapter 2 of *Predicting motion* that

$$F_r = -\frac{dE_{pot}}{dr}.$$

This means that the force-separation graph plotted in Figure 1.9 is *minus the gradient* of the potential energy-separation graph plotted in Figure 1.11. Notice, in particular, that the equilibrium separation of the atoms, r_0, corresponds to the *lowest* point on the potential energy graph, where the gradient of the graph, and hence the

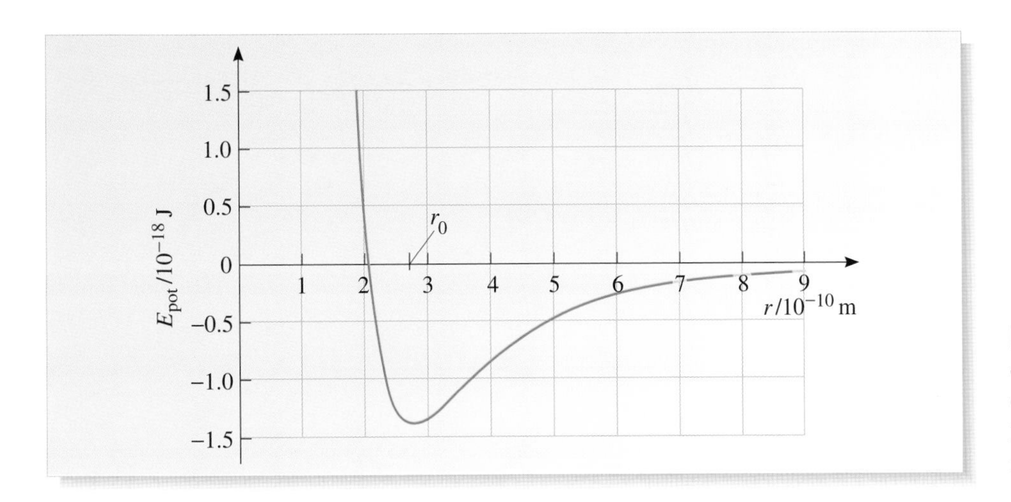

Figure 1.11 The potential energy–separation curve corresponding to the force–separation curve shown in Figure 1.9.

force, is zero. For separations smaller than r_0, the gradient of the potential energy graph is negative, corresponding to a positive radial force component and a repulsive force. For separations greater than r_0, the gradient of the potential energy graph is positive, corresponding to a negative radial force component and an attractive force.

We are often interested in the minimum energy needed to take two atoms in a diatomic molecule, initially at the equilibrium separation, and move them infinitely far apart. This is known as the **binding energy** of the molecule. Taking the potential energy at the bottom of the potential energy curve to be $-E_b$ and the potential energy at infinite separation to be 0, we have

$$\text{binding energy} = 0 - (-E_b) = E_b.$$

In other words, the binding energy is the 'depth' of the potential energy well.

The above discussion referred to atoms in a diatomic molecule. However, similar considerations apply to other forms of matter. For example, similar forces exist between molecules, and between atoms in a solid. The situation is slightly more complicated for atoms in a solid because any given atom will experience forces due to all the atoms around it. However, the forces between atoms are quite short in range so the most important influences come from an atom's nearest neighbours. Thus, in a solid, we can again expect the typical spacing between atoms to be close to r_0 and the atoms will again move to and fro, making small vibrations about fixed equilibrium positions. We can also imagine separating all the atoms in a solid from their equilibrium positions until they are infinitely far apart. The energy required to do this is the binding energy of the solid.

Question 1.4 Two atoms are initially at their equilibrium separation in a diatomic molecule, with forces and potential energies as shown in Figures 1.9 and 1.11. Use these graphs to estimate the minimum force and minimum energy needed to split the molecule into completely separate atoms. ■

3 Atomic descriptions of the phases of matter

A water molecule is a tightly bound combination of one oxygen and two hydrogen atoms. Steam, liquid water and ice are just collections of these water molecules. It is intriguing that the same building blocks are capable of producing such distinct phases, and that one phase can change into another quite abruptly: at 99 °C we have liquid water, and at 101 °C gaseous steam. In this section we will examine the three familiar phases of matter — gas, liquid and solid — and try to relate their behaviour to that of the underlying atoms and molecules. We start by noting that the phases have distinctive properties:

Gases occur at high temperatures. They are not very dense and are able to flow, adopting the shape and volume of any empty container they are placed in. They can be compressed quite easily.

Liquids form at lower temperatures. They are much denser than gases. Although they are able to flow, they do not expand to fill the whole volume of any empty container. If the volume of the container is greater than the volume of the liquid, the liquid will generally occupy the bottom of the container, but if the volume of the liquid is greater than that of the container, it will overflow. Liquids resist being compressed.

Solids form at even lower temperatures. They have roughly the same density as liquids (slightly greater in most cases) and are not able to flow. In fact they are rigid — kick one end of a brick, and the whole brick will move forward, maintaining a practically constant shape. Rigidity is a special property of solids that distinguishes them from both liquids and gases.

These distinctions are generalizations. A gas can be pressurized and made more dense so that it behaves rather like a liquid. Putty could be thought of either as a soft solid or a thick liquid. Nevertheless, the distinction between gases, liquids and solids is usually clear enough, especially when there are clear transitions from one to another, as when ice melts or liquid water boils.

The precise phase that is found, whether gas, liquid or solid, depends on two competing influences — binding energy and kinetic energy. Binding energy measures the attraction between molecules. If this were the only influence, the molecules would huddle closely together, with nearest neighbours being close to their equilibrium separation. While binding energy causes molecules to cohere together, kinetic energy has the opposite influence. If the molecules move around very energetically, they can easily shake themselves free from one another. Any small clump of molecules that does manage to cohere will soon be split apart when it is bombarded by other high-energy molecules.

The role of temperature is significant. As the temperature rises, so does the average kinetic energy of the molecules and this, in very broad terms, explains why solids melt and liquids vaporize. At low temperatures, the binding energy is much greater than the kinetic energy, and the molecules cling together vibrating about fixed equilibrium positions: this is the rigid, solid phase. As the temperature rises, the molecules achieve the freedom to move past one another, although they still stay in close contact: this is the liquid phase. At higher temperatures, the molecules break loose and the much more tenuous gas phase is formed. This is not quite the end of the story, for at even higher temperatures, molecules split apart into separate atoms, and atoms disintegrate into electrons and nuclei. A fully ionized gas has very distinctive properties and is regarded as a new phase of matter — the **plasma** phase. Although plasmas are widespread in the Universe, in stars and in the interstellar medium, we shall not discuss them further, but will concentrate on the three phases most familiar in everyday life: gases, liquids and solids.

3.1 The gas phase

We have just seen that gases appear at high temperatures, where the effects of kinetic energy dominate those of binding energy. Since the forces between molecules play a relatively minor role in gases, we can expect all gases to behave in roughly similar ways. This makes the gas phase a good place to start, as it is really the simplest phase to discuss.

To have a gas phase, it is *not* necessary for the kinetic energy to exceed the binding energy. The disruptive effect of collisions is such that a gas will generally form if the average kinetic energy per atom is greater than 10% of the binding energy per atom. This is shown in Figure 1.12. The strong correlation between the binding energy per atom and the average kinetic energy of an atom when the boiling point is reached, confirms the idea that there is a competition between binding energy and kinetic energy. In the gas phase, kinetic energy has the upper hand.

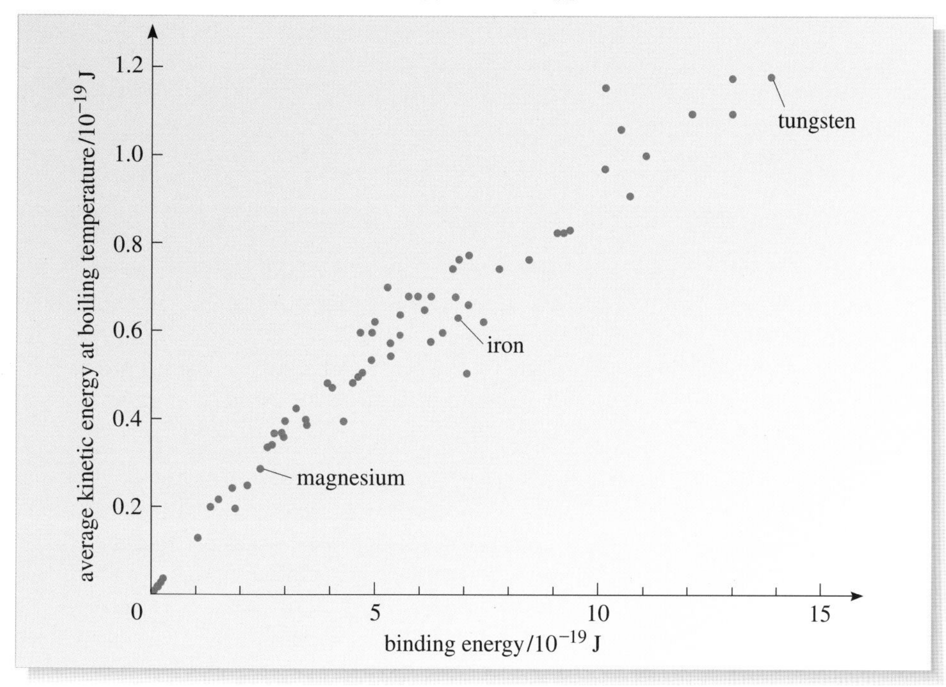

Figure 1.12 The average kinetic energy per atom for a range of selected elements at their boiling temperatures, plotted against the binding energy per atom. (To compare like with like, elements have been selected that vaporize into separate atoms. Thus elements such as nitrogen and oxygen, which have diatomic molecules in the gas phase, have been excluded.)

The density of a fixed mass of gas depends on the size of its container, but we can take the air in your living room as a typical example. The average spacing between molecules in air is about 3.3×10^{-9} m. While this may seem very small, it is much larger than the typical range of intermolecular forces so each molecule in a gas spends most of its time beyond the range of influence of others. You can picture the molecules in a gas as moving around freely, occasionally colliding with one another when their paths cross. This is why a gas expands to fill an empty vessel: the molecules experience no strong forces holding them together, so they just drift apart until they occupy the vessel more or less uniformly.

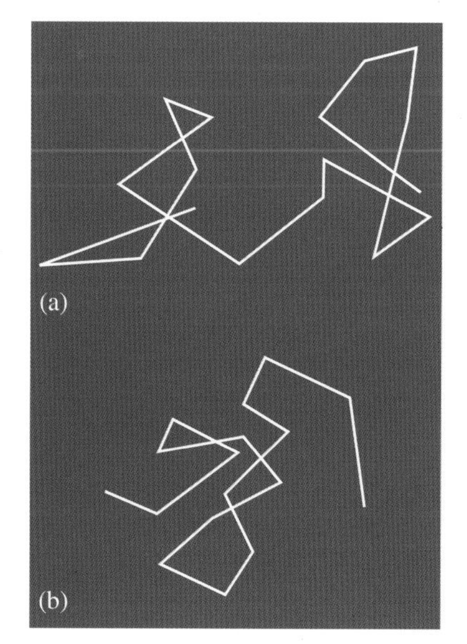

Figure 1.13 Typical paths followed by a gas molecule. (a) The path for a molecule of small radius. (b) The path for a molecule of larger radius. The number of molecules per unit volume is the same in both cases.

Figure 1.13a shows a typical path followed by a gas molecule. The path is highly erratic because each collision with another molecule involves an exchange of momentum and energy, and usually a sharp change in direction. A larger molecule would collide more frequently and its path would be even more erratic (Figure 1.13b) so, if the molecules were those of scents emerging from perfume bottles into still air, the smaller molecule would delight your nose first.

The random motion of a gas molecule is invisible, and can only be inferred indirectly. But if smoke particles are allowed to drift in air, and are observed under high magnification, their zigzag paths can be seen directly. This phenomenon (which is also observed in suspensions of particles in liquids) is called **Brownian motion** after its discoverer, the botanist Robert Brown. It is a direct consequence of collisions between molecules in the surrounding medium and the suspended particle. For a smoke particle, each zig and zag does not correspond to a single collision. The collisions occur at random, so that the smoke particle is continually jostled this way and that, but usually on a scale that is not visible. Occasionally, a number of collisions push the particle in the same direction, and it gains enough momentum to make a *visible* step forward. Historically, Brownian motion was important in establishing the reality of molecules and thus of atoms. Indeed, Einstein's 1905 paper on Brownian motion, and subsequent measurements carried out by Perrin in 1908, marked the end of all serious resistance to the atomic hypothesis.

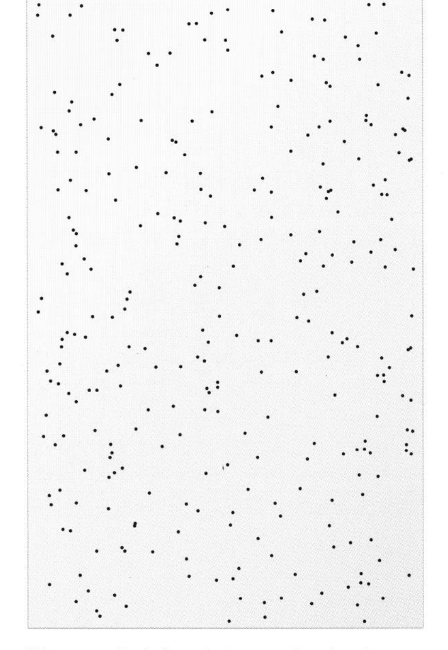

Figure 1.14 A 'snapshot' of molecules in a gas.

If you could take a snapshot of the individual molecules in a gas you would see an almost random arrangement, as shown in Figure 1.14. Randomness can be a tricky concept to understand. Looking at Figure 1.14, you can see that there are areas where molecules are quite closely packed, and areas where they are further apart. It is tempting to assume that this reveals a lack of randomness, but this is not true. If all the molecules were more or less uniformly spaced, that would be a *non-random* arrangement. Randomness implies a spread in particle spacings. It also implies a spread in particle speeds. If Figure 1.14 could be turned into a movie, you would see some molecules travelling very quickly and others moving very slowly. The word 'gas' is a corruption of the Greek *chaos*, which turned out to be an apt choice.

One way of describing the randomness in a gas is to say that there is no correlation between the positions of molecules in a gas. In other words, if I know that there is a molecule at a given place, that does not affect the chances of finding another molecule somewhere else. This is because the intermolecular forces between gas molecules have a negligible role. To visualize this idea more clearly, imagine choosing one particular molecule as a reference and counting the number of molecules, N, whose centres lie in a thin spherical shell centred on this molecule (Figure 1.15). If the spherical shell has radius r and thickness Δr, its volume is $4\pi r^2 \Delta r$, so the number of molecules *per unit volume* at a distance r from the chosen molecule is $N/(4\pi r^2 \Delta r)$. As the molecules move around, this quantity fluctuates, but we are interested in its value averaged over time:

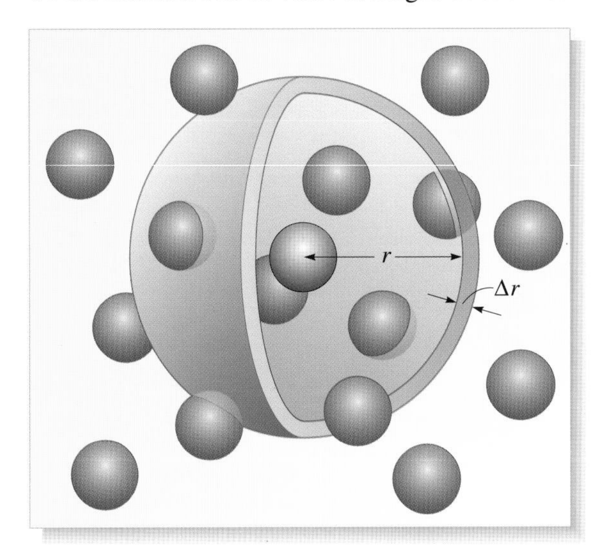

Figure 1.15 A spherical shell used to define the radial density function.

The **radial density function** $n(r)$ is defined as the time-averaged value of $N/(4\pi r^2 \Delta r)$, where N is the number of molecules whose centres lie in a thin spherical shell of radius r and thickness Δr, centred on a chosen molecule.

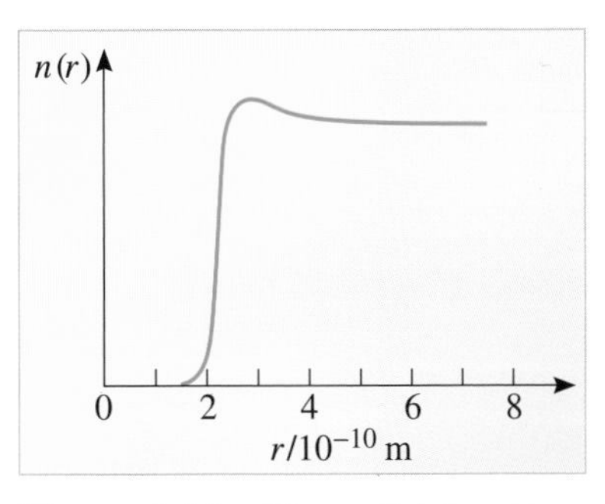

Figure 1.16 The radial density function of a gas.

The radial density function measures the average number of molecules per unit volume as we step outwards from a given molecule. Figure 1.16 shows the radial density function for a gas. For most separations this is practically a horizontal line, which confirms that one molecule has no influence on the position of another — the molecules are in random, uncorrelated positions. The only significant detail occurs at very low separations, where there is a dip in the radial density function. This is not surprising because the molecules repel one another if pressed too closely together — they cannot sit exactly on top of one another. Apart from the occasional collisions that occur when molecules get too close, the overall picture is one of independent molecules travelling around freely, with a range of speeds and kinetic energies.

3.2 The liquid phase

When a gas is cooled, the kinetic energy of its molecules falls until it reaches a point where collisions are no longer able to prevent the clumping together of molecules. At this point, a gas turns into a liquid.

The transformation of steam into water can be taken as a typical example. Rare clumps of molecules may form briefly in the steam, but are rapidly torn apart. As the temperature drops these clumps grow, persist for longer, and can be thought of as tiny droplets of water. Eventually, a few droplets become so large that, instead of decaying, they grow, enveloping molecules or other droplets that strike their surface. The steam then rapidly condenses into water.

You will have seen one of the key signatures of this process whenever you boil a kettle. Steam is actually colourless and transparent. The white haze you see forming a little way from the spout of the kettle is a suspension of water droplets formed as the steam condenses (Figure 1.17). These water droplets have grown to sizes that are comparable to, or larger than, the wavelength of visible light, so they are able to scatter light strongly. Exactly the same thing happens in clouds as water vapour contained in the atmosphere condenses.

Figure 1.17 Droplets of liquid water condensing from steam scatter light strongly when they grow to be larger than the wavelength of visible light.

Liquids are much denser than gases because their molecules are much closer together. Neighbouring molecules are within one another's spheres of influence, so you would expect the positions of the molecules to be much more correlated than for a gas. Figure 1.18 shows the radial density function for water at a variety of temperatures. In the lower graphs (4 °C, 25 °C), there are clear peaks and troughs which die away with distance. Each peak represents a surplus of molecules over the average, and each trough a deficit. These features show that the molecules of a liquid are arranged in a more orderly way than in a gas, but this order only persists over a short distance; beyond 4–6 molecular diameters from the chosen particle, the average density of particles is practically constant. Liquids are said to have **short-range order**. The upper graphs in Figure 1.18 show the radial density function for water at higher temperatures. Notice

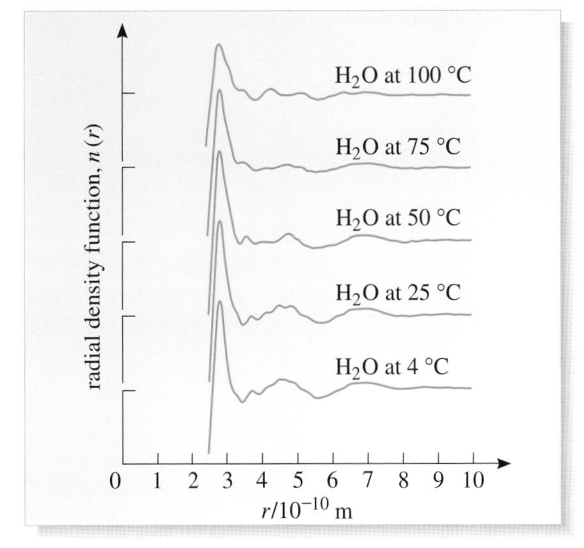

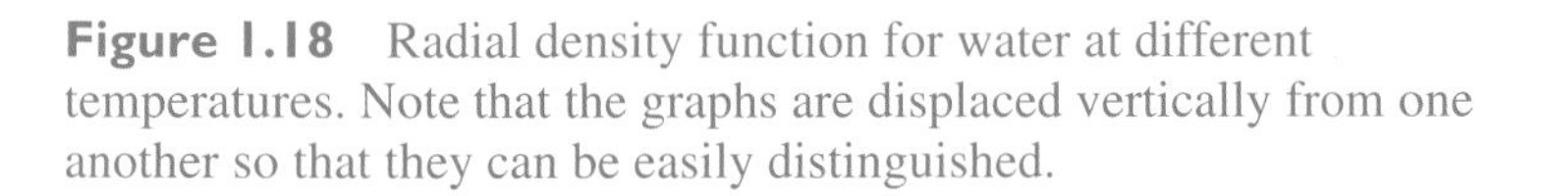

Figure 1.18 Radial density function for water at different temperatures. Note that the graphs are displaced vertically from one another so that they can be easily distinguished.

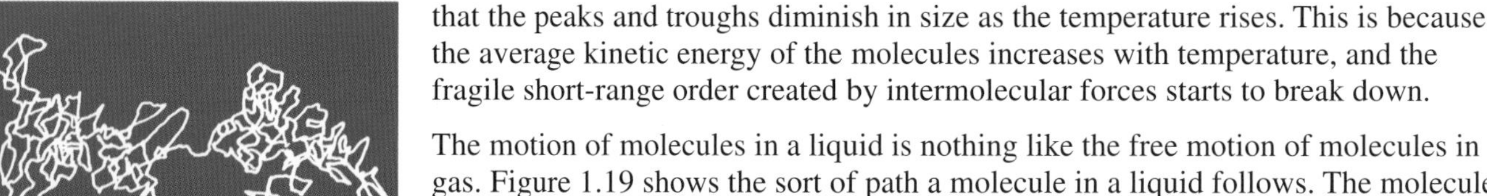

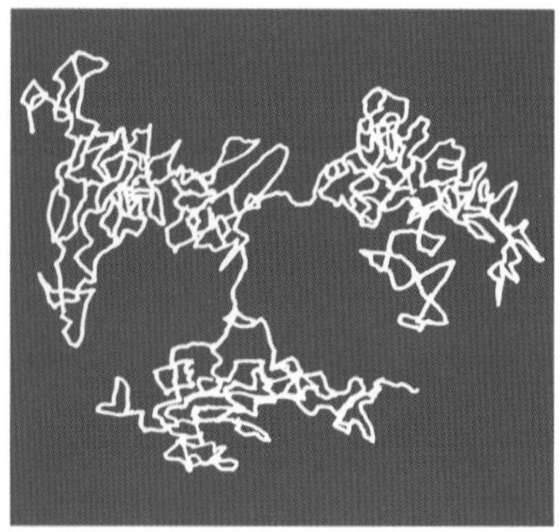

Figure 1.19 Path followed by a molecule in a liquid.

that the peaks and troughs diminish in size as the temperature rises. This is because the average kinetic energy of the molecules increases with temperature, and the fragile short-range order created by intermolecular forces starts to break down.

The motion of molecules in a liquid is nothing like the free motion of molecules in a gas. Figure 1.19 shows the sort of path a molecule in a liquid follows. The molecule spends much of its time jogging to and fro in a small region, but occasionally makes a longer jump into a neighbouring region. An analogy can be made with a country dance, where people are initially arranged into sets who dance together but, as the dance proceeds, some dancers move from one set to another. In a dance, these motions would be choreographed, but in a liquid the same effect is achieved in a more serendipitous way.

- Liquids are much harder to compress than gases. Use a microscopic model to explain why this behaviour is expected.

○ The molecules in a liquid are very close together so further compression invokes the strong repulsive force shown in Figure 1.9. The molecules in a gas can be brought closer together without invoking this force. ■

3.3 The solid phase

For all substances (except helium), liquids do not persist down to the lowest temperatures. Sooner or later, a temperature is reached at which the liquid turns into a solid. In a solid, each atom is tied to a specific equilibrium position. Although it vibrates to and fro about this position, an atom is very unlikely to move from its 'home' relative to its neighbours (Figure 1.20). That is why a solid is rigid. If you tap a solid gently it moves as a whole, rather than allowing the atoms to change their relative positions.

Figure 1.20 Movement of an atom in a solid.

In many cases, the solid is crystalline, either consisting of a single crystal, as in a gemstone, or a collection of microscopic crystals, as in most metals. In a crystal, the atoms oscillate around equilibrium positions that are *regularly ordered*. The atoms can be pictured like soldiers on parade, shuffling from one foot to another, but not moving from their position in the array. Figure 1.21 shows the equilibrium arrangement of water molecules in ice. Although it is rather difficult to see the details, you can discern that the whole structure is formed by repeating a simple pattern endlessly and periodically throughout space. This is rather like saying that a wallpaper pattern is produced by regularly spaced repetitions of a given motif (Figure 1.22).

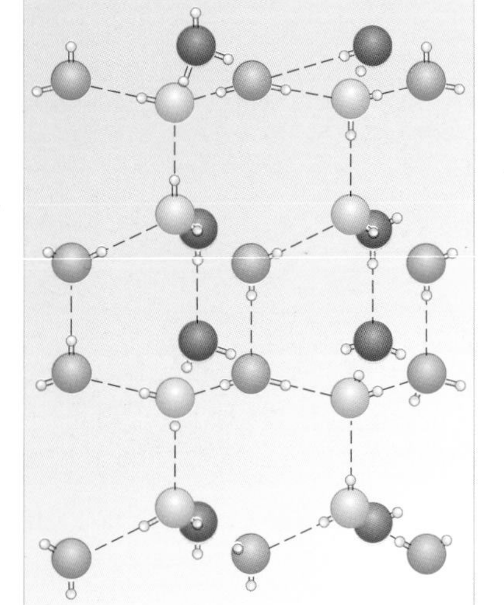

Figure 1.21 Arrangement of water molecules in ice.

Figure 1.22 A wallpaper pattern.

Once we know the structure of a crystal, and the immediate environment of a given atom, we can predict the positions of atoms many thousands of atomic spacings away. For this reason, a crystal is said to have **long-range order**, in contrast with liquids, which have only short-range order, and gases, which are disordered. The long-range order in a crystalline solid leads to a radial density function with very sharp peaks, extending far from the central atom (Figure 1.23).

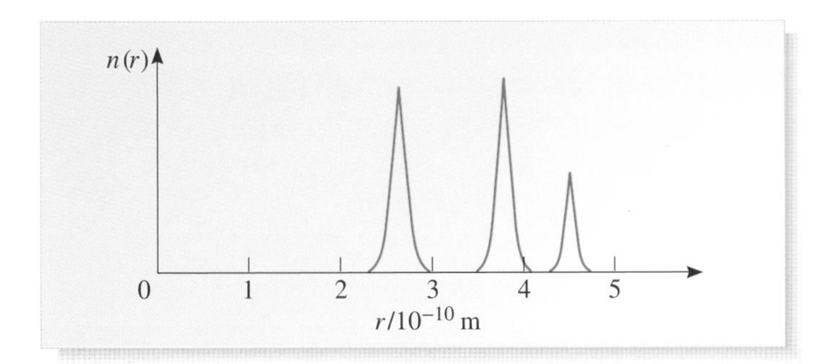

Figure 1.23 The radial density function in a solid.

Another distinctive feature of crystals is their symmetry. In ice, for example, the molecules pack together in an hexagonal arrangement. Intriguingly, this hexagonal symmetry has an influence on a much larger scale, in the beautiful form adopted by snowflakes (Figure 1.24). Also reflecting their underlying atomic arrangement, crystals cleave along lines of particular weakness, between one plane of atoms and the next. Even diamonds, the hardest of all substances, are cleaved in this way before being polished to produce gemstones (Figure 1.25). The study of crystals turns out to be full of fascinating detail. Ice, for example, has at least seven crystalline phases, with the water molecules arranged differently in each. In the case of steel, the existence of different phases is of major technological importance. Steel generally consists of a myriad of tiny crystals, so small that they can only be seen under the microscope (Figure 1.26). By arranging to have a suitable blend of crystals in different phases, it is possible to control the hardness and flexibility of different types of steel.

Figure 1.24 A snowflake.

Figure 1.25 A diamond crystal.

Figure 1.26 Microscopic crystals in steel.

4 Macroscopic variables

So far, we have looked at matter from an atomic perspective. For the rest of this chapter, we will take a more everyday view, on the scale of millimetres and above, rather than nanometres and below. The larger scale is often described as being **macroscopic**, while the smaller one is **microscopic**. Of course, the real interest lies in linking these two different scales: understanding the whole in terms of its parts.

To obtain a more quantitative description, we will introduce variables, such as density, pressure, temperature and internal energy, that are often used to describe matter on a large scale. Although you may be familiar with these concepts, it is worth reading on, to check that your understanding is in line with the technical meaning of these terms.

4.1 Density

The density, ρ, of a sample of matter is defined to be its mass per unit volume,

$$\rho = \frac{M}{V} \tag{1.2}$$

so the SI unit of density is kilogram per cubic metre (kg m^{-3}). The volume V should be chosen judiciously so that the sample of matter appears to be homogeneous. It must be large enough to avoid the grainy nature of matter on an atomic scale, and small enough to be unaffected by any large-scale variations. If the volume is chosen reasonably, the density will be quite insensitive to the exact choice.

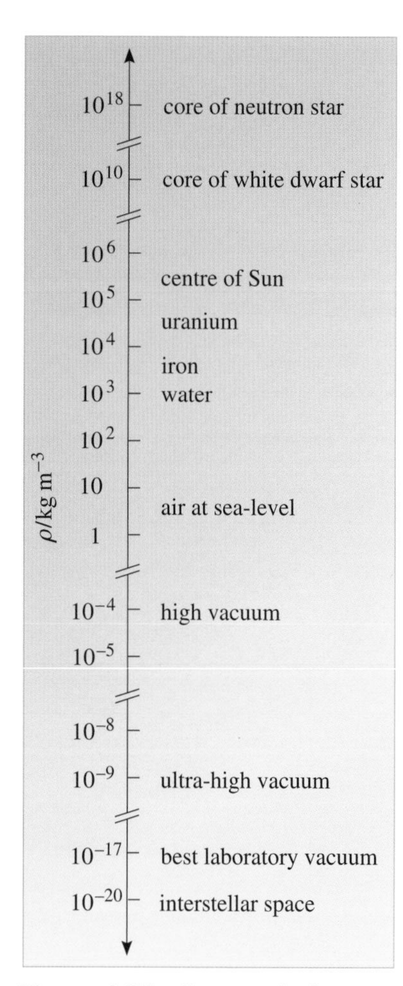

Figure 1.27 Some typical densities.

Some typical densities are shown in Figure 1.27. Of course, solids and liquids have much higher densities than gases, but even amongst solids there are large differences — the density of solid osmium is more than 40 times that of solid lithium. Under extreme conditions, the atomic nature of matter breaks down and very high densities are produced. Neutron stars, for example, are composed of matter with the density of atomic nuclei, roughly 10^{15} times greater than found in ordinary matter.

You will also come across the term **number density** for the number of molecules per unit volume. If each molecule has mass m, the number density is given by

$$\text{number density} = \rho/m. \tag{1.3}$$

4.2 Pressure

An asteroid, in empty space, would experience no pressure from its surroundings. But the matter we are familiar with on Earth is generally at some non-zero pressure, if only because the weight of the atmosphere presses down on it.

The upper surface of a flat roof, with dimensions 10 m × 10 m say, experiences a downward force due to the pressure of the Earth's atmosphere. The magnitude of this force is about 10^7 N, equivalent to the weight of a 1000 tonne object! This sounds alarming, but the space beneath the roof also contains atmosphere which exerts a force of 10^7 N on the lower surface of the roof, in an upward direction. So the roof will be slightly compressed, but it should not collapse on your head.

The large forces applied by the Earth's atmosphere only become apparent when they are unbalanced. This is what happens when a jam jar is sealed with a partial vacuum inside it. A more spectacular demonstration was given in 1657 by Otto von Guericke, one-time mayor of Magdeburg, Germany. Von Guericke placed two large copper hemispheres together to form a sphere (Figure 1.28). The hemispheres had been carefully made so that they fitted snugly together. He then pumped some of the air

Figure 1.28 Otto von Guericke's demonstration of the large forces that are applied by atmospheric pressure.

out of the sphere, and challenged people to pull it apart into separate hemispheres. To the amazement of the assembled crowd, no one could, nor could two teams of eight horses. The pressure inside was very small but the atmospheric pressure of about 10^5 newtons per square metre provided a large inward force, pressing the hemispheres together.

Pressure is quantified in terms of the force exerted on a flat surface. If a uniform force presses down on a surface, in a direction perpendicular to the surface, the pressure is defined to be

$$P = \frac{F}{A}$$

where F is the magnitude of the force and A is the area of the surface.

If the force $\boldsymbol{F}$ is not perpendicular to the area (Figure 1.29), the pressure is defined in terms of $F_\perp$, the component of the force that is perpendicular to the surface:

$$P = \frac{F_\perp}{A} = \frac{F\cos\theta}{A}. \tag{1.4}$$

The area is taken to be small enough for the force to be uniform across it. In the limiting case where the area is taken to be very small, the pressure is defined at a given point.

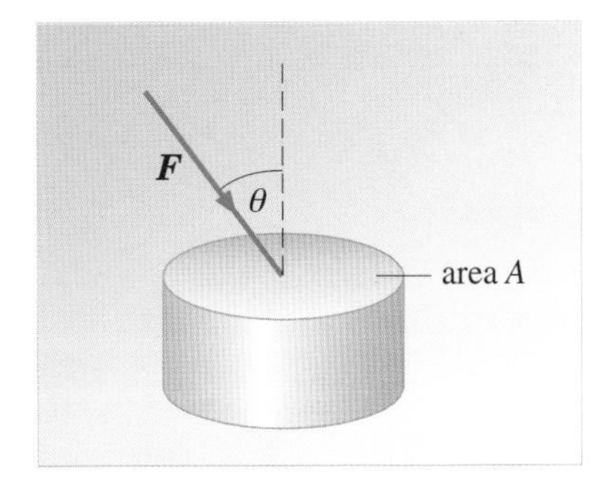

Figure 1.29 Definition of pressure.

Unlike force, pressure is a scalar quantity. In SI units, it is measured in newtons per square metre ($\mathrm{N\,m^{-2}}$). This unit is called a **pascal** (Pa) in honour of the French scientist Blaise Pascal (1623–1662) who gave an early definition of pressure and investigated its variation with altitude. In the context of these investigations, Pascal was fortunate in his sister's choice of husband. Just four years after the invention of the barometer, he asked his brother-in-law to carry a barometer up a mountain and observe any change in the pressure recorded. So Pascal's brother-in-law was the first

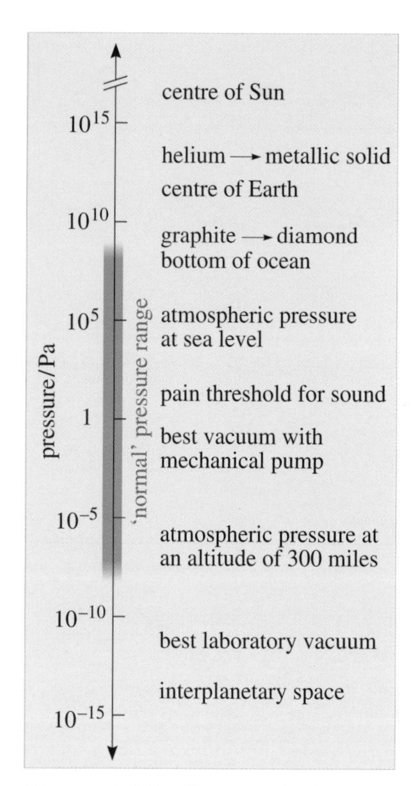

Figure 1.30 Some typical pressures.

to observe that pressure decreases with altitude, a topic we return to in Chapter 4. Normal atmospheric pressure at sea-level is around 10^5 Pa, but Figure 1.30 shows a range of values that are encountered in other contexts.

Sometimes, the pressure of a system is quoted as an overpressure. This is the amount by which the pressure of the system exceeds a standard value (such as normal atmospheric pressure). Tyre pressures, for example, are generally quoted in this way. When performing calculations in physics, though, it is usually advisable to convert overpressures into actual pressures.

Question 1.5 A woman wearing high-heel shoes pivots in such a way that, for a brief instant, all her weight is on one heel. If the woman has a mass of 56 kg and the heel has a cross-sectional area of 0.49 cm^2, what is the pressure exerted by the heel on the floor? ■

4.3 Temperature

We are used to ambient temperatures that vary over a relatively rather narrow range. In Britain, for example, we would regard 30 °C as a very hot day, and −10 °C as a bitterly cold one. Although some parts of the world are hotter than others, large temperature differences are moderated by winds and ocean currents which transport energy from hotter regions to cooler ones.

We know that our bodies work best within a narrow range of temperature and we do our best, with houses, clothing, ice creams and air conditioning to stay within that range. When these artificial protection mechanisms let us down, our bodies take over by shivering or sweating. In a way, we all act as thermometers. Our skin contains special nerve endings, or receptors, which are sensitive to temperature (Figure 1.31). Between about 15 °C and 45 °C, the active receptors are those that respond to cold or warmth; outside this range, pain receptors start to function. Thus, we literally have a 'feel' for the meaning of temperature.

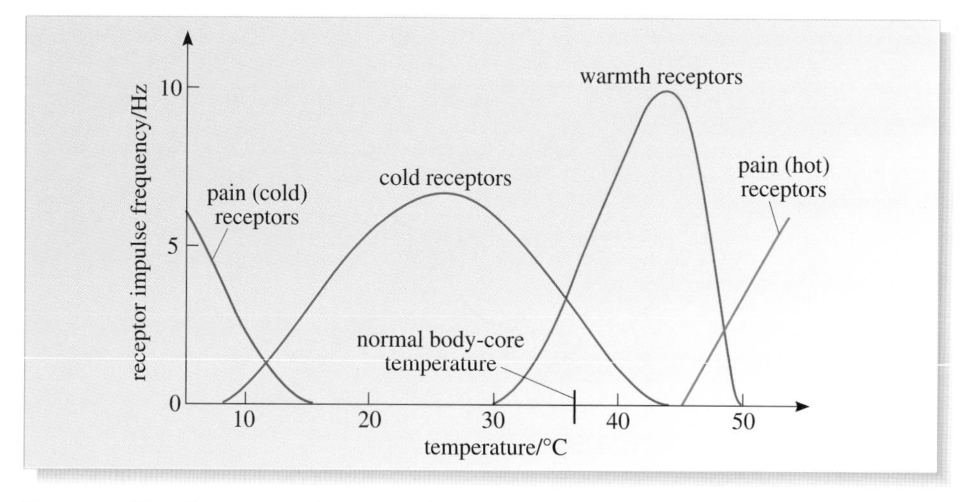

Figure 1.31 Four types of receptors in the skin respond to temperature. The frequency with which each receptor transmits impulses to the central nervous system varies with temperature. Note that the response curves for the cold receptors and warmth receptors overlap around the core body temperature (~ 37 °C).

The ability to sense temperatures is essential to our survival, but temperature is more than a vague, subjective feeling. It is a scientific variable with a precise meaning. Perhaps the most important point to emphasize is the distinction between temperature and heat. In the early days there was much confusion between these two terms, but they refer to very different things.

If two bodies, at different temperatures, are brought into contact, energy flows from the warm body to the cool body (Figure 1.32a). In the process, the temperature of the warm body falls and the temperature of the cool body rises. The energy transfer continues until both bodies are at the same temperature, when further change ceases (Figure 1.32b). The bodies are then said to be in **thermal equilibrium**.

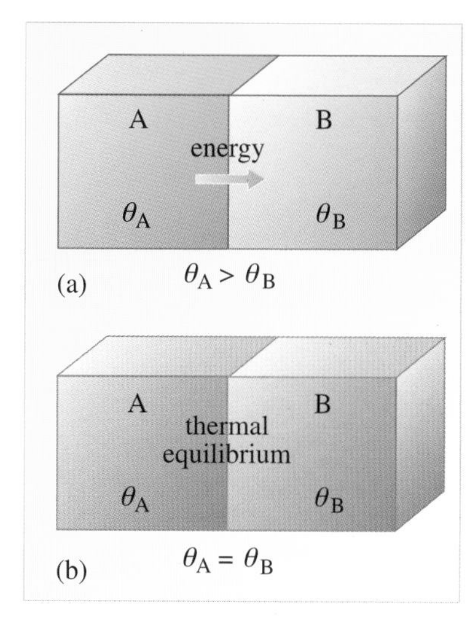

Figure 1.32 Temperature differences control the flow of energy (heat) from one body to another. (a) So long as the temperature of body A is greater than that of body B, energy will flow from A to B. (b) Thermal equilibrium is achieved when both bodies have the same temperature.

> The energy transferred from a warm body to a cool body, as a result of a difference in temperature, is known as **heat**, and this type of energy transfer is known as **heating**. If no other energy transfers take place, the energy lost by the warm body is equal to the energy gained by the cool body, ensuring that energy is conserved overall.

It is tempting, but WRONG, to think of temperature in the same way as energy. The first person to realize this was the Scottish physicist, Joseph Black. In 1760, Black placed a number of different objects, in turn, in contact with a hot object. For the same loss in temperature of the hot object, he found that different objects, of the same mass, were warmed to different temperatures. Such results could never be interpreted in terms of a flow of temperature from one body to another.

So what is temperature? We will arrive at many different answers to this question during this course, but the following statement makes an excellent starting point:

> **Temperature** is a label which determines the direction of heat flow between two bodies placed in contact with one another. Heat flows from the body with the higher temperature to the body with the lower temperature, and keeps on flowing until, in thermal equilibrium, both bodies are at the same temperature.

Several points can be lifted out of this statement:

1 Heat always flows in a definite direction. We define temperature in such a way that heat flows from a high-temperature body to a low-temperature body.

2 So long as there is a temperature difference between two bodies in contact, or within a single body, heat will flow and the system will not be in a state of equilibrium.

3 When the temperature is uniform, a state of thermal equilibrium is reached in which no more heat is transferred.

So far, we just know that temperature is a label that determines the direction of heat flow. For the moment, we will use the symbol θ to denote temperature, so if body 1 has temperature θ_1 and body 2 has temperature θ_2, with $\theta_1 > \theta_2$, heat will flow from body 1 to body 2. To proceed further, we must say something about how the labelling is carried out. In other words, we must decide how to *calibrate* a thermometer so that quantitative readings can be taken. This turns out to be a subtle issue but, to get started, we can consider an ordinary mercury-in-glass thermometer.

Following the instructions in Box 1.1, you could produce a calibrated thermometer which could be used to read off any temperature (in degrees Celsius) between 0 °C and 100 °C.

Box 1.1 Instructions for calibrating a thermometer

Suppose you had a mercury-in-glass thermometer before it had been calibrated. Because mercury expands when heated you will see the length of mercury in the column increasing as the temperature increases. To calibrate the thermometer, you could then take the following steps:

1 Plunge the thermometer into a beaker of ice, melting under normal atmospheric pressure. Allow the thermometer to settle down and reach the same temperature as the melting ice, then mark the position of the end of the mercury column as 0 °C.

2 Immerse the thermometer in water, boiling under normal atmospheric pressure. Allow the thermometer to settle down and reach the same temperature as the boiling water, then mark the position of the end of the mercury column as 100 °C.

3 Construct a linearly graduated scale between these two calibration points. This can be done as follows:

Suppose l_0 and l_{100} are the lengths of the mercury column when the thermometer is in thermal equilibrium with melting ice and boiling water respectively. *By definition*, these correspond to 0 °C and 100 °C on our temperature scale, and this is shown by the positions of the red dots on Figure 1.33. A linearly graduated scale is formed by drawing a straight line between these two points and taking readings from it. So, when the length of the mercury column is midway between l_0 and l_{100}, the temperature is taken to be 50 °C. More generally, when the length of the mercury column is l, the temperature is taken to be

$$\theta = 0\,^{\circ}\mathrm{C} + \frac{l - l_0}{l_{100} - l_0} \times 100\,^{\circ}\mathrm{C}. \tag{1.5}$$

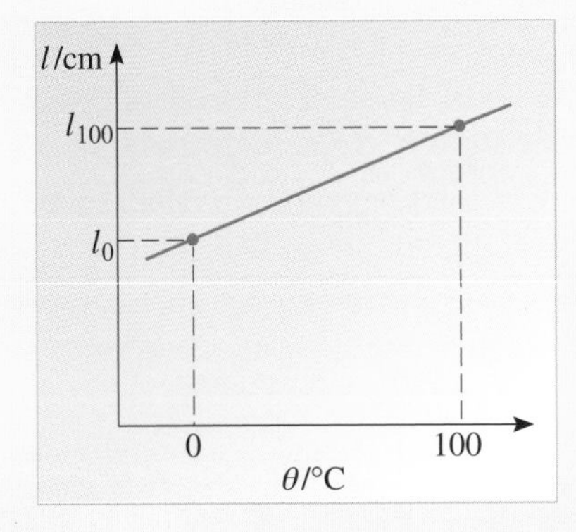

Figure 1.33 Forming a temperature scale.

Question 1.6 The length of mercury in the column of an uncalibrated mercury-in-glass thermometer is found to be 3.3 cm when it is held in contact with melting ice and 6.9 cm when it is immersed in boiling water. Estimate the temperature, measured by this thermometer on the Celsius scale, when the length of mercury in the column is 4.2 cm. ■

4.4 Internal energy

Energy has an important role to play in explaining the properties of matter, but care is needed, because some contributions to the energy have no influence on the properties of matter. For example, a steel bar may rotate about an axis. This motion certainly contributes to the total energy of the bar, but has a negligible effect on the bar's material properties, such as its strength or flexibility. Similarly, taking the bar to the top of a mountain will affect its gravitational potential energy, but will not affect the bar's ability to conduct electricity or heat. For present purposes, we can restrict attention to samples of matter that are macroscopically at rest, and far from all external influences, such as gravity. Under these circumstances, the total energy of the sample is referred to as its **internal energy**.

In microscopic terms, we can picture individual atoms or molecules moving around within the sample. These particles have kinetic energy and mutual potential energy. The internal energy can then be thought of as the sum of all these atomic or molecular contributions. (This energy is sometimes referred to as **thermal energy**, but should never be called heat, or heat energy, as this would clash with the careful definition of heat given earlier.)

As always, we must settle on a choice for the zero of potential energy. One common convention is to take the potential energy to be zero when the individual atoms or molecules are infinitely far apart. This implies that internal energies can be negative, as well as positive, depending on the precise combination of kinetic and potential energies. Alternatively, we can take the potential energy to be zero when the atoms are in their equilibrium positions in the solid phase. The choice is arbitrary and unimportant because internal energy *itself* cannot be measured. However, we can measure *changes* in internal energy. This is because energy is a conserved quantity so, if we keep track of the energy entering or leaving a system, we know by how much its energy has changed.

We can distinguish two ways of increasing the internal energy of a system. The first, called heating, has been described above. **Heat** is any transfer of energy that occurs because of a temperature difference, the energy flowing from a high-temperature region to a low-temperature region. Thus, if you place a frying pan on top of a glowing hot plate, the frying pan will be *heated*, according to our strict definition of the term. The term **work** is used for all energy transfers that are not classified as heat. Work often involves mechanical processes, such as compressing a gas, stirring a liquid, or scouring a solid. It may also involve electrical processes, as when the filament in a light bulb is made to glow by passing an electrical current through it. Calculations of heat, work and changes in internal energy form part of the subject of **thermodynamics** which will be discussed in Chapter 3 of this book.

4.5 Molar quantities

Many properties of matter depend on how large a sample is under consideration. It is therefore a good idea to specify a standard quantity of matter, relative to which all measured values are quoted. The simplest choice is to take one kilogram as this standard quantity. However, another choice, based on the concept of a mole, is equally important.

A mole contains a definite number of 'basic particles' (atoms or molecules). To see how it is defined, consider a pure substance composed of identical basic particles of mass M_r amu. (These could be molecules of relative molecular mass M_r or atoms of relative atomic mass M_r.)

The abbreviation for mole is mol.

Then one **mole** of the substance is defined to be a sample of mass $M_r \times 10^{-3}$ kg. This is a macroscopic quantity of matter, such as you might hold in your hand.

Note that the definition of a mole only applies to pure substances. We cannot talk of a mole of concrete, for example, because concrete is a mixture of several different chemicals, of different relative molecular masses. The factor 10^{-3} is an historical curiosity. It occurs because the definition of the mole dates back to a time when grams were regarded as fundamental units of mass, rather than kilograms, but it would be too risky to make changes now, given the widespread use of moles in medicine.

Avogadro's number, N_A, is defined as the number of basic particles in a mole. Since each basic particle has mass M_r amu, we can write:

$$\text{mass of 1 mole} = M_r \times 10^{-3}\,\text{kg} = N_A \times (M_r\,\text{amu})$$

so

$$N_A = \frac{10^{-3}\,\text{kg}}{1\,\text{amu}} = \frac{10^{-3}\,\text{kg}}{1.6605 \times 10^{-27}\,\text{kg}} = 6.022 \times 10^{23}.$$

The enormous size of Avogadro's number shows that ordinary samples of matter contain a vast number of molecules or atoms, giving the illusion that matter is continuous. Note that Avogadro's number is independent of the substance under consideration.

One mole of any pure substance contains 6.022×10^{23} basic particles. That is why a mole is a useful way of specifying a standard quantity of matter. Many properties of matter are expected to depend on the number of basic particles involved, so patterns are likely to emerge more clearly if measurements are expressed *per mole* of the substance.

Although it may seem like splitting hairs, it is conventional to distinguish between the number of particles *in one mole*, and the number of particles *per mole* in any given sample. The former is Avogadro's number, $N_A = 6.022 \times 10^{23}$. Of course, this is just a pure number, with no units. By contrast, the number of particles *per mole* has the units of mol^{-1}. To keep track of units, including moles, it is sensible to introduce a new symbol, N_m, for the number of particles per mole. This quantity is called **Avogadro's constant**, and is defined by

$$N_m = N_A\,\text{mol}^{-1}.$$

Having defined the mole, we can express other molar quantities, as values *per mole*. The **molar mass**, M_m, is the mass per mole so, by definition,

$$M_m = M_r \times 10^{-3}\,\text{kg mol}^{-1}.$$

The **molar volume**, V_m, is the volume per mole. This is related to the molar mass via the density,

$$V_m = M_m/\rho.$$

For a gas at room temperature and atmospheric pressure, the molar volume is of the order of a cubic foot per mole. A typical value for most solid elements is more like a cubic inch per mole.

The **molar internal energy**, U_m, is the internal energy per mole. This depends on the substance involved and the zero chosen for potential energy, but a typical room temperature value is of the order of 10^4 to 10^5 joules per mole.

- Lead has a relative atomic mass of 207. How many lead atoms are there in (a) one mole of lead and (b) one kilogram of lead?

❍ (a) By definition, one mole of lead contains Avogadro's number (6.022×10^{23}) of atoms.

(b) One mole of lead has a mass of 207×10^{-3} kg and contains 6.022×10^{23} atoms. Thus 1 kg of lead contains

$$6.022 \times 10^{23} \times \frac{1.0 \times \text{kg}}{207 \times 10^{-3}\,\text{kg}} = 2.9 \times 10^{24} \text{ atoms.} \quad ■$$

4.6 Response functions

A useful way of gauging the properties of a substance is to make a small change in one quantity and observe how some other quantity responds. For example, we could increase the temperature slightly, and see how much the volume increases; or we could increase the pressure slightly and see how much the internal energy changes.

Thermal expansion provides a familiar example. Consider a sample of matter that initially has volume V, pressure P and temperature θ. If this sample is heated at constant pressure, it will generally expand. The change in volume, ΔV, is therefore positive, and would be expected to be proportional to the change in temperature, $\Delta\theta$, and to the initial volume, V. We can therefore write

$$\Delta V = \alpha V \Delta\theta \tag{1.6}$$

where α is a coefficient of proportionality, which depends on the substance that is being heated. This quantity is known as the (isobaric) **expansivity** of the material, where isobaric means 'at constant pressure'. The factor of V on the right-hand side of Equation 1.6 ensures that α is independent of V, and so characterizes the *substance* under study (e.g. copper or steel), no matter what the size of the sample.

A second example is provided by compression. If you squeeze a sample of matter slightly, increasing the pressure, whilst keeping the temperature constant, there will be a small decrease in the sample's volume (it squashes). The change in volume is therefore negative, and would be expected to be proportional to the change in pressure, ΔP and to the initial volume, V. This implies that

$$\Delta V = -\beta V \Delta P \tag{1.7}$$

where β is a coefficient of proportionality, which depends on the substance that is being squeezed and is known as the (isothermal) **compressibility** of the substance. Isothermal means 'at constant temperature'. Rubber has a large value of β because it is very squashy, while steel has a much smaller value. The isothermal compressibility depends on a number of factors, but tables exist for a variety of materials under a range of conditions, so that designers and technologists can choose an appropriate material for a particular use.

Similar definitions can be made for other pairs of stimuli and responses. The coefficients of proportionality that occur in this way are called **response functions**. In the limiting case of small changes, Equations 1.6 and 1.7 give

$$\alpha = \frac{1}{V}\frac{\mathrm{d}V}{\mathrm{d}\theta} \quad \text{(at constant pressure)} \tag{1.8}$$

$$\beta = -\frac{1}{V}\frac{\mathrm{d}V}{\mathrm{d}P} \quad \text{(at constant temperature).} \tag{1.9}$$

All response functions can be defined in terms of derivatives describing how rapidly one quantity varies with respect to another, supplemented by conditions specifying which quantities are held constant during the change.

Question 1.7 When a metal block is heated by 5 °C at constant pressure, it expands. The pressure acting on the block is then increased at constant temperature until the bar contracts back to its initial volume. During the expansion, the expansivity is $3.6 \times 10^{-6}\,°C^{-1}$, and during the contraction the compressibility is $3.0 \times 10^{-11}\,Pa^{-1}$. By how much was the pressure increased? ■

5 Macroscopic equilibrium in ideal gases

Some macroscopic variables (such as pressure and temperature) help us to specify the state of a given sample of matter. Others (such as expansivity and compressibility) help us to quantify the inherent differences between one material and another. Either way, as soon as we start to look closely at these variables, a vast range of questions emerges. Usually, when we change the value of one macroscopic variable, many other variables will change, and we would like to understand these changes. With understanding comes power, so we can hope that an understanding of the properties of matter will allow us to control the behaviour of materials and shape them for our own ends.

When confronting a difficult question — such as trying to understand the properties of matter — it is a good idea to consider a particularly simple situation, to see if that can be understood first. For this reason, we shall concentrate first on *gases* which are *in equilibrium*, that is, in a settled, unchanging state.

At first sight, equilibrium may seem dull. The still air in a sealed room may seem less exciting than the air in a tornado or a furnace. Perhaps so, but equilibrium itself contains a great mystery, which deserves careful thought. The mystery can be stated in very general terms:

Any equilibrium state of a macroscopic sample of matter can be characterized by an amazingly small number of macroscopic variables — often just two or three are sufficient.

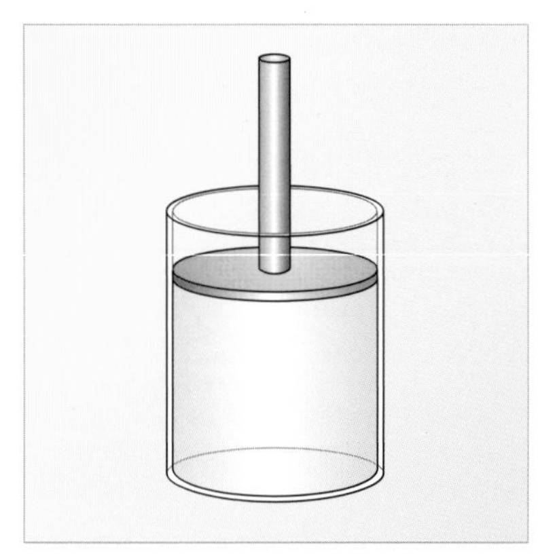

Figure 1.34 A sample of oxygen gas, studied under controlled conditions.

To take a definite case, consider a fixed quantity of oxygen gas, contained in a cylinder with a closely fitting but adjustable piston (Figure 1.34). The volume of the gas is easily controlled by adjusting the position of the piston: sliding the piston in reduces the volume and sliding it out increases the volume. The temperature of the gas can be controlled by immersing it in a large *thermal reservoir* (sometimes called a 'heat bath') which has a known, fixed temperature. If the gas is colder than the thermal reservoir, heat will flow into the gas and it will get warmer; if the gas is warmer than the reservoir, heat will flow out of the gas and it will get cooler. When the gas has settled down to a state of equilibrium, its temperature is the same as that of the thermal reservoir. Now for the surprise:

Having fixed the temperature and volume of our sample of oxygen, *all other macroscopic properties of this sample are determined.* The pressure, internal energy, compressibility, density and all other properties have definite values which can be reproduced whenever the sample has the same volume and temperature.

The surprising thing about this situation is that it is so simple. A typical macroscopic sample of matter contains 10^{18} molecules or more, each of which is described by three position coordinates and three velocity components so, from a microscopic point of view, a vast number of variables is needed to specify the gas completely. Yet, from a macroscopic point of view, all the equilibrium properties are determined by the volume and temperature. This extreme contrast between the microscopic and macroscopic descriptions is the essence of the mystery described above. How can we get away with such a simple macroscopic description of equilibrium states when the underlying microscopic description is so complicated?

The answer to this question will emerge in the next chapter, where you will see that the macroscopic properties of a gas are determined by the most likely, average behaviour of its molecules. In other words, the gulf between the macroscopic and microscopic viewpoints will eventually be bridged with help from statistics.

Figure 1.35 Robert Boyle (1627–91) was an Irish chemist who did much to advance the concept of a chemical element and the atomic hypothesis. Using a vacuum pump, Boyle demonstrated that Galileo was correct in asserting that all falling objects accelerate downwards at the same rate, and went on to perform the experiments which led to Boyle's law.

5.1 Experiments on gases

It is possible to perform experiments on gases, to see how different variables such as pressure, volume, temperature and internal energy are related to one another. An experiment of this type was carried out by Newton's contemporary, Robert Boyle (Figure 1.35).

Boyle's law

Boyle knew of Pascal's work and was stimulated to investigate the effects of pressure on what he called 'the spring of the air'. In 1662, he observed 'not without delight and satisfaction', that the more a gas is compressed, the more its pressure rises. By taking a fixed mass of gas, and maintaining it at a fixed temperature, he was able to establish the following law:

Boyle's law

If a fixed mass of gas is held at a constant temperature, its pressure is inversely proportional to its volume.

We can express this in terms of a proportionality:

$$P \propto \frac{1}{V}$$

or as an equation:

$$PV = \text{constant} \qquad \text{(fixed quantity, constant temperature)}$$

where the constant on the right-hand side depends on the amount (i.e. number of moles) of gas and on the temperature. Figure 1.36 shows what this equation means, in graphical terms.

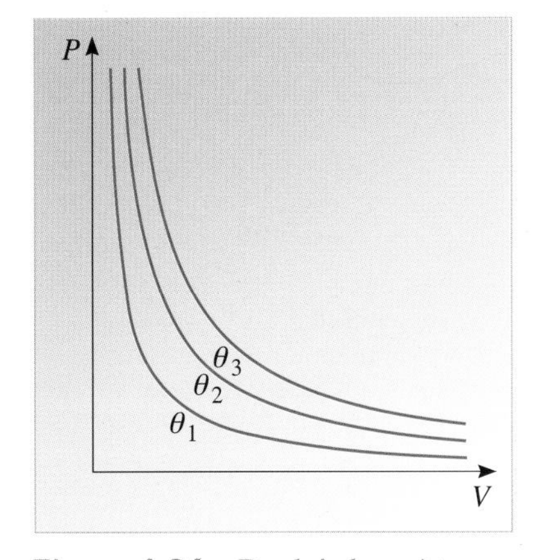

Figure 1.36 Boyle's law. At constant temperature, pressure is inversely proportional to volume. The precise shape of the curve depends on the fixed temperature: in this figure, $\theta_1 < \theta_2 < \theta_3$.

Boyle needed skill and care to arrive at this conclusion. Normally, when you compress a gas, it warms up (you may have noticed the temperature rise that occurs when you pump up a cycle tyre). Boyle was able to obtain a simple result only because he had the patience to let the compressed gas cool down to its initial temperature. This was good scientific practice: in order to investigate how one variable depends on another, it is generally a good idea to ensure that other variables, which could influence the result, are held constant.

According to Boyle's law, each doubling of the pressure leads to a halving of the volume, so we can compress a gas to a very small volume indeed, provided we apply a high enough pressure. This suggests that Boyle's law has its limitations. At very

Figure 1.37 A hot air balloon. Heating the air inside the balloon causes it to expand, and the reduction in density produces buoyancy.

high pressures, the molecules will be squeezed together so closely that the strong repulsive forces shown in Figure 1.9 will come into play. These forces will help the gas to resist further compression, so that further doublings of the pressure will not produce halvings of the volume, and deviations from Boyle's law will be observed. Nevertheless, Boyle's law works well for most ordinary gases, provided the pressure is not too high. In particular, it works well for air at room temperature and atmospheric pressure.

Charles's law and the absolute temperature scale

Ballooning, which became popular in the second half of the eighteenth century, provided strong motivation for more detailed studies of gases (Figure 1.37). In 1783, Jacques and Robert Charles completed the first manned ascent in a hydrogen balloon. As his fame spread, Jacques Charles (1746–1823) studied the properties of gases more thoroughly. One of the favoured means of ascent was in hot air balloons, so he studied the expansion that occurs when a gas is heated, arriving at the following conclusion:

Charles's law

If a fixed mass of gas is heated at constant pressure, the increase in volume is proportional to the increase in temperature.

So, if we plot a graph of volume against temperature at constant pressure, we will get a straight line (Figure 1.38). As with Boyle's law, there are deviations at very high pressures, but Charles's law works well enough for gases under normal conditions, and becomes exactly true in the limiting case of low pressures.

Figure 1.38 prompts an intriguing question. Since gases expand when they are heated, they must contract when they are cooled. What would happen if we carried on cooling a low-pressure gas indefinitely? If we simply extrapolate the straight lines in Figure 1.38 to very low temperatures, we get the strange situation shown in Figure 1.39. At a temperature of −273.15 °C the extrapolated volume for any gas is zero, and at temperatures lower than this, it is actually negative! Such a state of affairs is clearly impossible, so Charles's law cannot apply at temperatures below −273.15 °C. The reason for this turns out to be simple, and profound: *temperatures lower than −273.15 °C do not exist.*

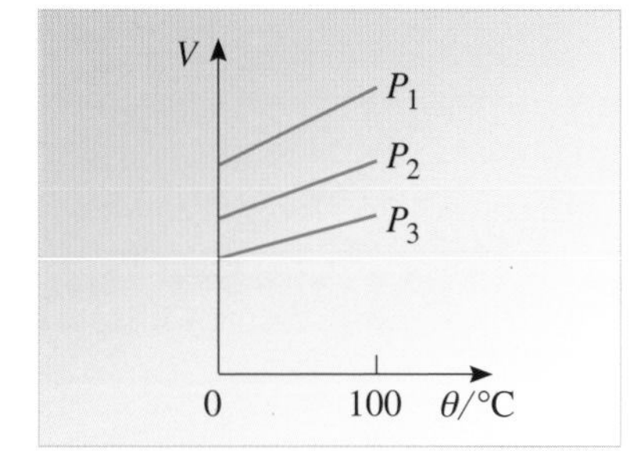

Figure 1.38 (a) Charles's law between 0 °C and 100 °C. At constant pressure, the volume of the gas is linearly related to the temperature. The slope of the line depends on the fixed pressure: in this figure, $P_1 < P_2 < P_3$.

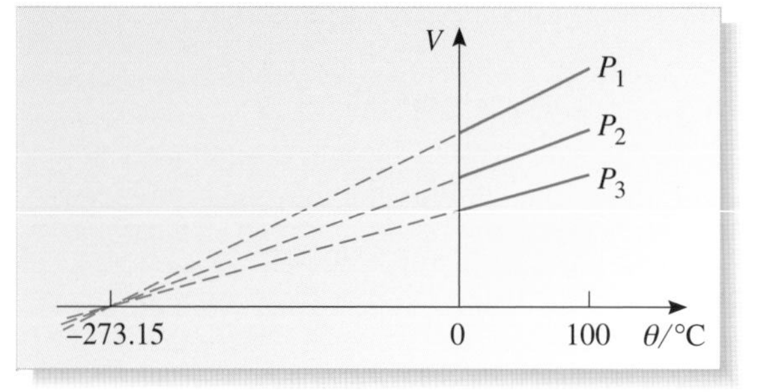

Figure 1.39 Charles's law extrapolated to very low temperatures.

The absolute zero of temperature

The lowest conceivable temperature is −273.15 °C. This is known as the **absolute zero** of temperature; it can be approached, but never quite reached. No system in a state of equilibrium can have a temperature below absolute zero.

5.2 The absolute temperature scale

The idea that –273.15 °C is the lower limit for temperature suggests that it would be useful to introduce a new temperature scale in which this lowest conceivable temperature is taken to be *zero*. This new scale is called the **absolute temperature scale**.

So far, we have used the symbol θ to represent a temperature on the Celsius scale, and the symbol °C for the Celsius degree. Thus, the temperature of boiling water has been specified as $\theta = 100$ °C. To avoid confusion, it is a good idea to use different symbols for our new scale. We shall use the symbol T to represent a temperature on the absolute temperature scale and we shall also introduce a new unit, the **kelvin**, abbreviated to K, for the new scale. The numerical value of a temperature on the absolute scale is found by adding 273.15 to the numerical value on the Celsius scale, so in terms of symbols

$$T/\text{K} = \theta/°\text{C} + 273.15. \qquad (1.10)$$

You have seen that the lowest conceivable temperature on the Celsius scale is $\theta = -273.15$ °C. Using Equation 1.10, the corresponding value on the absolute scale is

$$T/\text{K} = -273.15\ °\text{C}\,/\,°\text{C} + 273.15 = 0,$$

so the absolute zero of temperature on the absolute scale is $T = 0$ K, as required.

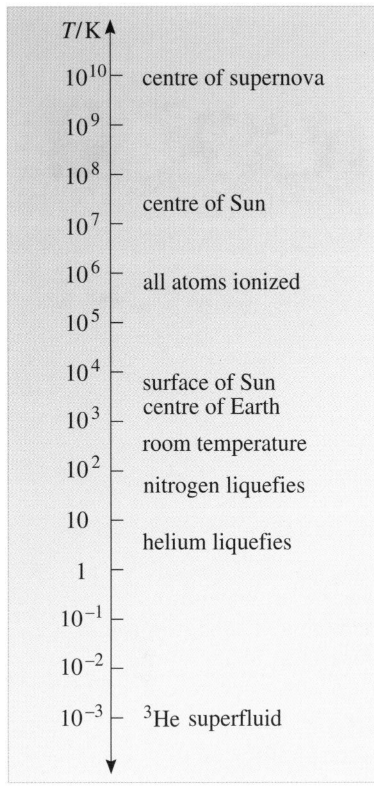

Figure 1.40 A range of temperatures.

- ● What is the temperature of boiling water on the absolute scale?
- ❍ Using Equation 1.10,

$$T/\text{K} = 100\ °\text{C}/\,°\text{C} + 273.15 = 373.15.$$

So the temperature of boiling water is $T = 373.15$ K. ■

If a temperature on the Celsius scale increases by 1 °C, the corresponding temperature on the absolute scale increases by 1 K. This means that quantities we defined earlier in terms of θ can be easily re-expressed in terms of T. For example, the isobaric expansivity can be written as:

$$\alpha = \frac{1}{V}\frac{\mathrm{d}V}{\mathrm{d}T} \quad \text{(at constant pressure).} \qquad \text{(Eqn 1.8)}$$

In spite of this close relationship, it is important to preserve the distinction between °C and K, so that it is clear that a temperature quoted as 300 °C refers to the Celsius scale, while one quoted as 300 K refers to the absolute scale. Figure 1.40 shows a range of temperatures found on Earth and in the Universe, all expressed on the absolute scale. Normal room temperature is around 295 K.

Finally, returning to Charles's law, Figure 1.41 shows the linear relationship between volume and temperature, expressed in terms of the absolute temperature scale. The straight line now passes through the origin, corresponding to the proportionality

$$V \propto T$$

or the equation

$$V = \text{constant} \times T \qquad \text{(fixed quantity, constant pressure)}$$

where the constant depends on the amount (i.e. number of moles) of gas and on the pressure.

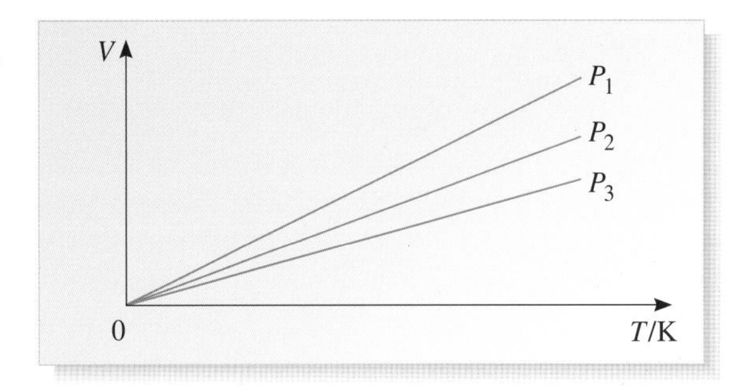

Figure 1.41 Charles's law expressed in terms of the absolute temperature scale.

5.3 The equation of state of an ideal gas

Boyle's law and Charles's law were important milestones in understanding the properties of gases. Charles's law, especially, pointed the way to the absolute zero of temperature and the absolute temperature scale. However, it is not necessary to remember these laws individually, because they are both special cases of a more powerful statement.

Suppose that n moles of a gas are in equilibrium, at pressure P, volume V and absolute temperature T. Then, it is found that

$$PV = nRT \tag{1.11}$$

where $R = 8.314\,\mathrm{J\,K^{-1}\,mol^{-1}}$ is called the **universal gas constant** or the **molar gas constant**.

Equation 1.11 works well for most ordinary gases, but deviations appear when the pressure becomes too high. We can simplify the discussion by introducing the concept of an **ideal gas**. An ideal gas is one with negligible forces between its molecules, and which obeys Equation 1.11 *exactly*. Any equation that links the equilibrium values of P, V and T is called an **equation of state**, so Equation 1.11 is known as the **ideal gas equation of state**.

The behaviour of a real gas becomes completely indistinguishable from that of an ideal gas in the limit of low pressure. This is understandable — as the pressure decreases, the molecules move further apart (on average), which reduces the importance of intermolecular forces. But, even under normal conditions, most real gases are approximately ideal, and Equation 1.11 provides an excellent description of their behaviour.

The fact that Equation 1.11 provides such a good description of gases, especially at low pressures, is used to construct the **ideal gas temperature scale**. This is done by *defining* temperature in such a way that Equation 1.11 is automatically satisfied at very low pressures. More explicitly, imagine taking measurements of the pressure P and volume V of a fixed mass of a gas at a fixed temperature. For each pressure, the quantity PV/nR can be calculated. As the pressure becomes smaller and smaller, the value of PV/nR is found to approach a limiting value. This limiting value is *defined* to be the temperature on the ideal gas temperature scale. In terms of symbols, we can write

$$T = \lim_{P\to 0}\left(\frac{PV}{nR}\right)$$

This is much more satisfactory way of defining temperature than using the expansion of a mercury column. There are more fundamental ways of defining temperature, based on statistical mechanics or thermodynamics but, for practical purposes, they are all equivalent to temperature defined on the ideal gas scale.

The ability of an equation to give a concise summary of many facts is well illustrated by the ideal gas equation of state. By considering this equation under special circumstances, we can recover all the facts about gases discussed earlier, and more besides.

1 Boyle's law For a fixed number of moles at a constant temperature, the right-hand side of Equation 1.11 is a constant, so we recover Boyle's law:

$$PV = \text{constant.}$$

2 Charles's law According to Equation 1.11,

$$\frac{V}{T} = \frac{nR}{P}.$$

So, for a fixed number of moles at a constant pressure, we recover Charles's law:

$$\frac{V}{T} = \text{constant}.$$

3 Avogadro's hypothesis According to Equation 1.11,

$$n = \frac{PV}{RT}.$$

So, if two samples of different ideal gases are maintained at the same temperature T and pressure P, and occupy the same volume V, they must contain the same number of moles, n. Since a mole contains a definite number of molecules (6.022×10^{23}), the two samples must also contain the same number of molecules, in agreement with Avogadro's hypothesis.

4 Universal behaviour Because the ideal gas equation of state contains a constant, R, which is the same for all ideal gases, many properties of ideal gases are *universal* (i.e. independent of whether the gas is hydrogen, oxygen or whatever). This can be illustrated by calculating the (isobaric) expansivity of an ideal gas. The equation of state of an ideal gas gives

$$V = \frac{nRT}{P}$$

so, at constant pressure, the rate of increase of volume with temperature is

$$\frac{\mathrm{d}V}{\mathrm{d}T} = \frac{nR}{P}.$$

Using Equation 1.8, the (isobaric) expansivity is then

$$\alpha = \frac{1}{V}\frac{\mathrm{d}V}{\mathrm{d}T} = \frac{nR}{PV} = \frac{1}{T},$$

where the last equality follows from Equation 1.11. We conclude that:

All ideal gases at the same temperature have the same (isobaric) expansivity, irrespective of their molecular composition.

This fact was known to Charles, and must have been of interest to him as he planned future ballooning exploits.

Question 1.8 We stated earlier that 1 mol of any gas, under normal conditions of pressure and temperature, occupies a volume of about a cubic foot. Use the ideal gas equation of state to establish this fact.

Question 1.9 A sample of ideal gas occupies a volume of $5.00\,\mathrm{m}^3$ and is at a pressure of 1.00×10^5 Pa and a temperature of 300 K. (a) How many moles of the gas are there in the sample? (b) How many molecules are there in the sample? (c) If the gas is made of hydrogen molecules, with a relative molecular mass of 2.02, what is the density of the gas? ■

Alternative ways of writing the equation of state

The ideal gas equation of state can be written in many different, but fundamentally equivalent, ways. The most common variant is to express it in terms of the number of molecules, N, rather than the number of moles, n. Remembering that the number of molecules per mole is given by Avogadro's constant, N_m, the total number of molecules is

$$N = nN_m.$$

The equation of state can therefore be written as

$$PV = \frac{N}{N_m}RT$$

$$\text{or} \qquad PV = NkT \tag{1.12}$$

where we have introduced a new constant,

$$k = \frac{R}{N_m} = 1.381 \times 10^{-23}\,\mathrm{J\,K^{-1}},$$

known as **Boltzmann's constant**. You will see Boltzmann's constant appearing in many equations of physics — especially where temperature is involved.

A third way of expressing the equation of state is also worth mentioning. If we have a gas composed of N molecules, each of mass m, the total mass of the gas is

$$M = Nm$$

and the density of the gas is

$$\rho = \frac{M}{V} = \frac{Nm}{V}.$$

Combining this with Equation 1.12 gives

$$\rho = \frac{mP}{kT}. \tag{1.13}$$

This form of the equation of state would be of direct interest to balloonists as the density of the gas within the balloon needs to be less than that of the surrounding air to produce buoyancy. The equation suggests three different strategies for achieving this:

- choose a gas with a low molecular mass;
- use a sealed container containing gas at very low pressure;
- heat the gas in a balloon that is not sealed.

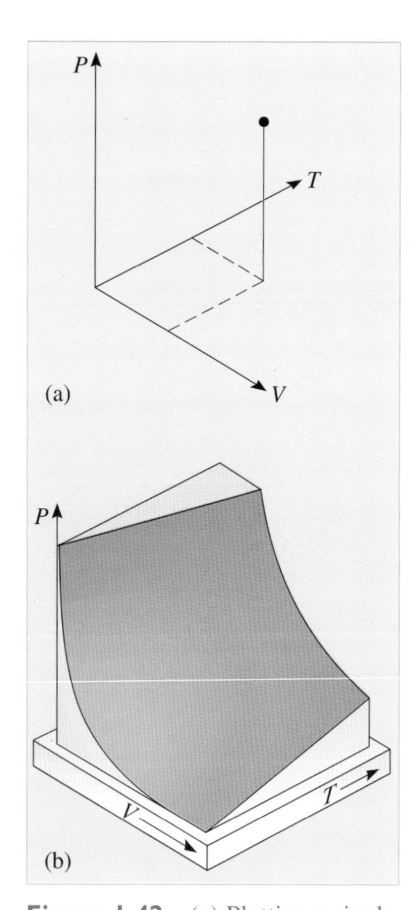

Figure 1.42 (a) Plotting a single point on the *PVT* surface. (b) The *PVT* surface of an ideal gas. The surface is normally drawn for one mole of a substance.

The PVT surface — visualizing the ideal gas equation of state

A useful way of visualizing the equation of state for a fixed quantity of ideal gas is to display it as a 'three-dimensional graph', with T and V plotted along two horizontal axes and P plotted along a vertical axis. Figure 1.42a shows how any single point is plotted. The set of all points generated in this way produces a continuous surface, known as the ***PVT* surface for an ideal gas** (Figure 1.42b).

Each point on the *PVT* surface corresponds to a combination of pressure, volume and temperature values that can be achieved in a fixed quantity of an ideal gas in equilibrium. Conversely, any set of pressure, volume and temperature values that does not lie on this surface does *not* satisfy Equation 1.11, and so *cannot* be achieved in the fixed quantity of ideal gas in equilibrium.

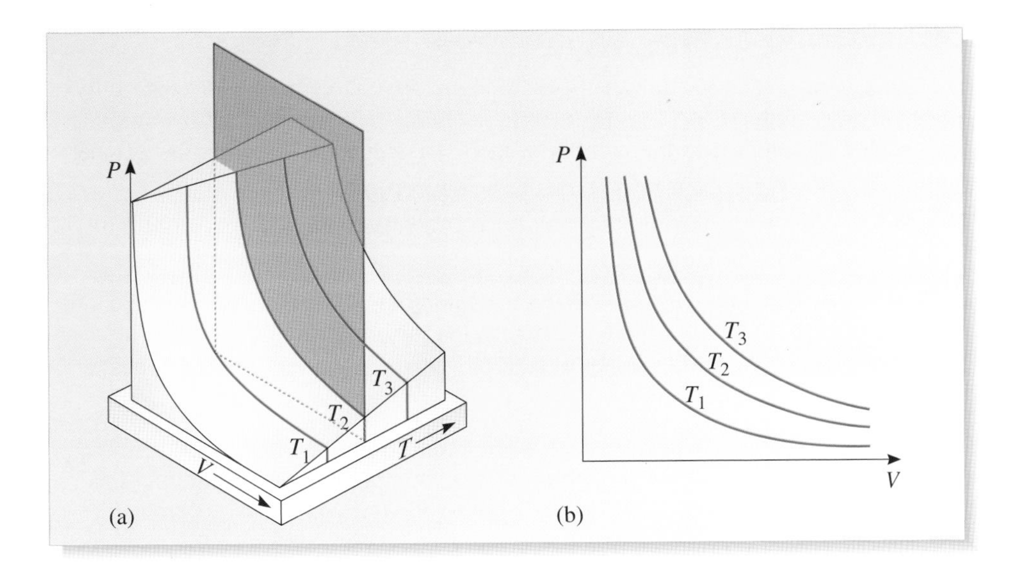

Figure 1.43 (a) Slicing through the *PVT* surface of an ideal gas at constant temperature T_2. The cut surface reveals (b) the dependence of pressure on volume at constant temperature (Boyle's law).

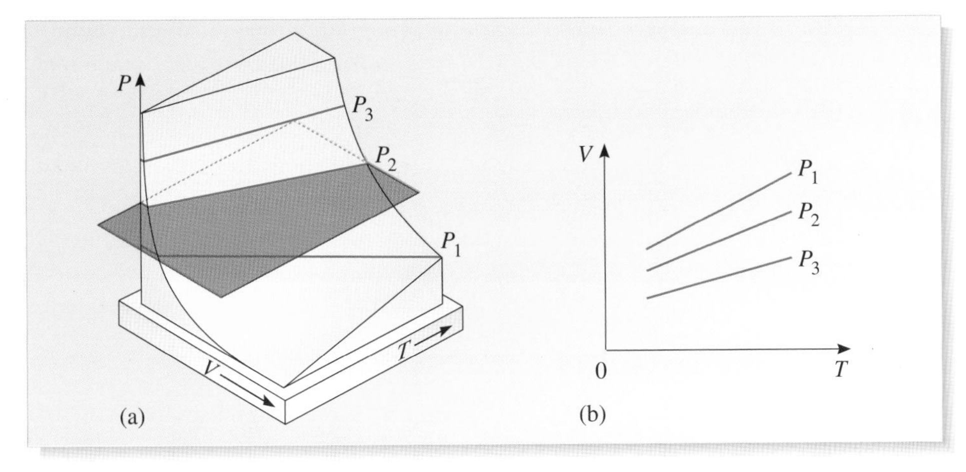

Figure 1.44 (a) Slicing through the *PVT* surface of an ideal gas at constant pressure P_2. The cut surface reveals (b) the dependence of volume on temperature at constant pressure (Charles's law).

Figure 1.43 shows a slice through the *PVT* surface at constant temperature. The cross-section revealed is identical to Figure 1.36 for Boyle's law. Figure 1.44 shows a slice through the *PVT* surface at constant pressure. This gives a cross-section identical to Figure 1.41 for Charles's law.

The *PVT* surface can also be used to visualize *processes*, such as compression or heating. Provided the process is gentle, the system will have time to adjust to the changing conditions, and can be thought of as being in equilibrium at each instant. Changes that are gentle enough for this to be a fair description are said to be **quasi-static**. Because quasi-static processes are effectively successions of equilibrium states, they can be represented by paths that lie on the *PVT* surface.

Figure 1.45 shows two paths that occur when a gas is compressed by a certain amount. The red path a–b shows what happens if the temperature is held constant during the compression, while the blue a–c path shows what happens when the gas is allowed to warm up as it compresses. The two paths are quite different, and the pressure at point c is greater than the pressure at point b. This illustrates the care that is needed when describing processes: it is not sufficient to say that a gas is compressed: we must also know how the temperature varies (or does not vary) during the compression.

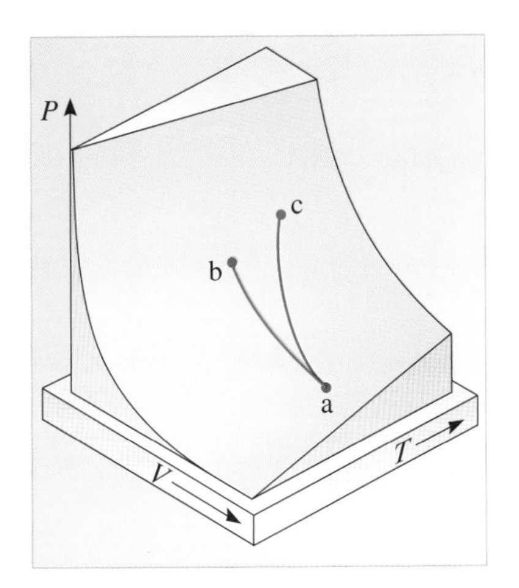

Figure 1.45 Two paths across a *PVT* surface representing quasi-static processes.

5.4 Internal energy equations of ideal gases

You have seen that the equation of state relates the pressure of a gas to other variables — the number of moles, the volume and the temperature. We can also ask whether the internal energy of a gas is linked to these variables. For example, you might expect the internal energy of a gas to be proportional to the number of moles of the gas, and to increase with temperature. To say any more, we need to turn to experiments for guidance.

James Prescott Joule (1818–1889) conducted a number of experiments on the nature and effect of heat which did much to establish the general principle of the conservation of energy. At the same time, he investigated the factors that affect the internal energy of gases.

Joule found that the internal energy of a gas is proportional to the number of moles and increases with temperature. However, at fixed temperature, he found that the internal energy is practically independent of pressure and volume. The phrase 'at fixed temperature' is significant because, when a gas is compressed, its temperature will generally rise, leading to an increase in internal energy. But if the gas is allowed to cool back to its initial temperature, the internal energy returns to its initial value, with the gas at a smaller volume than before. So, *at a fixed temperature*, the internal energy of the gas is unaffected by the volume change. This result can be interpreted by recalling that intermolecular forces play a negligible role in gases.

In a solid, the individual molecules are close together and interact strongly. Reducing the volume of a solid brings the molecules even closer together, and is accompanied by large changes in potential energy. In a gas, though, the molecules are so far apart that bringing them a little closer together makes practically no difference to their mutual potential energy. Provided the temperature remains fixed, the kinetic energy of the molecules also remains constant, so the total internal energy does not depend on the volume of the gas.

It is again convenient to summarize our discussion by using the concept of an ideal gas, in which molecular interactions have a negligible effect.

> In equilibrium, the internal energy of an ideal gas is independent of pressure and volume, and can be written as
>
> $$U = nF(T) \qquad (1.14)$$
>
> where n is the number of moles of the gas and $F(T)$ is some function of the absolute temperature. This result is known as **Joule's law of ideal gases**.

In some ways, Joule's law is simpler than the ideal gas equation of state, but the result is less universal because the function $F(T)$ depends on the type of gas under investigation, especially on the number of atoms per molecule.

As always, when discussing energy, it is important to specify a zero of energy. We will take the zero of energy to correspond to a situation where the molecules are at rest, an infinite distance apart. With this convention, the internal energy of n moles of any monatomic gas is accurately described by the equation

$$U = \tfrac{3}{2}nRT \qquad (1.15a)$$

where R is the universal gas constant introduced earlier and T is the absolute temperature.

Diatomic gases, such as nitrogen and oxygen, have higher internal energies than monatomic gases at the same temperature. Their internal energies depend on temperature in a complicated way, but around room temperature a typical diatomic molecule has

$$U = \tfrac{5}{2}nRT. \qquad (1.16a)$$

The simple relationships expressed by Equations 1.15a and 1.16a intrigued physicists in the nineteenth century, and they invested much effort to explain them. You will see some of their arguments in the next chapter. For the moment, you can accept Equations 1.15a and 1.16a as experimental results.

These equations can also be rewritten in terms of the number of molecules, N, rather than the number of moles, n. When this is done, the internal energy equation for a monatomic ideal gas becomes

$$U = \tfrac{3}{2}NkT \qquad (1.15b)$$

and that of a typical diatomic gas at room temperature is

$$U = \tfrac{5}{2}NkT \qquad (1.16b)$$

where k is Boltzmann's constant introduced earlier. As with the equation of state, we can visualize these results graphically, by plotting a three-dimensional graph of internal energy against pressure and temperature to produce the *UPT* surface of an ideal gas (Figure 1.46). This is simply a plane sloping upwards as temperature increases, but not sloping as P varies, in accordance with Joule's law. The slope is greater for diatomic gases than for monatomic gases.

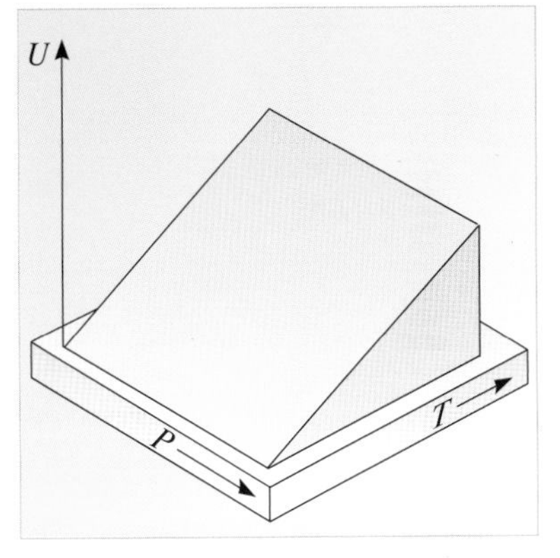

Figure 1.46 The *UPT* surface for an ideal gas.

Absolute zero and kinetic energy

Equations 1.15 and 1.16 suggest that the internal energy of an ideal gas vanishes as the temperature falls to absolute zero. Because intermolecular forces have a negligible effect in an ideal gas, the internal energy contains no contributions from potential energy, and we can say that the kinetic energy of the molecules becomes zero at absolute zero.

This turns out to be a general statement, valid for all phases of matter, throughout classical physics. The lower the temperature falls, the slower the molecules move until, at absolute zero, they stop moving and have zero kinetic energy. This helps to explain why absolute zero is a fundamental limiting temperature. The process of cooling slows the molecules down, but once the molecules have come to rest, no further cooling is possible.

Question 1.10 Oxygen exists usually as a diatomic molecule and has a relative molecular mass of 32. How much energy is needed to increase the temperature of 1.6 kg of oxygen gas from 0 °C to 100 °C?

Question 1.11 Helium is a monatomic gas. Suppose 0.2 kg of helium gas initially has a volume of 3.2×10^{-3} m^3, at a temperature of 0 °C. The gas is then heated at constant pressure and its internal energy increases by 6.5×10^3 J. By how much do the temperature and volume of the gas increase? ■

5.5 A puzzle resolved

We started the chapter with a puzzle, claiming that turning on an electric fire does not increase the internal energy of the air in a room. We can now investigate this statement, to see why it is true, and why the glowing fire nevertheless provides a comforting experience.

The key to the solution is to combine the equation of state with the internal energy equation. Air is mostly nitrogen and oxygen, both diatomic gases, so it is appropriate to use Equation 1.16 to describe the internal energy. Combining Equations 1.12 and 1.16b gives

$$U = \tfrac{5}{2}PV. \tag{1.17}$$

The volume of air in the room is clearly unaffected by lighting the fire (i.e. the volume of the room does not change). The pressure of air in the room is unaffected too, remaining at atmospheric pressure. The pressure remains constant because the room is not completely sealed — tiny gaps underneath doors or between floorboards, for example, allow air to escape, and the pressure to equalize with that outside. We can be sure that this happens because it is easy to open the door of the room from the outside, showing that there is no extra pressure inside the room to press against.

With both the pressure and the volume remaining constant, Equation 1.17 shows that the internal energy of the air in the room remains constant too, as we claimed. However, something has changed. Rearranging Equation 1.12 gives

$$N = \frac{PV}{kT}$$

so, as the temperature increases, the number of gas molecules in the room decreases. This agrees with our description of air escaping through tiny gaps. The average energy per molecule is

$$\frac{U}{N} = \tfrac{5}{2}kT$$

which certainly increases with temperature, but the number of air molecules in the room decreases so the total internal energy remains unchanged. The fire nevertheless does its job, partly because heat is radiated directly in the form of infrared radiation, but also because the molecules that remain in the room have increased average energies, and so are more effective in transferring energy to us.

6 Macroscopic descriptions of the phases of matter

6.1 The generic *PVT* surface and phases of matter

You have seen that a *PVT* surface is useful in visualizing the equation of state of an ideal gas. It is even more useful when considering different phases of matter. You should not, of course, assume that the ideal gas equation of state applies to liquids or solids — there will be quite different equations of state between pressure, temperature and volume for these phases, and the precise relationship will vary from substance to substance. Nevertheless, each phase of each substance will have a definite relationship linking pressure, temperature and volume for a fixed quantity of substance, so we can plot a *PVT* surface encompassing a wide range of temperatures and volumes, beyond any one phase. Each substance has its own characteristic *PVT* surface. Rather than look at any particular case we shall consider a so-called **generic *PVT* surface**, which illustrates the general features expected for a typical substance, and allows us to point out the important features without getting immersed in specific detail.

Figure 1.47 shows a generic *PVT* surface. This surface shows the conditions under which a substance is in a particular phase, or combination of phases. For example,

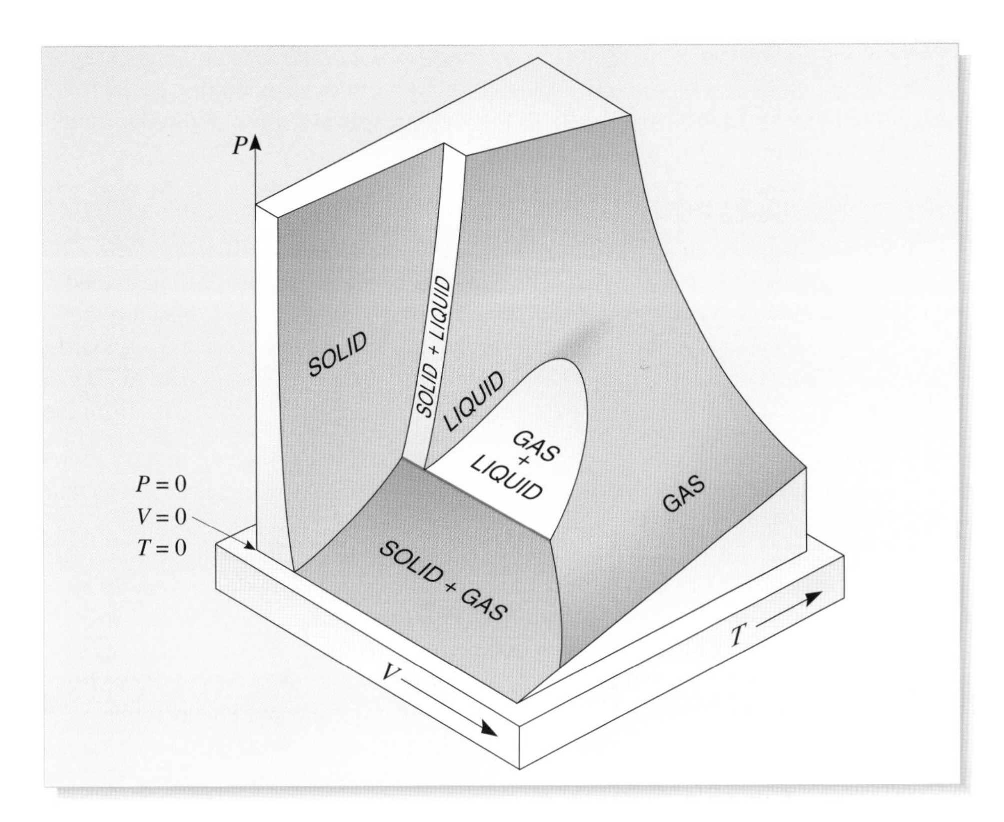

Figure 1.47 A generic *PVT* surface.

when the temperature and volume are high, it is in the gas phase; when the temperature and volume are small, it is solid. This is exactly as you would expect from everyday observations of matter. At intermediate temperatures and volumes, the substance may exist as a liquid, or as a combination of different phases in equilibrium with one another. As you can see from Figure 1.47, each of the possible combinations of phases — solid+liquid, solid+gas, and gas+liquid — can coexist in equilibrium. Along the line marked in red on Figure 1.47, all three phases coexist in equilibrium. This very special state of affairs only occurs at one specific temperature and pressure: in the case of water at 273.16 K and 6.1×10^2 Pa.

One way of exploring the *PVT* surface is to consider a quasi-static process that takes us on a journey across the surface. Figure 1.48 shows a process, drawn in red and marked A_1, A_2, A_3, A_4, that occurs at constant temperature. At the start of the process (point A_1) the system is in the gas phase, at high volume and low pressure. As the system is compressed, the volume decreases and the pressure rises. When the point A_2 is reached, liquid starts to appear, and the system becomes a mixture of two phases — liquid at the bottom of the container and gas above. Further compression can be accomplished without an increase in pressure and this steadily converts the gas into liquid, until at point A_3 the system is entirely in the liquid phase. In the liquid phase, compression is much more difficult, so the pressure has to be increased very substantially in order to reduce the volume a little more to reach point A_4.

Figure 1.48 shows another route between A_1 and A_4, marked in blue, involving a very different experience. Following a route like this, you would notice no abrupt transition between gas and liquid, and there is no intermediate mixture of the two phases. You would just notice the gas getting denser, behaving more and more like a liquid. The route follows a path that requires high temperatures and pressures, so is not very familiar in everyday life, but its existence suggests that liquids and gases are not so different after all, as one phase can merge imperceptibly into the other.

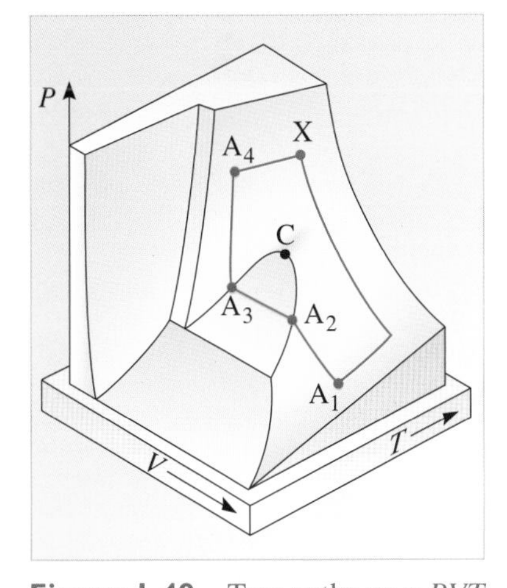

Figure 1.48 Two paths on a *PVT* surface. The red path occurs at constant temperature and involves clear transitions between the gas and liquid phases. The blue path occurs at higher pressures and temperatures, and encounters no sharp phase transitions.

There is no route that passes gradually from a liquid to a crystalline solid, because the long-range order that characterizes a crystal cannot appear gradually. In the liquid phase, this order is absent; and it is the sudden appearance of this order that marks the arrival of a crystalline solid phase.

The point marked C on Figure 1.48 is known as the **critical point**, and its temperature and pressure are called the critical temperature and the critical pressure. For water, the critical temperature and pressure are 647 K and 2.2×10^7 Pa. By convention, the liquid phase occupies the region shown in Figure 1.47, which only extends up to the critical temperature. So, to liquefy a gas you need to cool below the critical point. If the gas is already below the critical temperature (e.g. point A_1 in Figure 1.48), a liquid will be produced by reducing its volume. But if the gas starts out at a point like X in Figure 1.48, further cooling is essential. Before this understanding had been reached, many fruitless attempts were made to liquefy gases like oxygen. The critical temperature of oxygen is 155 K, so no amount of compression at room temperature (300 K) will produce a liquid. The first step in producing liquid oxygen is to cool the gas to below 155 K. The hardest gas to liquefy proved to be helium, which has a critical temperature of 5.2 K. It was not until the early years of the twentieth century that liquid helium was finally produced. Once this was done, a number of amazing low-temperature phenomena were discovered, including superflow (the ability of helium to flow without any viscosity) and superconductivity (the ability of electric currents to flow without any electrical resistance.) On at least seven occasions, the Nobel Prize has been awarded for work that revealed or helped to explain these phenomena and their consequences.

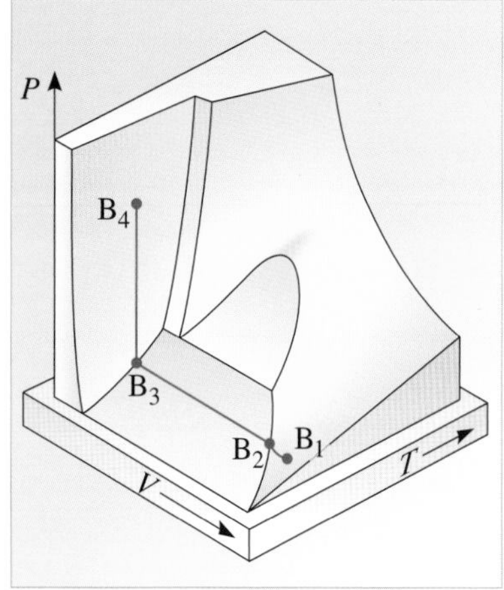

Figure 1.49 A path on a *PVT* surface.

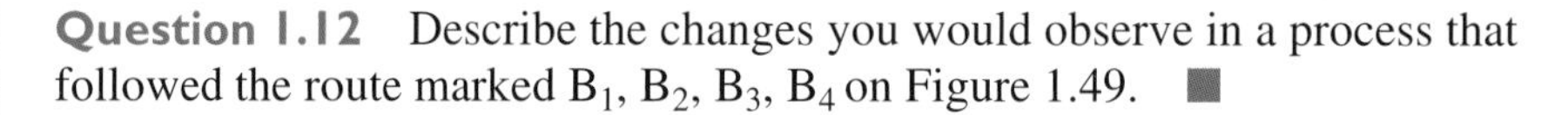

Question 1.12 Describe the changes you would observe in a process that followed the route marked B_1, B_2, B_3, B_4 on Figure 1.49. ■

6.2 The generic *UPT* surface and latent heats

Phase transitions and latent heats

For an ideal gas, a rise in internal energy goes hand-in-hand with a rise in temperature. But this is not always the case. As a substance changes from a phase found at low temperatures to a phase found at higher temperatures, the internal energy can increase, whilst the temperature remains fixed. For example, energy is required to convert ice at 0 °C into water at 0 °C. The energy input does not cause an increase in temperature, and it does not cause molecules to move faster, but it does increase their potential energy, and so helps break down the rigid crystalline structure of ice.

The amount of energy needed to convert *one kilogram* of a substance from one phase to another is called the *specific latent heat* of that phase transition. Similarly, the amount of energy that must be supplied to one mole of a substance to convert it from one phase to another is called the *molar latent heat* for that phase transition. The melting of a solid to a liquid is characterized by a **molar latent heat of melting**, and the vaporization of a liquid to a gas is characterized by a **molar latent heat of vaporization**. Table 1.2 shows the values of these quantities for a number of common substances. In general, the latent heats of vaporization are much larger than those of melting. The values are quoted under normal conditions of temperature and pressure. In fact, the latent heat of vaporization decreases as we approach the critical point.

The latent heat of vaporization of water helps to explain why you feel cold when you step out of a bath or shower, even when your bathroom is warm. As the water on your skin evaporates, energy must be supplied to convert it into gaseous form. This energy comes from the warmth of your body, and its loss explains why you feel cold. When a substance converts back from a high temperature phase to a low temperature phase, the latent heat is released. This explains why steam is so dangerous: as it condenses on your skin, a mole of steam liberates 4×10^4 J of energy, roughly seven times greater than the energy liberated by a mole of liquid water cooling from boiling point to room temperature.

Table 1.2 Latent heats of some common substances.

	Molar latent heat of melting (kJ mol^{-1})	Molar latent heat of vaporization (kJ mol^{-1})
helium	0.02	0.08
oxygen	0.44	6.82
water	5.9	40.7
aluminium	10.7	294
platinum	19.7	511

Ideas about latent heats and phase transitions can be visualized by plotting out a *UPT* surface over a range of pressures and temperatures, covering a number of different phases. Figure 1.50 shows a **generic *UPT* surface**, which illustrates some features expected for a typical substance. The important point to notice is the large vertical steps, which correspond to latent heats.

For example, the red path marked S_1, …, S_6, shows a process that occurs in a closed system at constant pressure. The system begins in the solid phase S_1. At S_2 the solid starts to melt, forming a mixture of solid and liquid. During the melting the temperature remains fixed, but the internal energy increases. At S_3 the system is entirely in the liquid phase. The latent heat of melting is given by the height of the vertical step from S_2 to S_3. The internal energy then continues to rise until vaporization begins at S_4. The system is then in a mixed liquid–gas phase. During vaporization the internal energy increases significantly, but the temperature remains fixed. Finally, at S_5, the system is entirely in the gas phase. The latent heat of vaporization is the height of the vertical step from S_4 to S_5. The internal energy of the gas continues to rise on the way to S_6 as the temperature increases.

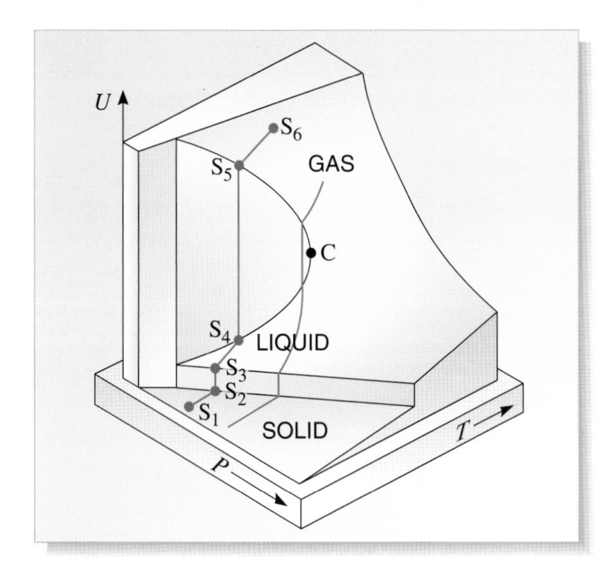

Figure 1.50 A generic *UPT* surface drawn for a fixed quantity of a substance. The red and blue paths show processes that involve phase transitions.

Similar effects are observed along the blue path in Figure 1.50, which is closer to the critical point, C. The latent heat of vaporization, revealed again by the height of the vertical step, is smaller in this case than on the red path, and would be zero for any path that passed through the critical point. At the critical point, the distinction between a liquid and a gas has become meaningless.

7 Closing items

7.1 Chapter summary

The atomic viewpoint (Sections 2–3)

1. Each element has a distinctive type of atom, with a definite number of protons in the nucleus and an equal number of electrons outside the nucleus. Different isotopes have different numbers of neutrons in the nucleus. They have similar chemical and physical properties, but slightly different masses. Images of atoms can be obtained using powerful electron microscopes or scanning tunnelling microscopes.
2. Atoms experience attractive forces when far apart, and repulsive forces when close together. The equilibrium separation of two atoms in a diatomic molecule occurs when the interatomic force vanishes, and the potential energy–separation curve reaches its minimum value. The binding energy of a diatomic molecule is the minimum energy needed to separate two atoms initially at the equilibrium separation.
3. The separation between atoms in a diatomic molecule oscillates simple harmonically with a frequency that depends on the masses of the atoms and the gradient of the force–separation curve at the equilibrium separation.
4. According to Avogadro's hypothesis, equal volumes of different gases, at a given temperature and pressure, contain the same number of molecules. This allows a scale of relative atomic and molecular masses to be devised. The atomic mass unit (amu) is 1/12 the mass of one atom of the carbon-12 isotope. The relative atomic mass of an atom is a dimensionless number formed by dividing the mass of the atom by 1 amu. A similar definition is made for relative molecular mass.
5. The different phases of matter are the result of a competition between kinetic energy and binding energy. Gases are found when the average molecular kinetic energy is much more than 10% of the average molecular binding energy. As the system is cooled, the average molecular kinetic energy falls, and molecules cohere together to form liquid and solid phases.
6. The radial density function measures the average number of molecules per unit volume as we step outwards from a given molecule. This function reveals any correlations that exist between the positions of molecules in a given phase. In the gas phase, the molecules are distributed almost at random. Liquids show some short-range order, extending transiently over a few molecules, but crystalline solids have a long-range order, extending indefinitely over the whole crystal.
7. In a gas, molecules follow zigzag and highly erratic paths. In a solid, molecules oscillate around fixed equilibrium positions. Liquids are an intermediate case: molecules oscillate for a while in the local environment of their neighbours, but they soon move on to join another group of molecules.

The macroscopic viewpoint (Sections 4–6)

8 The density of a sample of matter is $\rho = M/V$, where M is the mass and V is the volume of the sample.

9 The pressure acting on a small element of area A is the perpendicular component of the force pressing down into the area, divided by the area: $P = F_{\perp}/A$.

10 Energy transferred from a warm body to a cool body, as a result of a difference in temperature, is classified as heat. Energy transferred by non-thermal means is classified as work.

11 Temperature is a label that determines the direction of heat flow: heat flows from a body with a higher temperature to a body with a lower temperature, and keeps on flowing until both bodies are at the same temperature.

12 The internal energy of a sample of matter is the same as its total energy, provided the sample is at rest, and is far from external influences, such as gravity.

13 If a pure substance consists of molecules of relative molecular mass M_r, one mole of the substance has mass $M_r \times 10^{-3}$ kg. A mole of any substance contains Avogadro's number (6.022×10^{-23}) of molecules.

14 Response functions tell us how rapidly one macroscopic quantity varies in response to changes in another. Examples include the isobaric expansivity and the isothermal compressibility.

15 The lowest conceivable temperature is −273.15 °C. This is known as the absolute zero of temperature. According to classical physics, molecules stop moving at absolute zero, and so have zero kinetic energy. On the absolute temperature scale, absolute zero is taken to be zero kelvin (0 K).

16 An ideal gas is one in which the molecular interactions play a negligible role. Most gases under normal conditions are approximately ideal. An ideal gas obeys the ideal gas equation of state:

$$PV = nRT$$

where P is the pressure, V the volume, n the number of moles, T is the absolute temperature and R is the universal gas constant. A useful alternative form of this equation is

$$PV = NkT$$

where N is the number of molecules and k is Boltzmann's constant. The ideal gas equation incorporates Boyle's law ($P \propto 1/V$ at constant n and T), Charles's law ($V \propto T$ at constant n and P) and Avogadro's hypothesis.

17 The internal energy of an ideal gas is independent of pressure and volume, and depends only on the number of moles and the temperature. This is Joule's law of ideal gases. The internal energy of any monatomic gas is

$$U = \tfrac{3}{2}nRT = \tfrac{3}{2}NkT.$$

The internal energy of a diatomic gas is a more complicated function of temperature, but close to room temperature typical diatomic gases obey

$$U = \tfrac{5}{2}nRT = \tfrac{5}{2}NkT.$$

18 The equations of state of the phases of a substance can be visualized by plotting the equilibrium values of P, V and T as points in three-dimensional space. The set of all such points forms the PVT surface of the substance. The PVT surface can be divided into regions corresponding to different phases or combinations of phases. Liquids only exist below a critical temperature so, to liquefy a gas, it is necessary to cool to below the critical temperature.

19 A quasi-static process is one that occurs so gently that the system can be thought of as being in equilibrium at each instant. Quasi-static processes can be represented by paths that lie on the *PVT* surface.

20 Energy is generally needed to convert a phase found at low temperatures to a different phase. During the phase transition, the energy input does not cause a rise in temperature, but is associated with the change in phase, which takes place at a fixed temperature. The amount of energy needed to melt a solid is called the latent heat of melting, and that needed to vaporize a liquid is the latent heat of vaporization. Latent heats appear as abrupt steps on a *UPT* surface.

7.2 Achievements

Now that you have completed this chapter, you should be able to:

A1 Understand the meaning of all the newly defined (emboldened) terms introduced in this chapter.

A2 Use the terms atom, molecule, isotope, ion, atomic number and mass number correctly.

A3 Use force–separation and potential energy–separation curves to interpret interactions between atoms, and to calculate the frequency of vibration and binding energy of a diatomic molecule.

A4 Describe the gas, liquid and solid phases of matter, from macroscopic and microscopic viewpoints.

A5 Explain the meaning of a radial density function.

A6 Define and use the terms density and pressure.

A7 Define internal energy, heat and work, and interpret temperature as a label that determines the direction of heat flow.

A8 Define and use the concept of a mole, and molar quantities.

A9 Define the concept of an ideal gas.

A10 State the equation of state of an ideal gas and recognize various alternative formulations. Describe the experimental origins of this equation and use it to calculate the equilibrium properties of ideal gases.

A11 State Joule's law for ideal gases. Recall and use internal energy equations for monatomic and diatomic gases.

A12 Interpret the changes that are observed in processes defined by paths on *PVT* or *UPT* surfaces.

A13 Describe the role of latent heats in phase transitions between solids, liquids and gases.

7.3 End-of-chapter questions

Question 1.13 Distinguish between gas, liquid and crystalline solids in terms of typical features in (a) the radial density function and (b) the motion of molecules.

Question 1.14 An empty room contains 30 cubic metres of air, at normal room temperature and atmospheric pressure. About 20% of the molecules in the air are diatomic oxygen molecules. Estimate the total number of oxygen atoms in the air of the room.

Question 1.15 In the summer, at 30 °C, a tyre is filled with air at an overpressure of 1.9×10^5 Pa (i.e. the pressure in the tyre is 1.9×10^5 Pa greater than atmospheric pressure (1.0×10^5 Pa). Estimate the overpressure of air in the tyre in the middle of winter, when the temperature is −20 °C. You may assume that no air escapes from the tyre and that the volume of the tyre remains constant.

Question 1.16 1.5 kg of nitrogen gas is initially at atmospheric pressure and room temperature. How much energy must be supplied to this sample to double its volume at constant pressure? (Nitrogen is a diatomic gas, with relative molecular mass 28.)

Question 1.17 Fifteen moles of a monatomic ideal gas are at a pressure of 1.8×10^5 Pa in a container with a volume of 0.25 m^3.

(a) What is the temperature of the gas?

(b) What is the average kinetic energy of a gas molecule?

Question 1.18 (a) Describe the changes you would observe in a process that followed a constant pressure route from point B_4 on Figure 1.49 to point A_4 on Figure 1.48. (b) How could the isobaric expansivity of the solid phase be deduced from the shape of the *PVT* surface? ■

Chapter 2 Microscopic models of gases

1 Introduction

> 'In truth there is nothing but atoms and empty space; all other things are merely opinion.'
>
> Democritus of Abdera, *c.* 420 BC.

Chapter 1 has already explored some of the properties of gases. You saw that both microscopic and macroscopic descriptions can be given.

From a *microscopic* viewpoint, the picture that emerges is one of independent molecules, a long way apart on average, moving around freely. The molecules follow highly irregular zigzag paths, changing their directions and speeds abruptly when they collide, but spending most of their time as freely moving particles, beyond the range of intermolecular forces.

From a *macroscopic* viewpoint, the picture that emerges can be summarized by a handful of laws that describe patterns of behaviour observed in equilibrium. These laws are neatly summarized by equations, such as

$$PV = NkT \qquad \text{(ideal gas equation of state)}$$

$$U = \tfrac{3}{2}NkT \qquad \text{(internal energy equation for a monatomic gas)}$$

$$U = \tfrac{5}{2}NkT. \qquad \text{(internal energy equation for a typical diatomic gas)}$$

On the one hand, we have a powerful image of what a gas *is*, in microscopic terms; on the other hand, we can state a number of laws that gases obey in practice. This is clearly not a satisfactory way to leave things. What is needed is a link between our microscopic image of a gas and the observed macroscopic properties. We must try to understand the whole gas in terms of the behaviour of its constituent molecules. This chapter will provide that link, by introducing a subject known as statistical mechanics, which is one of the cornerstones of science.

The basis of statistical mechanics was set out by Ludwig Boltzmann (1844–1906). Unfortunately, his ideas were poorly received, partly because of their originality and partly because of the tedious detail, obscurity and complexity of his publications. Around 1910, Einstein remarked, 'Boltzmann's work is not easy to read. There are great physicists who have not understood it.' That included Maxwell, noted for writing concise papers, who said '[Boltzmann] could not understand me on account of my shortness, and his length was and is an equal stumbling block to me.' Depressed by the lack of appreciation for his work, Boltzmann committed suicide while on holiday, just as his theories were on the threshold of acceptance.

Ludwig Boltzmann (1844–1906)

Boltzmann was born in Vienna in 1844, the son of a tax official. He moved around frequently in his career, holding professorships in Graz (twice), Vienna (four times), Munich and Leipzig. He attributed his restless nature to being born during a Mardi Gras ball. Although Boltzmann's papers were renowned for their difficulty, he was remembered by his students as a wonderful teacher. He was fond of using startling phrases such as 'gigantically small', and his lectures were full of scintillating anecdotes. Lise Meitner (who later discovered atomic fission) recalled that 'After each lecture it seemed as if we had been introduced to a new and wonderful world, such was the enthusiasm that he put into what he taught.' Another student said 'He never exhibited his superiority. The conversation took place quietly and the student was treated as a peer. Only later one realized how much had been learned from him.' Boltzmann's philosophy lectures became so famous that it was impossible to find a lecture hall large enough to hold the audience.

Another aspect of Boltzmann that is not always appreciated is his interest in inventions and technology. He argued that technology was the intellectual equal of science, and took a keen interest in modern developments. One of his convictions was that the aeroplane would prove to be superior to the airship.

Figure 2.1 Ludwig Boltzmann (1844–1906).

Statistical mechanics rests on concepts of chance and probability which had been developed by mathematicians from the seventeenth century onwards. Rather unusually in this course, we will begin by discussing these mathematical ideas, in order to clear the way for later discussions of the physics of matter. The mathematics required is not difficult, but the concept of probability is a subtle one, so it is worth checking that you understand it in the context of simple cases, like tossing coins or rolling dice, before attempting to use similar ideas in the context of atoms and molecules.

2 Chance and probability

> 'The true logic of this world lies in the calculus of probabilities.'
>
> James Clerk Maxwell.

2.1 Probability

We live in an uncertain world, and are obliged to make judgements about uncertain events. Very often, we might know that a number of outcomes are possible, but be unable to predict which one will actually happen. The whole economy of gambling is based on this fact. In the face of uncertainty, we are invited to make good use of our knowledge to increase our chances of winning.

We concentrate here on simple cases, where all the possible outcomes are judged to be equally likely. For example, when a coin is tossed, the possible outcomes are heads or tails, and there is no reason to expect that one is more likely than the other. When a die is rolled we could score '1', '2', '3', '4', '5' or '6', and would expect all of these outcomes to be equally likely (the die being taken as fair and unloaded).

Die is the singular of dice.

The **probability** of a given outcome is defined to be the fraction of times that outcome is expected to happen in the long run. For example, the probability of tossing heads is 1/2 and the probability of tossing tails is also 1/2. In general, a probability of 0 corresponds to impossibility, and a probability of 1 corresponds to inevitability. The interesting area lies between these two extremes, where outcomes are uncertain. The closer the probability is to 1, the more likely it is to happen.

● What is the probability of rolling '4' with a single unloaded die?

❍ A score of '4' is one of six equally-likely outcomes, so is expected to happen one time out of six, in the long run. Its probability is therefore 1/6. ■

For a coin there are two alternative outcomes, each with probability 1/2. For a die there are six alternative outcomes, each with probability 1/6. In both cases, the sum of all the probabilities is 1. This can be expressed as a completely general rule:

The normalization rule for probabilities

The sum of the probabilities of all the alternative outcomes is equal to 1.

The concept of probability is a theoretical one, based on our best estimate of the chances of something happening. Strictly speaking, no experiment measures a probability: what is measured is the fraction of times an outcome occurs in a *finite* set of attempts: this is known as the **fractional frequency** of the outcome. In the long run, the measured fractional frequency is expected to approach the theoretical probability. Nevertheless, when dealing with uncertain outcomes there can be no cast-iron guarantees. You could toss a fair coin 100 times and get heads on every single toss.

There are many misconceptions about rare events. Some people are reluctant to accept that they happen by chance, and look for rational, or irrational, 'explanations'. Sometimes this will be appropriate — perhaps the coin was double-headed — but it is important to realize that unusual things can *and do* happen entirely by chance — they are just very unlikely. To explain a coincidence as being due to chance may sound feeble, but so many things are happening around us that it is wise to remember that chance will turn up unlikely events all the time. There is a fundamental difference between noticing that *some* unlikely event has happened, and predicting that *a given* unlikely event will happen.

One common misconception is related to the (correct) expectation that everything should even out in the long run. Knowing this, many people believe that chance must actively conspire to help bring this about. If the first 100 tosses of a coin have produced 60 heads and 40 tails, they believe that they are owed a surplus of tails, to balance things out. This is not true. A coin has no memory so, on each toss, heads and tails are equally likely, irrespective of any previous history. Nevertheless, in an *extremely* long run, the imbalance in heads and tails is expected to become negligible. For example, in the next two million tosses we would expect there to be about a million heads and about a million tails, and these numbers will swamp the 60–40 imbalance of the first 100 tosses.

So far, we have looked at the probabilities of individual events, such as those associated with tossing a single coin or rolling a single die. Pioneering mathematicians, Pierre de Fermat (1601–1665) and Blaise Pascal (1623–1662), realized that such probabilities provide the raw material for tackling a much wider range of questions. For example, suppose we toss a coin *and* roll a die. What is the probability of tossing heads *and* rolling '4'? The answer is found by *multiplying* the probability of getting heads by the probability of getting '4' to obtain

$$\frac{1}{2} \times \frac{1}{6} = \frac{1}{12}.$$

More generally, we have:

The multiplication rule for probabilities

If a number of outcomes occur independently of one another, the probability of them all happening together is found by *multiplying* their individual probabilities.

For example, if you roll nine dice, the probability of scoring '6' on every one of the dice is

$$\tfrac{1}{6}\times\tfrac{1}{6}\times\tfrac{1}{6}\times\tfrac{1}{6}\times\tfrac{1}{6}\times\tfrac{1}{6}\times\tfrac{1}{6}\times\tfrac{1}{6}\times\tfrac{1}{6}=\tfrac{1}{6^9}\approx 10^{-7}\,.$$

Well, I suppose you knew it would be unlikely! If everyone in the country had one go at this, perhaps a handful would succeed: the odds are therefore comparable to those for winning the National Lottery jackpot.

Probabilities can also be added together. This is appropriate when there are a number of alternative (mutually exclusive) outcomes, such as rolling either a '4' or '5' on a single throw of a die.

The addition rule for probabilities

If a number of alternative outcomes are mutually exclusive, the probability of getting *one or other* of these outcomes is found by *adding* their individual probabilities.

For example, the probability of rolling a '4' or '5' on a die is

$$\tfrac{1}{6}+\tfrac{1}{6}=\tfrac{1}{3}.$$

The normalization rule, quoted earlier, can therefore be interpreted as expressing the obvious fact that we are *certain* to get one or other of the possible outcomes.

2.2 Average values

In a dictionary, the word 'average' is said to mean typical or normal. This conveys a fair impression, but is too vague for scientific purposes. Let's consider the familiar example of rolling a die, and let n be the score obtained. We know that the possible scores are 1, 2, 3, 4, 5 or 6, each of which has a probability of 1/6. The predicted average score is given the symbol $\langle n\rangle$ and is obtained by multiplying each possible value by its probability and then adding together the results:

$$\langle n\rangle=\tfrac{1}{6}\times 1+\tfrac{1}{6}\times 2+\tfrac{1}{6}\times 3+\tfrac{1}{6}\times 4+\tfrac{1}{6}\times 5+\tfrac{1}{6}\times 6=3.5\,. \qquad (2.1)$$

More generally, if a quantity x has possible values x_1, x_2, … x_N, with probabilities p_1, p_2, … p_N, its **predicted average value** is defined to be

$$\langle x\rangle = p_1x_1+p_2x_2+\ldots p_Nx_N\,. \qquad (2.2a)$$

This can also be written in a more abbreviated form as

$$\langle x\rangle=\sum_{i=1}^{N}p_ix_i \qquad (2.2b)$$

but both these formulae carry the same message: the average value is found by multiplying the possible values by their corresponding probabilities and then adding together all the terms.

We can also define the **measured average value** by multiplying each possible value by its *fractional frequency* and adding together all the terms. In the long run, the fractional frequencies are expected to approach the corresponding probabilities, so the measured average value is expected to approach the predicted average value. Often, we will simply use the term **average value**, and allow the context to show whether this average is based on theoretical probabilities or measured fractional frequencies.

As a further example, let's calculate the average value of n^2, the *square* of the score on the die. The possible values of n^2 are 1, 4, 9, 16, 25 and 36, each of which has the same probability of 1/6. So Equation 2.2 gives

$$\langle n^2 \rangle = \tfrac{1}{6}\times 1 + \tfrac{1}{6}\times 4 + \tfrac{1}{6}\times 9 + \tfrac{1}{6}\times 16 + \tfrac{1}{6}\times 25 + \tfrac{1}{6}\times 36 = 15.2. \qquad (2.3)$$

One point may surprise you. Squaring Equation 2.1 gives $\langle n \rangle^2 = 12.25$, which is clearly *not* the same as $\langle n^2 \rangle$.

In general, the square of the average of a quantity is *not* the same as the average of the square of the quantity.

An illustration of this fact is given by the average velocity of pedestrians walking along a straight pavement, with roughly equal numbers walking in either direction (Figure 2.2). If we choose our x-axis to point along the pavement, about half the pedestrians will have positive values of v_x and about half will have negative values of v_x. This means that $\langle v_x \rangle$ and therefore $\langle v_x \rangle^2$ will be very small. By contrast, v_x^2 is positive for all the pedestrians, so $\langle v_x^2 \rangle$ will be much larger than $\langle v_x \rangle^2$.

Figure 2.2 Pedestrians walking along a pavement.

These examples show that some care is needed when using averages. In particular, it matters *when* the average is taken (before or after taking the square, in the above example). Beginners often take averages at a very early stage of their calculations. In the case of molecules moving around in a gas, for example, it is tempting to think of a 'typical' molecule, moving with the average speed. While this is acceptable for rough estimates, it is not really recommended. It is better to consider an *arbitrary* molecule, moving with an *arbitrary* speed, follow the calculation through to the end, and then take averages at the last possible moment.

In spite of this caution, many operations involving averages are quite straightforward. For example, we know that the translational energy of a molecule is

$$E_{\text{trans}} = \tfrac{1}{2}mv^2 .$$

As the molecule moves around and collides with others, its speed v varies, but its mass m remains constant. If we now consider the average translational energy, we can write:

$$\langle E_{\text{trans}} \rangle = \langle \tfrac{1}{2} m v^2 \rangle = \tfrac{1}{2} m \langle v^2 \rangle. \tag{2.4}$$

There is no problem in taking the constant $(\frac{1}{2}m)$ outside the average. Similarly, we know that the square of the speed can be written as

$$v^2 = v_x^2 + v_y^2 + v_z^2$$

and we can write

$$\langle v^2 \rangle = \langle v_x^2 + v_y^2 + v_z^2 \rangle = \langle v_x^2 \rangle + \langle v_y^2 \rangle + \langle v_z^2 \rangle. \tag{2.5}$$

There is no problem in treating the average of a sum as a sum of averages. What *cannot* be justified is replacing $\langle v_x^2 \rangle$ by $\langle v_x \rangle^2$, as we emphasized above.

It is sometimes convenient to introduce the *root mean square speed* of a particle, i.e. v_{rms}. This is defined in such a way that

$$v_{\text{rms}} = \sqrt{\langle v^2 \rangle}. \tag{2.6}$$

The root mean square speed is not the same as the average speed but, from Equation 2.4, it is related very simply to the average translational energy:

$$\langle E_{\text{trans}} \rangle = \tfrac{1}{2} m v_{\text{rms}}^2. \tag{2.7}$$

2.3 Fluctuations

One of the general ideas we shall need is that of a random fluctuation. For example, imagine repeatedly rolling an ordinary die. You have seen that the probability of rolling a '4' is 1/6, so if you rolled the die 60 times, you would expect to score '4' on about 10 occasions. If you rolled the die 6000 times, you would expect to score '4' on about 1000 occasions. These statements are deliberately expressed in rather guarded terms. When dealing with probabilities, there can be no cast-iron guarantees. With 6000 rolls of the die, the best estimate is that '4' will be obtained on 1000 occasions, but I would not be too surprised if you scored '4' on 980 occasions, since this is pretty close to 1000. By contrast, it would be astonishing if you scored '4' on 1400 occasions, since this would *not* be close to 1000.

It might strike you that this is dangerously vague, because it is not at all clear what criteria are being used: how close is close enough? A useful rule can be used to judge this.

The square root rule

Consider a random process which is repeated in exactly the same way a large number of times. Suppose that our best estimate (based on a rational assessment of the chances) is that an outcome O will occur N times. Then:

- We should not be surprised if O occurred somewhere between $N - \sqrt{N}$ *and* $N + \sqrt{N}$ times.
- We should be *astonished* if O occurred less than $N - 10\sqrt{N}$ times or more than $N + 10\sqrt{N}$ times. (In a typical case, there is a better chance of winning the National Lottery jackpot *three weeks in a row*!)

- Use the square root rule to justify the claim made above that, in 6000 rolls of a fair die, it would not be surprising to score '4' on 980 occasions, but it would be astonishing to score '4' on 1400 occasions.

- ❍ Since all six scores are judged equally likely, our best estimate for the number of times '4' is scored is $N = 6000/6 = 1000$. The square root of N is 32. Since 980 lies comfortably within the expected range from $N - \sqrt{N} = 968$ to $N + \sqrt{N} = 1032$, it is not at all surprising. By contrast, 1400 is greater than $N + 10\sqrt{N} = 1320$, so such a result would be astonishing. ■

2.4 Dicing with Boltzmann

I shall now describe an effect that can be observed using ordinary dice. So far as I know, it does not correspond to any popular game played with dice, so it may seem rather contrived. Nevertheless, the situation described reveals one aspect of probability that is essential for understanding the physics of matter. I do not know if Boltzmann played dice, but he would probably have enjoyed the game suggested below, because it rehearses a key argument that justifies statistical mechanics. We won't make any such connections yet, but will just concentrate on revealing an interesting phenomenon.

Suppose you have a set of four ordinary dice labelled A, B, C and D so that they can be distinguished from one another. If you roll all four dice together and add their scores, you could get any total between 4 (four '1's) and 24 (four '6's). Now imagine that you repeatedly roll the dice, and that you keep a record of those rolls for which the total score is 9. Each time a total score of 9 is obtained you note down the scores on the dice, but if the total score is not 9, the results are ignored and not recorded.

Please do not try rolling dice for yourself: it would take a ridiculously long time, probably drive you to distraction, and is quite unnecessary. It is tedious but not very difficult to list every combination of scores on the four dice that add up to a total of 9. This is done in Table 2.1, which shows that there are 56 combinations in all. Naturally, all of these combinations are equally likely.

Now, concentrate on the scores on just one of the dice. Any one of the four dice could be chosen, but we will choose die A. Table 2.1 shows that die A scores '1' in 21 combinations, '2' in 15 combinations, '3' in 10 combinations, '4' in six combinations, '5' in three combinations and '6' in just one combination. Since all 56 combinations are equally likely, we conclude that the probabilities of scoring '1', '2', '3', '4', '5' and '6' with die A are 21/56, 15/56, 10/56, 6/56, 3/56 and 1/56 respectively. So, there is a dramatic decrease in probability as the score on the given die rises. Let's be clear about this. If there were no restriction on the total score, and we considered every conceivable combination, individual scores of '1' and '6' would be equally likely. It is the restriction of the total score to a fixed value of 9 that tips the balance in favour of lower scores.

The probabilities we have just calculated are plotted as points in Figure 2.3, while the line in this figure is a curve representing an exponential decrease: notice that it provides a reasonable, though by no means exact, fit to the calculated probabilities. This observation can be pursued further by considering even more dice. It is then more difficult to list all the combinations, but a computer can do most of the work.

Table 2.1 All the ways in which the scores on four dice A, B, C and D can add up to a total of 9. There are 56 different combinations. In 21 combinations die A has score '1'; in 15 combinations it has score '2', in 10 combinations score '3', in six combinations score '4', in three combinations score '5', and in just one combination it has score '6'. Similar results apply to each of the other dice, but entries have been ordered in such a way that it is easier to read off this information for die A than for the other dice.

A	B	C	D
1	1	1	6
1	1	2	5
1	1	3	4
1	1	4	3
1	1	5	2
1	1	6	1
1	2	1	5
1	2	2	4
1	2	3	3
1	2	4	2
1	2	5	1
1	3	1	4
1	3	2	3
1	3	3	2
1	3	4	1
1	4	1	3
1	4	2	2
1	4	3	1
1	5	1	2
1	5	2	1
1	6	1	1
2	1	1	5
2	1	2	4
2	1	3	3
2	1	4	2
2	1	5	1
2	2	1	4
2	2	2	3
2	2	3	2
2	2	4	1
2	3	1	3
2	3	2	2
2	3	3	1
2	4	1	2
2	4	2	1
2	5	1	1
3	1	1	4
3	1	2	3
3	1	3	2
3	1	4	1
3	2	1	3
3	2	2	2
3	2	3	1
3	3	1	2
3	3	2	1
3	4	1	1
4	1	1	3
4	1	2	2
4	1	3	1
4	2	1	2
4	2	2	1
4	3	1	1
5	1	1	2
5	1	2	1
5	2	1	1
6	1	1	1

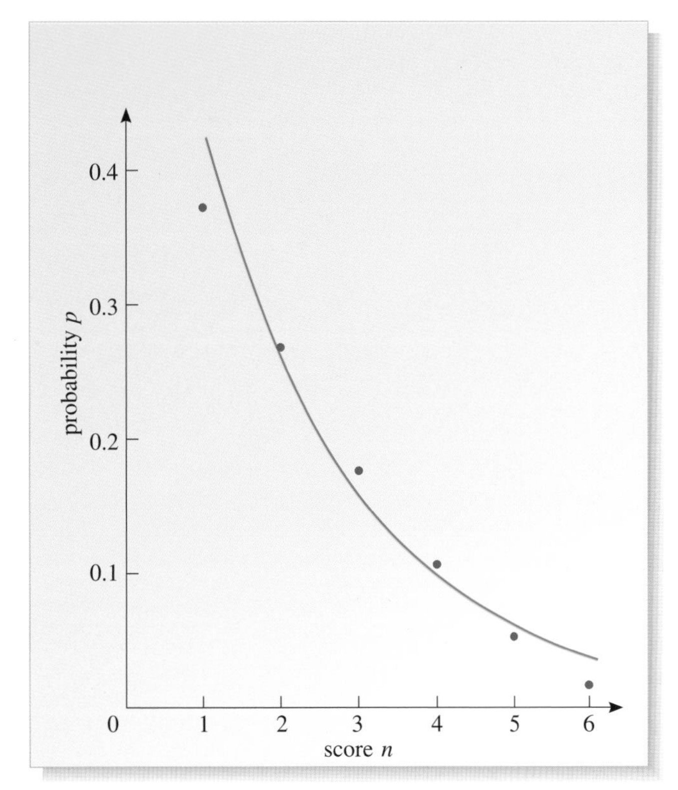

Figure 2.3 The probabilities of scores on a single die when it is rolled with three others and the total score on all four dice is restricted to 9.

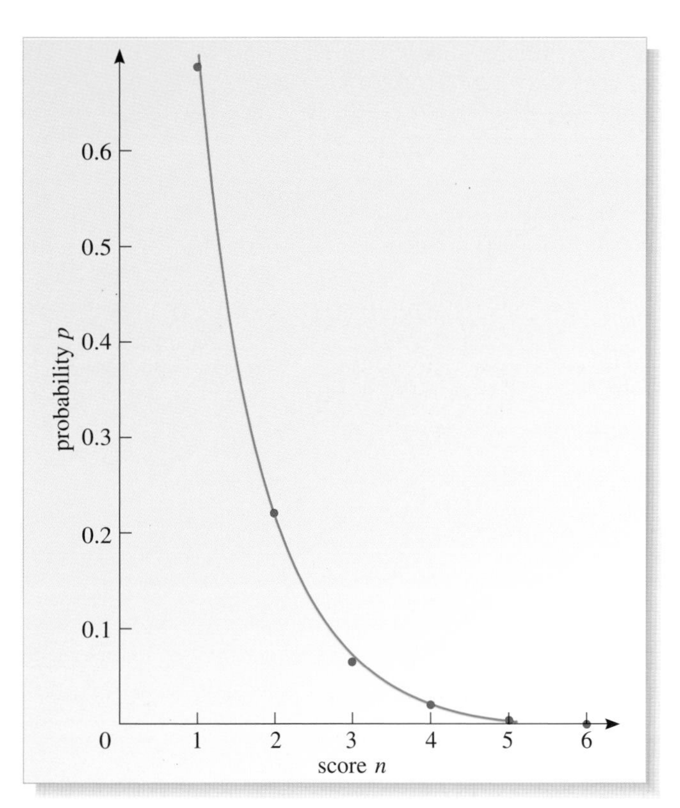

Figure 2.4 The probabilities of scores on a single die when it is rolled with 20 others and the total score on all 21 dice is restricted to 30.

Figure 2.4 is similar to Figure 2.3, but applies to a set of 21 dice, with the total score restricted to 30. The line represents an exponential decrease, and this time you can see that it fits the points very well. To summarize:

- If any number of dice are rolled with no restrictions in place, the probabilities of scoring '1', '2', '3', '4', '5' and '6' on a given die are all the same (1/6). If the total score on the dice is restricted to a fixed value, these probabilities can be strongly modified.
- You have seen examples where the probability decreases rapidly as the score on a given die increases. When a large number of dice are used, and the total score is not too high, this decrease is closely approximated by an exponential decrease.

Question 2.1 What is the predicted average score for die A in the game involving four dice described above, summarized in Table 2.1 and Figure 2.3?

Question 2.2 We are told that, if enough monkeys, with enough typewriters, hit the keys at random, they will eventually produce the entire works of Shakespeare. Never mind the entire works, let's just take the sentence: 'To be or not to be, that is the question.' Ignoring capital letters and punctuation but including spaces between words, this quote is made up of 39 independent outcomes, each of which could take any one of 27 values (26 letters or a space). (a) What is the probability that a single monkey will type this sentence with its first 39 keystrokes? (b) If a billion monkeys had each been typing 39 symbols every 10 seconds for the entire age of the Universe (approximately 15 billion years), would they have had a reasonable chance of succeeding by now? (*Note:* 1 billion = 10^9.) ■

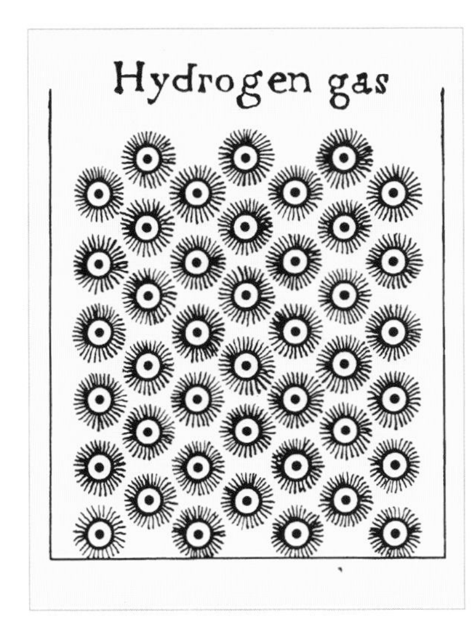

Figure 2.5 Dalton's picture of a gas, as published in his book *A New System of Chemical Philosophy*. The picture is misleading because it shows a regular arrangement of closely spaced atoms, and does not indicate their motion.

3 The pressure of a gas

3.1 Molecular bombardment

What is the cause of pressure in a gas? Had you asked this question in the year 1700 or even 1800, you would probably have received a rather strange answer. Pascal compared the atmosphere to 'a great bulk of wool, say 20 or 30 fathoms high, compressed by its own weight' and Boyle believed a gas was 'a heap of little bodies, lying one upon another, as may be resembled to a fleece of wool'. He discussed how the tiny gas particles were 'springy' like balls of wool, and could be squashed together by increased gas pressure in such a way that

$$PV = \text{constant.} \qquad \text{(Boyle's law)}$$

Even in 1810, Dalton was sketching pictures of a gas like that shown in Figure 2.5.

All these models were essentially static, and missed out the important idea that the molecules in a gas are in ceaseless and rapid motion. Theories which emphasize the importance of molecular motion are called kinetic theories. It was Daniel Bernoulli (1700–1782) (Figure 2.6) who developed the first convincing **kinetic theory of gases**. In 1738, he proposed that a gas consists of a large number of rapidly *moving* particles, which cause pressure by bombarding the walls of their container.

Each time a molecule rebounds from a wall, its momentum changes; it therefore experiences a force exerted by the wall. By Newton's third law, the molecule must exert an equal and opposite force on the wall. Corresponding to this force, there is a pressure. The pressure exerted by the gas is the result of the vast number of microscopic impacts. So many molecules bombard the wall that any fluctuations are unnoticed. The pressure, as detected on a macroscopic scale, therefore appears to be steady, in spite of the fact that the molecules arrive at the wall at random times, travel with different speeds, and approach from different directions. It is rather like rain falling on a tin roof so rapidly that all you experience is a steady roar.

Bernoulli used this model to derive Boyle's law, and also suggested that the dependence of pressure on temperature would be explained if the molecules moved more rapidly at higher temperatures. This was a most remarkable achievement, but Bernoulli achieved fame for other work rather than this. His first clear statement of the kinetic theory was so far ahead of his contemporaries that it was ignored and forgotten for over a century. It is almost as unfortunate to have an idea too soon as too late! (Eventually, Bernoulli's ideas were rediscovered by Joule, Maxwell and Waterston, of whom we shall have more to say later.)

Figure 2.6 Daniel Bernoulli (1700–1782) was one of a family of eminent Swiss mathematicians. He became interested in the properties of fluids and, in 1738, published his book *Hydrodynamica*. This included an explanation for the elastic properties of air, based on the existence of 'extremely small bodies agitated in very rapid random motion'. At the time, there was no understanding of the molecular composition of matter and this aspect of his work was largely ignored.

3.2 The simple gas model

The basic idea that the pressure in a gas is due to the bombardment of molecules is straightforward enough but is really only a starting point. In physics, it is regarded as essential to combine such an idea with the laws of physics, to deduce quantitative consequences, and then compare them with experiment.

Many books refer to the simple gas model as the ideal gas model. This can lead to confusion about what an ideal gas really is. Our preference is to use a different name for the model, and to assess later the extent to which the simple gas model reproduces the behaviour of an ideal gas.

In order to build a theory, we need to have a model. We will therefore begin by listing our simplifying assumptions about the behaviour of molecules in a gas, thereby defining the **simple gas model**, which we use throughout this section. Some of the assumptions may appear a bit unrealistic at first sight, but in fact the simple gas model provides a good description of ideal gases, and a good approximation to most real gases under typical conditions.

Assumptions of the simple gas model

1. The molecules are treated as structureless particles.
2. The molecules exert no forces on one another unless they collide.
3. Newton's laws of motion govern molecular dynamics, but gravity has a negligible effect.
4. All molecular collisions are elastic (this includes molecule–molecule collisions and collisions between molecules and the walls of the container).
5. The molecules are in ceaseless random motion.

From the outset, it is important to realize the spirit in which the simple gas model is proposed. Remember, above all, that it is a model. That is to say, it is a simplification of reality made *knowingly and deliberately* so that further analysis will be possible. In physics, a model is always simpler than the real world — it leaves out many messy details that would block further analysis. The secret of a good model is that it must retain the *essence* of the true situation, whilst ignoring all irrelevant, and troublesome, complications.

The rationale of using a model is sometimes misunderstood. Some people have been known to read the assumptions listed above, complain that molecules are not structureless particles, and then reject the whole model as being *wrong*. Such a criticism is wide of the mark. We all know that molecules have some finite size, but that is not the point. The model does not claim to be an exact summary of the real world. It admits to simplifying reality, but hopes to do so in a way that does not significantly affect the final conclusions that will be drawn from it — in our case, an expression for the pressure of a gas.

You might be suspicious about this process. What if one of the details neglected by the model turned out to be important after all? Well, that would mean that the model was not a good one. It would then disagree with observation or experiment, and be rejected: we would go back to the drawing board and devise a new model that did a better job. The process of obtaining a good model can be a protracted business but, once found, it is a precious thing. Many physicists have made their reputations by proposing new models which in spite of, *and because of*, their simplifications, have achieved their objective of deepening our understanding of the physical world. The simple gas model turns out to be a very good one indeed. Before using it, let's take a closer look at the assumptions that it makes.

Assumptions 1 and 2 reflect our discussion of the gas phase in Chapter 1. In air, at atmospheric pressure and room temperature, the average spacing between molecules is around 3×10^{-9} m. This is an order of magnitude larger than the size of a small molecule, and several times larger than the typical range of intermolecular forces, so Assumptions 1 and 2 seem reasonable as a starting point.

Assumption 3 underlines our decision to use Newtonian mechanics to analyse the colliding molecules. What else could we do? Well, we might suppose that quantum mechanics, or relativity, or some other theory would give a better description, but such complications need not detain us. As the success of the model will show, Newtonian mechanics is accurate enough for all present purposes.

You might wonder about neglecting gravity. It is true that molecules experience a downward gravitational force and therefore accelerate downwards between collisions (at $9.8\,\mathrm{m\,s^{-2}}$). But the molecules are moving very rapidly — at hundreds or even thousands of metres per second — so this acceleration causes no significant deviation in their trajectories: between collisions, the molecules travel along essentially straight-line paths. You might think that the force of gravity would cause the gas to settle near the bottom of its container, but this is a small effect too, and can be neglected provided the container is not too large. (The density of air is practically the same at floor and ceiling levels in a room, although it is larger at sea-level than on top of Mount Everest, as you will see in Chapter 4.)

Question 2.3 At room temperature, nitrogen molecules typically move with speeds around $500\,\mathrm{m\,s^{-1}}$. Suppose a nitrogen molecule in a vacuum system at very low pressure sets off horizontally at this speed and travels right across the chamber for a horizontal distance of 0.5 m. Calculate the vertical displacement of the molecule during this movement due to gravity. Is your result in agreement with Assumption 3? Is ignoring gravity a reasonable assumption for the gas in the air around us at atmospheric pressure? ■

Assumptions 4 and 5 The idea that the collisions of molecules are elastic is central in our analysis. It means that translational energy is conserved in all collisions and molecules neither break up nor stick together on collision. So, while collisions may (and generally do) redistribute translational energy between molecules, they cannot change the total translational energy of the gas.

As a result of collisions, it is reasonable to expect the molecules to acquire a *well-defined distribution* of translational energy (and also of speed). In equilibrium, the same distribution will be found in all parts of the gas so, while molecule–molecule collisions may change *which* molecules have a given energy, they will *not* change the overall pattern. This is very convenient because it means that molecule–molecule collisions have no influence on the pressure. For simplicity, we ignore them altogether, and imagine molecules drifting undisturbed from one side of the container to the other.

To take account of the spread in energies and speeds, we will consider averaged quantities. In fact, we calculate the *average* pressure (averaged over many molecular bombardments) and relate this to the *average* translational energy of the molecules.

3.3 Calculating the pressure

Having listed the assumptions that define our model of a simple gas, we can now put the model to work by deriving an expression for the pressure exerted by such a gas. Let's begin by considering a cubic container, with sides of length L, containing a *single* molecule of mass m and speed v. Suppose the molecule travels to and fro between the left and right walls, as shown in Figure 2.7. Because a collision between the molecule and a wall is elastic (and because the massive walls scarcely move

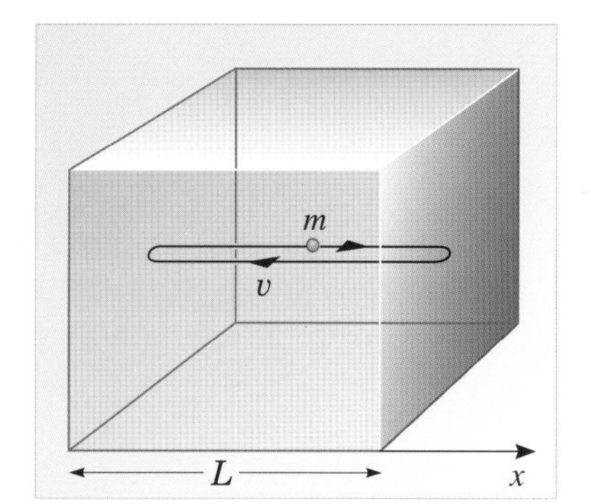

Figure 2.7 A molecule shuttles to and fro between two walls in a container.

when the molecule bounces off them), the translational energy of the molecule is the same before and after a collision with a wall. So, on colliding with a wall, the molecule reverses its direction of motion, but its speed remains unchanged. It shuttles to and fro between the walls at *constant* speed v.

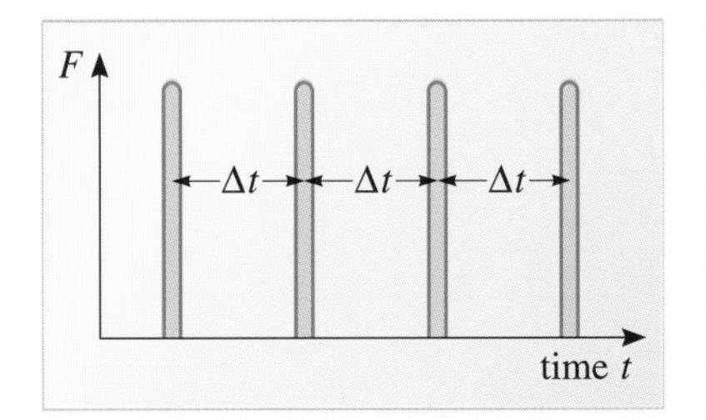

Figure 2.8 The magnitude of the force on the right-hand wall due to collisions with the molecule in Figure 2.7 is a series of sharp spikes occurring at regular time intervals of $\Delta t = 2L/v$.

We concentrate on the force exerted by the molecule on the right-hand wall. The force is only felt when the molecule collides with the wall and bounces off it. Most of the time the molecule is in flight somewhere inside the container, and exerts no force on the wall. Collisions with the right-hand wall take place at regular time intervals of $\Delta t = 2L/v$, so if we plot a graph of the magnitude of the force experienced by the right-hand wall against time, it looks something like Figure 2.8 — a series of sharp spikes. We are not interested in the force during each spike, but in the force averaged over an extended time.

During each collision with the right-hand wall, the momentum of the molecule changes from

$$p_{x,1} = mv$$

to

$$p_{x,2} = -mv$$

so the change in its momentum is

$$\Delta p_x = p_{x,2} - p_{x,1} = -2mv.$$

Averaged over a number of collisions, the rate of change of momentum of the molecule due to collisions with the right-hand wall is

$$\frac{\Delta p_x}{\Delta t} = \frac{-2mv}{2L/v} = -\frac{mv^2}{L}.$$

Since force is the rate of change of momentum, this is the average force exerted on the molecule by the right-hand wall. By Newton's third law, the average force exerted on the right-hand wall by the molecule is

$$F_x = \frac{mv^2}{L}. \tag{2.8}$$

The corresponding average force *per unit area* on the right-hand wall is

$$\frac{F_x}{L^2} = \frac{1}{L^2} \times \frac{mv^2}{L} = \frac{mv^2}{L^3}. \tag{2.9}$$

This is the contribution to the average pressure due to the chosen molecule. We must now scale this up to take account of all N molecules in the container, remembering that they have a variety of speeds and move in a variety of directions. The simplest way forward is to follow a suggestion originally made by James Joule.

Joule's classification

For the purposes of calculating the pressure, Joule asserted that it would make no difference if we imagined dividing the molecules into three classes: one-third moving to and fro along the x-axis, one-third moving to and fro along the y-axis and one-third moving to and fro along the z-axis.

If this classification is accepted, the total pressure on the right-hand wall can be found from Equation 2.9 by:

- taking its average (to allow for the spread in molecular speeds);

- multiplying this average by $N/3$ (because only one of Joule's three classes involves collisions with the right-hand wall).

We are therefore led to the following formula for the pressure:

$$P = \frac{N}{3} \times \frac{\langle mv^2 \rangle}{L^3}. \qquad (2.10)$$

Although we have concentrated on the right-hand wall, a similar argument could clearly be constructed for *any* of the walls, so this is actually the pressure *of the gas*. Remembering that the volume of the container is $V = L^3$, and the translational energy of a molecule is $E_{\text{trans}} = \frac{1}{2}mv^2$, we finally have

$$PV = \tfrac{2}{3} N \langle E_{\text{trans}} \rangle. \qquad (2.11)$$

The result is actually correct, but there is something slightly shaky about Joule's argument. Dividing molecules into three imaginary classes is an unproven simplification, which is not really part of the simple gas model. It might even be regarded as cheating in order to avoid some tricky mathematics!

Fortunately, a better argument can be found. Figure 2.9 shows a molecule moving in an *arbitrary* direction in the container. The molecule has a component of velocity v_x along the x-axis. It may bounce off a number of walls, but its round-trip time between its collisions with the right-hand wall is

$$\Delta t = \frac{2L}{|v_x|}.$$

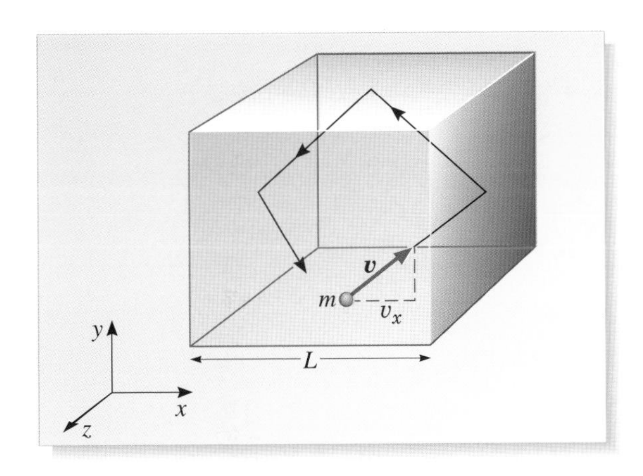

Figure 2.9 A molecule moving in an arbitrary direction in a container.

When the molecule strikes the right-hand wall, we will assume that the x-component of its momentum reverses but the y- and z-components of its momentum remain the same. The molecule therefore bounces off the wall at the same angle as it approached the wall — a process known as **specular reflection** (Figure 2.10). The change of momentum during such a collision has magnitude

$$|\Delta p_x| = 2m\,|v_x|.$$

Our previous derivation can then be repeated to give

$$\frac{F_x}{L^2} = \frac{m\,|v_x|^2}{L^3} = \frac{mv_x^2}{L^3}.$$

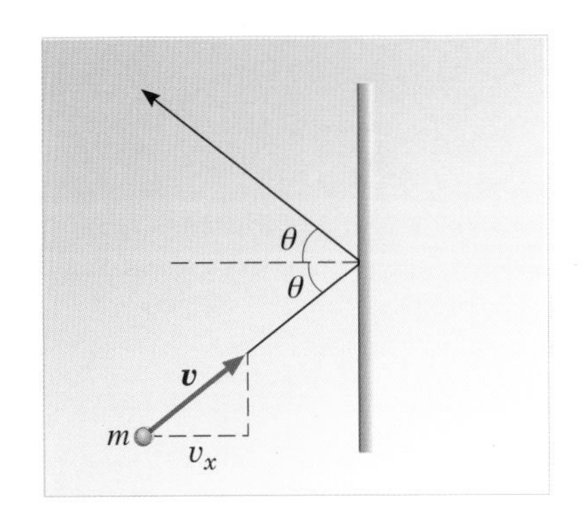

Figure 2.10 Specular reflection.

This formula applies for *any* molecule in the gas. To find the total pressure, we have to take an average (to allow for the spread in v_x) and multiply by the total number of molecules, N. We then obtain

$$P = N\frac{m\langle v_x^2 \rangle}{L^3}. \qquad (2.12)$$

Notice that we have not put angular brackets around the left-hand side of this equation. The quantity P calculated here is really an averaged quantity but, as we emphasized earlier, the fluctuations in pressure are completely negligible. The average pressure is effectively the same as the actual pressure that would be measured instant by instant.

Finally (and this is probably the step that Joule missed), we can use the fact that the directions of motion are random. No direction of motion is special so, averaging over all the molecules in the gas gives

$$\langle v_x^2 \rangle = \langle v_y^2 \rangle = \langle v_z^2 \rangle$$

and Equation 2.5 then allows us to show that

$$\langle v^2 \rangle = \langle v_x^2 \rangle + \langle v_y^2 \rangle + \langle v_z^2 \rangle = 3\langle v_x^2 \rangle.$$

Replacing $\langle v_x^2 \rangle$ in Equation 2.12 by $\langle v^2 \rangle/3$ and recalling that the volume of the container is $V = L^3$, and the translational energy of a molecule is $E_{\text{trans}} = \frac{1}{2}mv^2$, we obtain once more

$$PV = \tfrac{2}{3}N\langle E_{\text{trans}} \rangle. \qquad \text{(Eqn 2.11)}$$

This final result makes very good sense. One quick way of checking it is to compare the units on both sides.

● Show that both sides of Equation 2.11 have the units of energy.

❍ On the right-hand side, the units are obviously those of energy, since N has no units. On the left-hand side, the SI units of PV are $\text{N m}^{-2} \times \text{m}^3 = \text{N m} = \text{J}$, which are also the units of energy. ■

A much more demanding check is given by combining Equation 2.11 with the ideal gas equation of state

$$PV = NkT$$

to obtain $\langle E_{\text{trans}} \rangle = \frac{3}{2}kT$. (2.13)

Is this equation true? Well, in the case of monatomic gases, we can give an immediate answer. In Chapter 1, you saw that the internal energy of a monatomic gas is

$$U = \tfrac{3}{2}NkT. \qquad \text{(Eqn 1.15b).}$$

This means that Equation 2.13 is true provided that

$$U = N\langle E_{\text{trans}} \rangle$$

in other words, provided that the internal energy of an ideal monatomic gas is the total translational energy of all its molecules. This makes excellent sense because the molecules in an ideal gas have no mutual potential energy, and they are single atoms so would not be expected to vibrate or rotate. They behave just like the structureless particles envisioned in the simple gas model.

For diatomic and other gases, the situation is more complicated. Molecules in these gases *would* be expected to vibrate and rotate, and their internal energies would be expected to contain contributions due to these motions. We know that the internal energies of diatomic gases are greater than those of monatomic gases, but does Equation 2.13 continue to describe the average *translational* energy in any gas? This is something we will examine closely later; for the moment, let's just say the answer will turn out to be a resounding 'yes'.

On this basis, we can regard Equation 2.11 as a highly successful prediction. It provides a good illustration of the ways in which microscopic models, on the scale of atoms and molecules, can be used to explain macroscopic phenomena, at least in terms of average behaviour. Indeed, the model is so successful and appealing that it carries the danger of being misapplied. Please remember that the simple gas model applies only to gases! It is also possible to talk about the pressure in a liquid or a solid, but in these cases the pressure includes a contribution from interatomic forces, and is not just due to molecular bombardment.

Having obtained an expression for the pressure of a gas, we can now go back and examine one of the assumptions on which it relies in a little more detail. We assumed that the molecule always bounces back elastically after a collision with the wall. This assumption is hard to investigate unless the rate of molecular collisions is greatly reduced and controlled conditions are established. At very low pressures, studies have shown that the collisions between molecules are far from being elastic. In nearly every case, the molecules transfer all their energy and momentum to the wall and remain stuck on the surface for 'dwell times' of about a microsecond. This might not seem very long, but it is a million times longer than the time for a real elastic collision, which is comparatively rare. Yet fortunately our simple gas model gives the right answer! How is this possible?

Any surface in contact with a gas soon acquires a coating of molecules adhering to the surface. This coating is generally a slushy mess several molecules thick, like the layer of condensation on a window when you breathe hard onto it, but much thinner. If a gas molecule approaches the surface, it usually becomes incorporated in the coating and so sticks to the wall. It may remain there for many years! However, this does not lead to the build-up of a thicker and thicker coating, because from time to time molecules also leave the surface. A steady-state situation develops in which molecules arrive and leave at the same rate. Molecules arrive with random directions and speeds, and they leave with random directions and speeds. All the random effects average out, so the *overall* process of arrival and departure is exactly as if each molecule were reflected elastically and specularly.

You may have the idea that the pressure generated from the impact of tiny molecules cannot be very great. However, the molecules are moving very quickly (at hundreds or thousands of metres per second) and there is a huge number of impacts (around 10^{25} air molecules strike your hand every second) so the pressure can be huge. The air around you exerts a pressure of $10^5\,\mathrm{N\,m^{-2}}$, enough to cause great damage if not handled correctly. James Clerk Maxwell is best known as a theoretical physicist, but he also did some experimental work, including an investigation of frictional forces in gases at reduced pressures. During this work Maxwell wrote to a colleague, 'I made an erroneous estimate by rule of thumb as to the strength of a glass plate $\frac{1}{2}$ inch thick in consequence of which when exposed to a pressure of $\frac{3}{4}$ atmosphere it succumbed with a stunning implosion and set me back a month with regard to the friction of gases.' It is fortunate for the development of physics that Maxwell himself emerged relatively unscathed.

Question 2.4 What is the translational energy of all the molecules in the air that fills an empty room measuring 3 m × 4 m × 2.5 m? (Assume the air is at atmospheric pressure, 1×10^5 Pa.) ■

4 Distributions of speed and translational energy in gases

The simple gas model has already anticipated the fact that molecules in a gas have a range of energies and speeds, and this is reflected in the fact that Equation 2.4 involves the *average* translational energy. We now take a closer look at the way in which speed and translational energy are distributed among the molecules in a gas. This may seem to be a detail, but it is not. The problem of distributing energy among the molecules of a gas turns out to be of fundamental importance. Not only will it cement our understanding of gases, it will also shed new light on the meaning of temperature and provide an introduction to statistical mechanics.

4.1 Measuring the distribution of molecular speeds

A good place to start any investigation of this kind is experiment. Several experimenters have devised ingenious methods for measuring the distribution of molecular speeds. The apparatus for one method is illustrated schematically in Figure 2.11. A sample of metallic bismuth is put into an oven and heated until it vaporizes. The vapour is brought to a carefully controlled temperature, and then allowed to escape from the oven through a slit. A second slit, some distance from the first, allows only those molecules that are travelling in an almost parallel beam to carry on towards the drum. The drum has another narrow slit in its wall and is lined with a thin transparent film. It is capable of rotating at up to 100 revolutions per second. At the start of the experiment, the stationary drum is positioned so that all three parallel slits line up. The bismuth molecules in the beam therefore travel straight across the drum and strike the film in a line directly opposite the slit. They condense out on the film at this point, leaving a metallic deposit to provide a 'zero mark', a reference point for the molecular speed measurements.

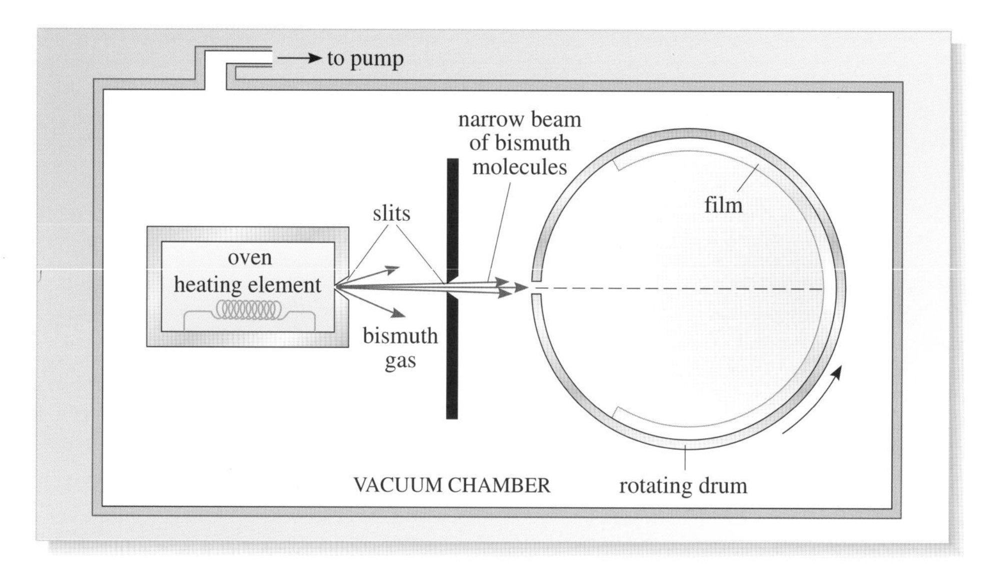

Figure 2.11 Rotating-drum apparatus used to measure the distribution of molecular speeds in bismuth gas.

Figure 2.12 shows how the measurements are made. The drum rotates at a constant angular speed, and as its slit cuts through the molecular beam it admits a sample group, or 'pulse', of molecules to the drum's interior. It takes a short time for this pulse of molecules to cross the drum, which is continually rotating. Hence the molecules do not strike the film at the zero mark, but at a distance from it which depends on the speed of rotation of the drum and on the time taken by the molecules to cross the drum. Because the molecules in the pulse have different speeds, individually they take different times to cross the drum. This results in the trace of the pulse being smeared out across the film.

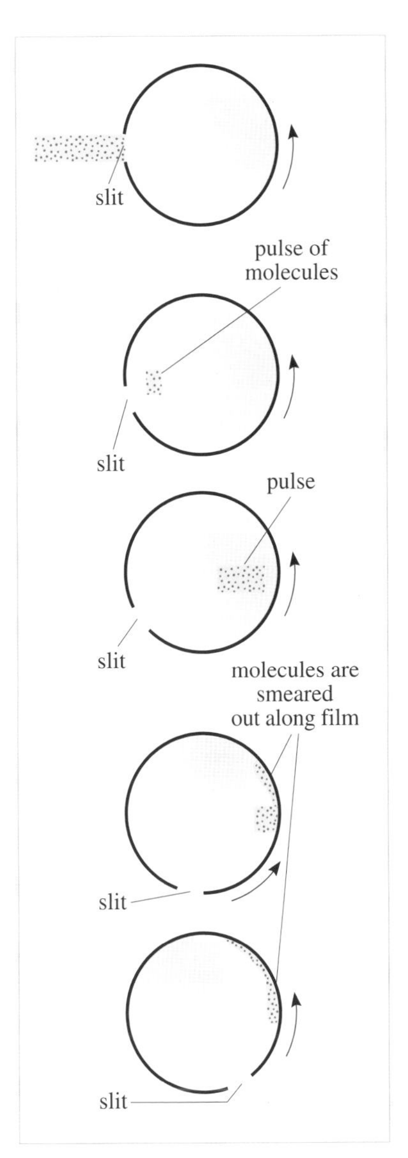

Figure 2.12 Rotating drum in operation: the pulse of gas molecules passes across the drum and leaves a smeared-out trace on the film.

After many similar pulses have been sampled in this way, the film can be studied more closely. The thickness of the deposit can be measured by viewing light transmitted through the normally transparent film, with the darkening proportional to the number of molecules deposited. Figure 2.13 shows the thickness of deposit for a typical experiment. The variation in thickness shows that the molecules have a wide range of speeds. The darkest part of the trace is produced by molecules travelling at the **most probable speed**, v_{mp}. Of all the speeds that might be measured (to within a given accuracy), this is the single most likely value.

Question 2.5 In a rotating-drum experiment to measure the speed distribution of bismuth molecules, a drum of diameter 0.30 m is rotated at 100 revolutions per second. The results shown in Figure 2.13 are obtained, with the darkest part of the trace occurring 95.5 mm away from the zero mark. Estimate the most probable speed, v_{mp}, of the bismuth molecules. ■

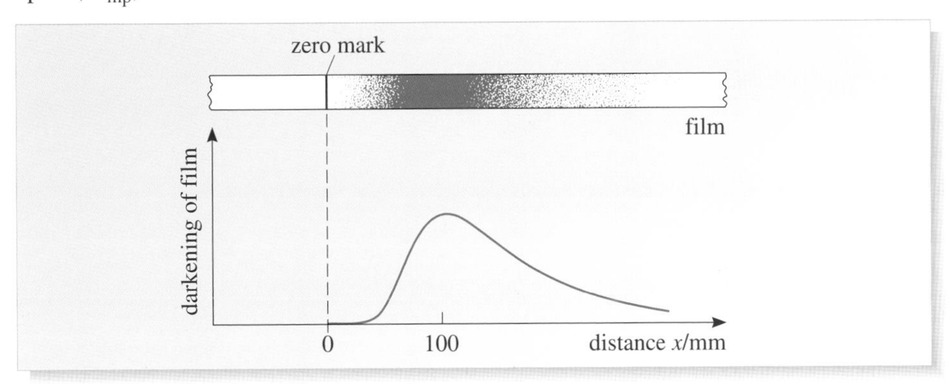

Figure 2.13 Data from a rotating-drum experiment.

The quantities plotted on Figure 2.13 are those directly measured in the rotating-drum experiment. However, data from this type of experiment can be presented in a more useful way by noting that:

- the degree of darkening of the film is proportional to the number of molecules that have condensed on to it;
- the position at which a molecule strikes the film (relative to the zero mark) is proportional to the time it has taken the molecule to cross the drum, and this is *inversely* proportional to its speed.

Therefore, it is sensible to replot the data of Figure 2.13 in terms of the fraction of molecules having speeds within a chosen small range. Figure 2.14 shows the results of such an analysis. Note that these results are presented in the form of a **speed histogram**. It is not useful to ask: 'how many molecules have a speed of $500\,\mathrm{m\,s^{-1}}$?'. If you mean *exactly* $500\,\mathrm{m\,s^{-1}}$, the answer almost certainly is *none*. It is much more sensible to ask a question such as: 'how many molecules have a speed between

500 m s^{-1} and 505 m s^{-1}?'. Each histogram bar covers a *range* of speeds along the horizontal axis, and the height of the bar is the fraction of molecules that have speeds in this range. All the bars have the same width, Δv, so the overall shape of the histogram gives a good impression of the pattern of speed distribution among the molecules. Again, the most probable speed corresponds to the peak of the histogram.

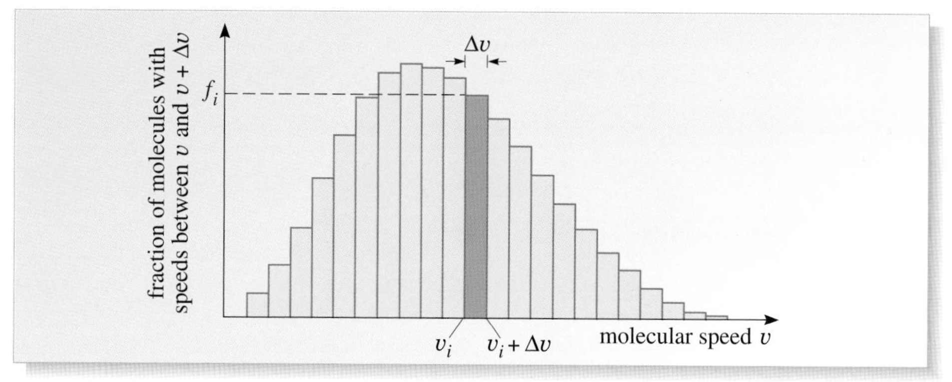

Figure 2.14 A histogram showing the distribution of molecular speeds in a given gas at a given temperature. The height of any histogram bar is the fraction of molecules that have speeds in the range covered by the width of the bar. Since all bars have the same width, the most likely speed ranges correspond to the tallest bars.

The size of Δv is chosen with two conflicting demands in mind. It should be large enough to ensure that large numbers of molecules contribute to each bar of the histogram. This prevents random fluctuations from changing the shape of the histogram from one moment to the next. However, we should also choose Δv to be small enough to get plenty of separate speed ranges along the speed axis, so allowing fine discrimination between one range of speeds and another.

In the notation shown in Figure 2.14, a typical histogram bar covers a speed range from v_i to $v_i + \Delta v$ and has height f_i. Because the height of a histogram bar is equal to the *fractional frequency* of molecules with speeds in the range of the width of the bar, the sum of the heights of all the bars must be equal to 1. Thus,

$$\sum_i f_i = 1. \tag{2.14}$$

The average speed of the molecules can be defined in the same way as any other average. *Provided the histogram bars are very narrow*, all the molecules associated with the histogram bar with speeds between v_i and $v_i + \Delta v_i$ can be taken to have the same speed, v_i. The average speed is then given by multiplying each speed v_i by the corresponding fractional frequency f_i and adding together all the results. This gives

$$\langle v \rangle = \sum_i f_i v_i \,. \tag{2.15}$$

If the speed histogram were exactly symmetrical on either side of its peak, the average speed would be the same as the speed at the peak of the histogram (which you have seen is the most probable speed). But, if you look closely at the speed histogram, you will see that it is not quite symmetrical: the distribution has a tail at higher speeds, which skews the histogram towards higher speeds: as a consequence, the average speed is slightly greater than the most probable speed. In fact, in all cases,

$$\langle v \rangle = \frac{2}{\sqrt{\pi}} v_{\text{mp}} \approx 1.13 v_{\text{mp}} \,. \tag{2.16}$$

The root mean square speed is even higher:

$$v_{\text{rms}} = \sqrt{\tfrac{3}{2}}\, v_{\text{mp}} = 1.22 v_{\text{mp}} \,. \tag{2.17}$$

4.2 Analysing the experiment in terms of translational energy

So far, we have analysed the motion of molecules in a gas in terms of their speeds. There is an alternative way of representing the same information, in terms of translational energy. For molecules of a given mass, m, any particular speed corresponds to a definite translational energy

$$E_{\text{trans}} = \tfrac{1}{2}mv^2 .$$

To keep the notation as simple as possible, we will drop the subscript 'trans', and use the symbol E to represent the translational energy. Do not be misled by this: diatomic and more complicated gases have contributions to the total energy due to vibration and rotation, which are not included in E. When we want to refer to the total energy, including all such contributions, we will use the symbol E_{tot}.

The analysis carried out for speed can now be repeated in terms of translational energy. Figure 2.15 shows the translational energy histogram. A typical bar of this histogram extends along the horizontal axis from E_i to $E_i + \Delta E$, and its height h_i is the fraction of molecules with translational energies in this range. The shape is quite different from the speed histogram in Figure 2.14. Notice that the energy histogram is much more asymmetric, and has a much more pronounced tail; this can be explained with the aid of Figure 2.16.

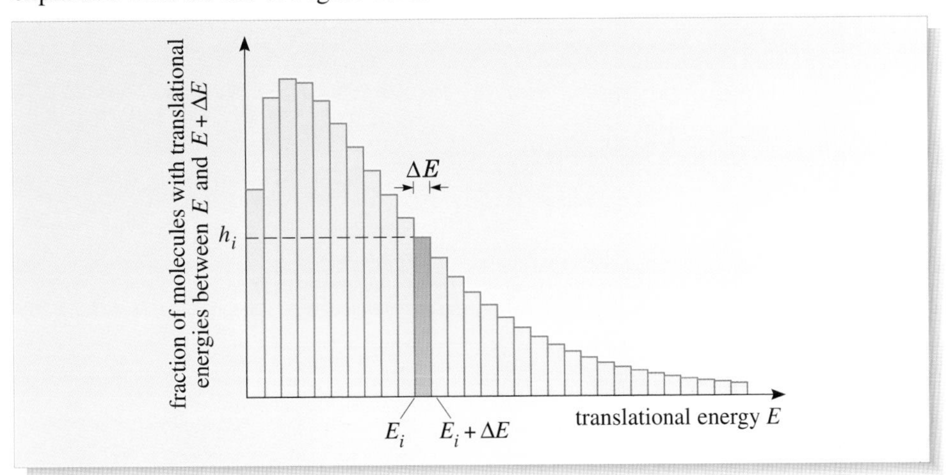

Figure 2.15 Translational energy histogram.

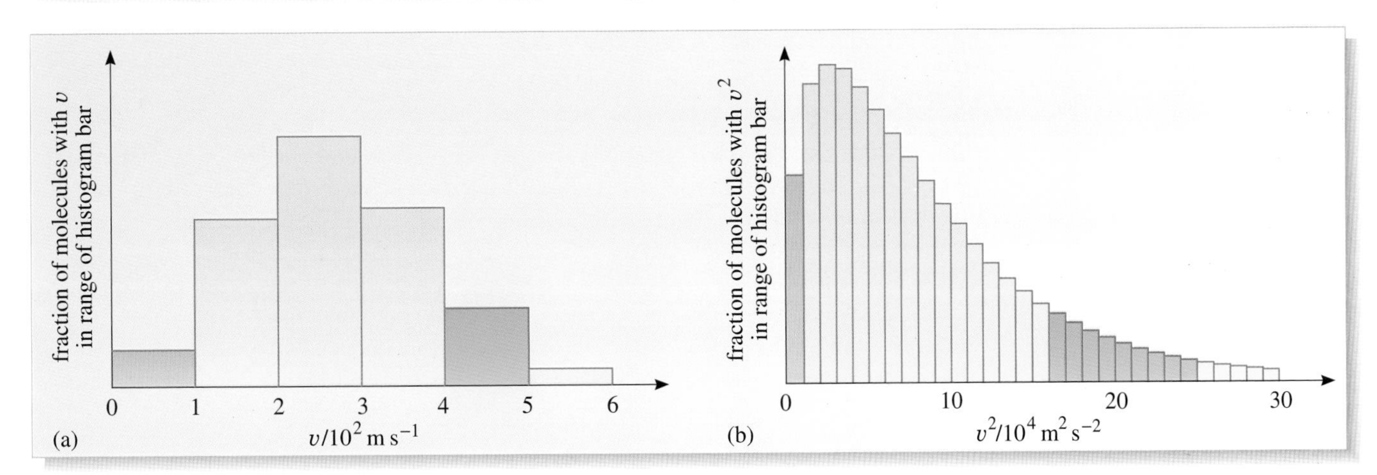

Figure 2.16 Why the histogram for translational energy has a much more extended tail than the histogram for speed.

Figure 2.16a shows a speed histogram with two bars picked out. The green bar extends from $0\,\mathrm{m\,s^{-1}}$ to $100\,\mathrm{m\,s^{-1}}$, while the red bar extends from $400\,\mathrm{m\,s^{-1}}$ to $500\,\mathrm{m\,s^{-1}}$. Figure 2.16b shows the corresponding distribution for v^2. Note that the green bar in Figure 2.16a corresponds to one bar in Figure 2.16b ($0\,\mathrm{m^2\,s^{-2}}$ to $1 \times 10^4\,\mathrm{m^2\,s^{-2}}$), while the red bar in Figure 2.16a corresponds to nine bars in Figure 2.16b ($16 \times 10^4\,\mathrm{m^2\,s^{-2}}$ to $25 \times 10^4\,\mathrm{m^2\,s^{-2}}$). Thus, the tail of Figure 2.16a is considerably stretched out in Figure 2.16b. Since translational energy is proportional to v^2, the shape of the energy histogram is similar to that shown in Figure 2.16b, and that is why it has an extended tail.

Apart from its detailed shape, the energy histogram has much in common with the speed histogram. For example, the sum of the heights of all the histogram bars is equal to 1. That is,

$$\sum_i h_i = 1. \tag{2.18}$$

The most probable translational energy of the molecules, E_{mp}, is the energy at which the histogram has its peak value and, provided the histogram bars are very narrow, the average translational energy of a molecule is

$$\langle E \rangle = \sum_i h_i E_i \,. \tag{2.19}$$

Because the energy distribution is highly asymmetrical, the average translational energy is much greater than the most probable translational energy. In fact,

$$\langle E \rangle = 3E_{\mathrm{mp}} \,. \tag{2.20}$$

4.3 What do the distributions of speed and translational energy depend on?

You have seen that experiments reveal how speed and translational energy are distributed amongst the molecules in a gas. So far, we have given a brief summary of an experiment and looked at ways in which its results can be analysed — in terms of histograms for speed and translational energy, average values, most probable values etc. But these results inevitably lead to other questions:

- What factors influence the shapes of the histograms?
- Do the histograms fluctuate in time?
- What physical principles explain the shapes of the histograms and their fluctuations?

Answers to the first two questions can be provided by experiments and/or computer simulations, but the third question is an exceptionally deep one. Do you have any idea what the answer might be? Perhaps not. Great physicists like Maxwell, Boltzmann and Gibbs struggled for many years with this question, and it took even longer for the scientific community to accept their answers. Section 5 will describe these developments, which led to the creation of a branch of physics known as statistical mechanics. You will see that more is at stake than the behaviour of molecules in a gas. Statistical mechanics provides a deeper understanding of the meaning of temperature, and its methods can be applied very widely to particles in the early Universe and atoms in solids, as well as to the molecules in a gas.

To begin, we must concentrate on the first two questions to collect as much information as possible about the distribution of speed and energy in gases. This will be done with the support of computer simulations.

Open University students should leave the text at this point and carry out the multimedia activity *Sharing out the energy in gases*. This should take about an hour. When you have finished, turn to the Appendix to Chapter 2 for a summary, and then continue with the main text. Do not even think of looking at the appendix before your multimedia investigations are complete! Readers who cannot access the computer simulations should go directly to the appendix, and then continue with the main text.

Question 2.6 Compare the histograms for molecular speed and molecular translational energy shown in the Appendix to Chapter 2 (Figures A1 and A2). Identify the main similarities and main differences between these histograms. ■

4.4 Distribution functions

This section is about a detail of presentation rather than an issue of fact. This does not mean it is unimportant, though — the terminology introduced here will be used extensively later on.

The speed distribution function

Although a histogram gives a simple way of summarizing the distribution of speed in gases, it suffers from one disadvantage. The sum of the heights of all the histogram bars is equal to 1, so the height of each bar depends on the number of bars chosen. If Δv in Figure 2.14 were halved, for example, there would be twice as many bars and we would expect the height of each bar to halve. This is unfortunate because different researchers might choose to present their results in different ways, which could lead to confusion. The problem is overcome by moving one step beyond histograms.

We define the **speed distribution function**, $f(v)$ as follows:

$$f(v) = \frac{\text{height of histogram bar for speeds betweeen } v \text{ and } v + \Delta v}{\text{width of histogram bar, } \Delta v}.$$

Provided the histogram bars are narrow (i.e. Δv is small), this function is independent of Δv. Whereas the height of the histogram bar halves when Δv halves, $f(v)$ remains unchanged.

In a speed histogram, the height of any histogram bar tells us the fraction of molecules whose speeds lie within the width of the bar. Thus, the speed distribution function $f(v)$ tells us the fraction of molecules that lie within a given speed range, *per unit speed range*. The meaning of the speed distribution function can be understood by multiplying it by a small speed range, Δv, and stating that

$$f(v)\,\Delta v = \text{fraction of molecules with speeds between } v \text{ and } v + \Delta v. \qquad (2.21)$$

Although this is slightly cumbersome (we have to remember to multiply $f(v)$ by a small speed range, Δv), the convenient point is that this equation is valid for *any* small speed range Δv.

Figure 2.17 is a graph of $f(v)$. It looks just like a smoothed-out version of the histogram, but with the vertical axis calibrated in the units of $(\text{speed})^{-1}$. This means that the product $f(v)\,\Delta v$ has no units, which is reasonable, since the right-hand side of Equation 2.21 also has no units. Functions like $f(v)$ are known as

distribution functions, because they tell us how a quantity is *distributed* amongst the particles of a system. The distribution function shown in Figure 2.17 is therefore called the *speed* distribution function.

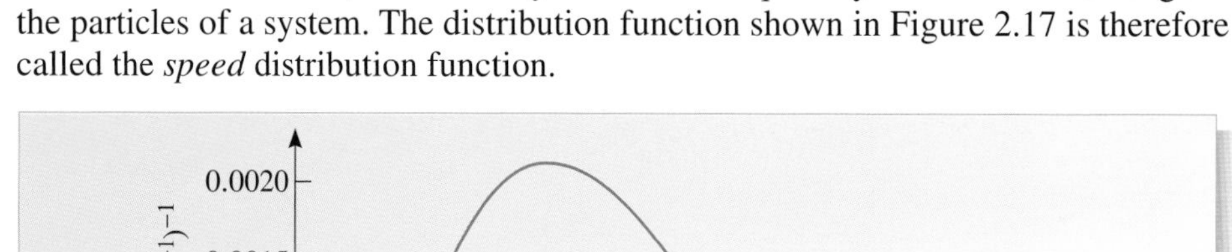

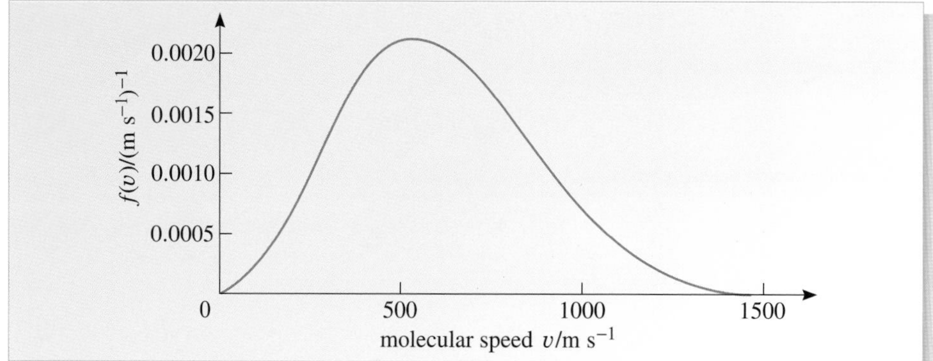

Figure 2.17 A graph of the speed distribution function.

As it stands, Equation 2.21 assumes that the speed range, Δv, is *small*. More generally, the fraction of molecules in any given speed range can be found from the area under the speed distribution graph for this speed range (Figure 2.18). In general,

$$\left(\begin{array}{l}\text{area under speed distribution graph}\\ \text{between } v_1 \text{ and } v_2\end{array}\right) = \left(\begin{array}{l}\text{fraction of molecules with}\\ \text{speeds between } v_1 \text{ and } v_2.\end{array}\right) \quad (2.22)$$

Recalling that the area under a graph can be written in terms of a definite integral, we can also express this as

$$\int_{v_1}^{v_2} f(v)\mathrm{d}v = \text{fraction of molecules with speeds between } v_1 \text{ and } v_2. \quad (2.23)$$

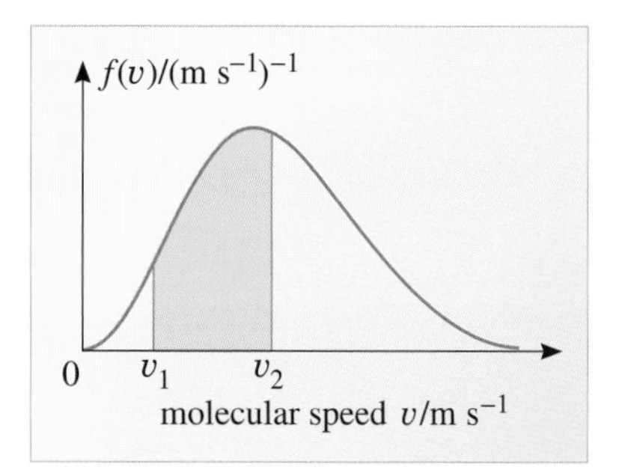

Figure 2.18 The area under the speed distribution graph between v_1 and v_2 (shaded area) is equal to the fraction of molecules with speeds between v_1 and v_2.

As a special case, suppose we are interested in the fraction of molecules that have speeds between zero and infinity. What fraction is that? Of course *all* the molecules have a speed somewhere in this range, so the fraction is 1. We can therefore conclude that

$$\text{total area under speed distribution graph} = 1 \quad (2.24\text{a})$$

or, equivalently,

$$\int_0^\infty f(v)\,\mathrm{d}v = 1. \quad (2.24\text{b})$$

You might want to know the fraction of molecules that have speeds greater than a fixed value ($1000\,\mathrm{m\,s^{-1}}$, say). A question like this can be answered by looking at the area under the tail of the distribution. For example, with the distribution curve shown in Figure 2.19, the shaded area (corresponding to speeds greater than $1000\,\mathrm{m\,s^{-1}}$) is roughly one-third the total area under the curve, so roughly one-third of the molecules would have speeds in excess of $1000\,\mathrm{m\,s^{-1}}$ in this case. (Of course, this is not a general result, because different gases at different temperatures have different distribution functions, but it illustrates how such information can be obtained.)

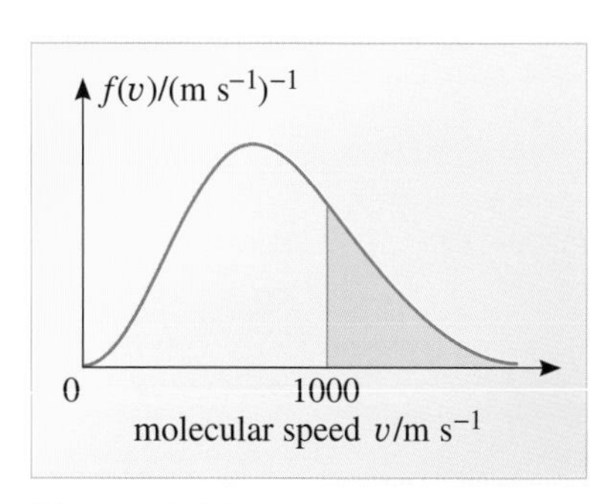

Figure 2.19 The fraction of molecules with speeds greater than $1000\,\mathrm{m\,s^{-1}}$ is given by the shaded area under the tail of the speed distribution graph. This is roughly one-third of the total area under the graph, which is equal to 1. So roughly one-third of the molecules have speeds greater than $1000\,\mathrm{m\,s^{-1}}$ for this distribution function.

The notation of definite integrals can also be used to give an expression for the average speed in terms of the speed distribution function:

$$\langle v\rangle = \int_0^\infty v f(v)\,\mathrm{d}v. \quad (2.25)$$

As usual, we do not expect you to remember or use this formula; we present it just to make the general point that much useful information, including the average speed, can be extracted from the distribution function using standard mathematical techniques.

The translational energy distribution function

Exactly the same considerations apply to the spread of translational energy in gases. As before, we can use the energy histogram to construct an energy distribution function. We introduce the **translational energy distribution function** $g(E)$ (sometimes called the energy distribution function for short). This is defined in such a way that for any small energy range, ΔE,

$$g(E)\,\Delta E = \text{fraction of molecules with translational energies between } E \text{ and } E + \Delta E \tag{2.26}$$

Figure 2.20 is a graph of the energy distribution function, $g(E)$, plotted against E. It looks just like a smoothed-out version of the energy histogram, but with the vertical axis calibrated in the units of (energy)$^{-1}$.

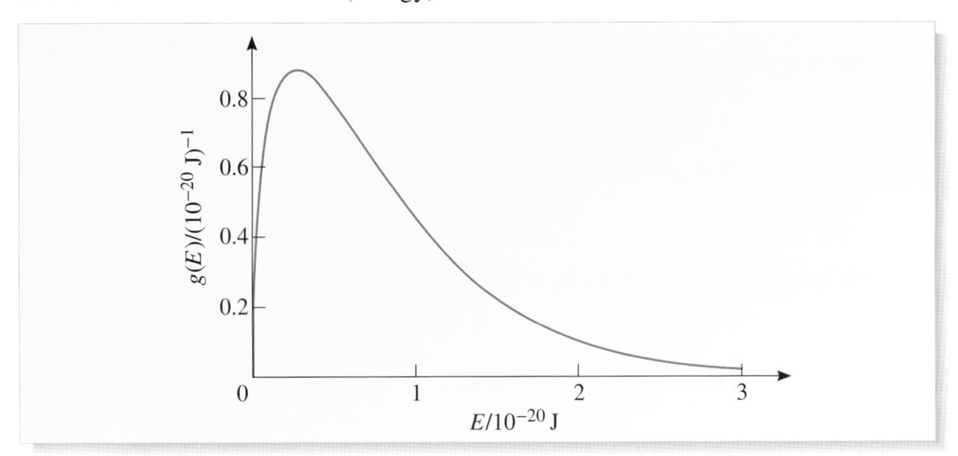

Figure 2.20 A graph of the translational energy distribution function at a given temperature.

Because the translational energy distribution function is defined in a similar way to the speed distribution function, it has similar properties. In particular, Equations 2.22–2.25 can be adapted by simply replacing $f(v)$ by $g(E)$ and speed by translational energy throughout. Try this now in the following exercises. Be sure to look up the answers if you are in any doubt.

In the following three questions, E, E_1 and E_2 are translational energies and $g(E)$ is the translational energy distribution function in a gas.

Question 2.7 (a) What physical meaning can be given to the area under the graph of $g(E)$ against E, between E_1 and E_2?

(b) What physical meaning can be given to the integral $\int_{E_1}^{E_2} g(E)\,\mathrm{d}E$?

Question 2.8 (a) What is the total area under the graph of $g(E)$ against E?

(b) What is the value of $\int_0^{\infty} g(E)\,\mathrm{d}E$?

Question 2.9 What physical meaning can be given to the integral $\int_0^{\infty} Eg(E)\,\mathrm{d}E$? ■

The distribution functions defined above refer to the fraction of molecules with speeds or translational energies in a given range. To find the *number* of molecules with speeds in a given range, it is necessary to multiply these distribution functions by the total number of molecules N in the gas. Thus,

$$Nf(v)\,\Delta v = \text{number of molecules with speeds between } v \text{ and } v + \Delta v$$

and $\quad Ng(E)\,\Delta v =$ number of molecules with translational energy between E and $E + \Delta E$.

Some texts define:

$$F(v) = Nf(v)$$

$$G(E) = Ng(E)$$

and call $F(v)$ and $G(E)$ the distribution functions of speed and translational energy. The final book of this course (*Quantum physics of matter*) will use the function $G(E)$, but it will not be needed here.

5 Statistical mechanics

'The whole is simpler than its parts.'

J. Willard Gibbs.

5.1 From chance to practical certainty

You have seen that computer simulations reveal definite patterns of speed and translational energy in gases (refer to the Appendix for Chapter 2 for details). A gas, in equilibrium at a given temperature, has a definite distribution of molecular speeds. The molecules do not all have the same speed, but we can say what fraction of the molecules have speeds lying within any specified range. Temperature is a crucial parameter that determines the overall shape of the distribution. If a given sample of gas is compared at two different temperatures, then at the higher temperature the sample will have:

- a higher speed for the peak of the distribution;
- a less prominent peak and a broader distribution.

The mass of the molecule also has an important effect on the speed distribution. For two different gases at the same temperature, the speed distribution for the sample with molecules of lower mass will have a less prominent peak, occurring at a higher speed, and the distribution will be broader. Thus, *for the speed distribution*, lowering the mass has a similar effect to increasing the temperature.

Temperature is also important in determining the distribution of translational energy. If a given gas is compared at two different temperatures, the translational energy distribution for the sample at the higher temperature will have a less prominent peak, occurring at a higher energy, and the distribution will be broader. However, the role of mass is very different. The mass of the molecule has no effect on the distribution of translational energy.

These are some of the facts we would like to understand. First, let me remind you why the situation is a difficult one. Newtonian mechanics led to the belief that the entire future of a system could be predicted, once its initial conditions have been specified. Proud boasts were made, including that of Laplace who claimed he would be able to predict the entire future of the Universe, if he were given sufficient data about its present state. 'Does God have any role in your Universe?', Napoleon is said to have enquired, but Laplace was quite clear: 'I have no need of that particular hypothesis!' Yet reality never matched the boast. No one could predict the entire future of the Universe, then or now. In fact, even a flask full of gas proved far too complex to analyse via the Newtonian scheme of using forces to predict accelerations. A cubic centimetre of air contains about 2×10^{19} molecules of gas, each changing its motion about 5×10^{9} times per second via collisions. How could we ever hope to understand such a system?

If 'understanding' means being able to predict the motion of every single molecule, this is far beyond our capabilities. The brilliant idea of statistical mechanics is to concentrate instead on the average or most likely behaviour of the gas, using plausible assumptions about the likelihood of different possibilities. This avoids the need for detailed Newtonian calculations and actually turns the burden of an enormous number of molecules into an asset: the large number of molecules helps to ensure the reliability of the statistical estimates. At first there was great resistance to this approach, which was considered feeble and lacking in rigor, but shortly after 1900 the full power and precision of these methods became apparent.

How reliable could it be to apply statistical methods to gases? One indication can be obtained by considering a cubic centimetre of air, which contains about 2×10^{19} molecules. Imagine plotting a speed histogram with a thousand bars, and suppose that the smallest bar is at least one-thousandth the height of the largest bar. Then even this smallest bar corresponds to more than 2×10^{13} molecules. Using the square root rule outlined in Section 2, we should be astonished if the number of molecules associated with this bar were to vary by as much as $10 \times \sqrt{2 \times 10^{13}} = 4.4 \times 10^{7}$. This may sound like a large fluctuation, but it is tiny compared to the (more than 2×10^{13}) molecules expected: expressed as a fraction of the height of the bar, it is only two parts in a million.

Fluctuations in the *fractional* number of particles associated with a histogram bar become smaller as the total number of particles increases.

Here, we have looked at a very testing case: a small volume of gas, a histogram containing many bars, and we have concentrated on one of the smallest bars that could be usefully plotted. Even so, the fluctuations are negligible, and they will be even smaller for larger volumes of gas or larger histogram bars, closer to the peak of the distribution. The fact that the fluctuations in a gas are so small is very encouraging. It suggests that it will be good enough to use concepts of chance and probability to estimate the most likely shapes of the speed and energy histograms: these estimates should be reliable because the random fluctuations will be very small. In other words, a sort of alchemy is wrought: uncertain statistical estimates are converted into practical certainties, thanks to the vast number of molecules involved.

Box 2.1 Black skies, blue skies and red sunsets

In spite of the general rule that fluctuations are unimportant in gases, there is one notable exception. At a height of nearly 100 km from the surface of the Earth, the air density is reduced and there are only about 10^{20} molecules per cubic metre. Light is sensitive to the number of gas molecules it meets in a distance of one wavelength, which for blue light is about 4×10^{-7} m. In a cube with that dimension as an edge, there are only about 10 ± 3 molecules. The fluctuation in the number is therefore around $\pm$ 30%.

So, as a blue light wave travels through the atmosphere, it encounters random changes of air density. These cause the wave to be deflected. Though each deflection is quite small, many separate deflections combine to produce a large effect. Most of the blue light is scattered out of the direct sunlight beam after travelling about 50 km through the upper atmosphere. Red light has a longer wavelength than blue light, so a cube with the dimension of a wavelength as an edge contains more molecules, leading to a smaller percentage fluctuation. It is mainly for this reason that red light is scattered ten times less than blue light.

Because sunlight is scattered, daylight comes from all directions rather than in a beam from the Sun. If it were not for this scattering, the sky would be black and any place shadowed from direct sunlight would be intensely dark, which is what happens on the Moon. Blue light is scattered more than red light, so the sky is blue. The scattering is even larger during a sunset because the sunlight passes through a greater thickness of atmosphere, and that is why the setting Sun appears to be red. So, the next time you enjoy a blue summer sky or the glory of a sunset, remember that you are seeing an effect of statistical physics! See Figure 2.21.

Figure 2.21 (a) There is no atmosphere on the Moon and so the sky is black. (b) On Earth, the beautiful blue sky is caused by fluctuations in the atmosphere. (c) A red sky at night is the physicist's delight. (It is worth noting that there have been disputes about whether the blueness of the sky is due to scattering by fluctuations in the density of air (as outlined in Box 2.1) or to scattering by water droplets and dust particles (as Lord Rayleigh believed). This was settled in 1911 by Einstein, who showed that the phenomenon is explained by fluctuations and went on to deduce a value of Avogadro's number from his theory.)

5.2 Defining configurations in a gas

When the rules of chance are applied to dice, two basic steps are carried out: the possible outcomes are identified, and probabilities are assigned to them. For rolling a die, this is easy enough: the possible outcomes are '1', '2', '3', '4', '5' and '6', and these are assigned equal probabilities of 1/6.

In the case of molecules in a gas, the first step of identifying the possible outcomes means specifying the possible microscopic states of the gas. In broad terms, this is done by stating the position and velocity of each molecule in the gas. We can imagine labelling all the molecules A, B, C, ... etc. so that one molecule can be distinguished from another. (This is not really a practical proposition, of course, given the vast number of molecules in a gas, but we are just assuming that such labelling makes sense *in principle*.) Next, we need to specify the state of each of the labelled molecules. In Newtonian physics, the state of a particle can be defined by giving its position and velocity vectors. But position and velocity components are *continuous* variables, so they have an *infinite* number of possible values. This is not very helpful — how can we assign sensible probabilities if there are an infinite number of possible values?

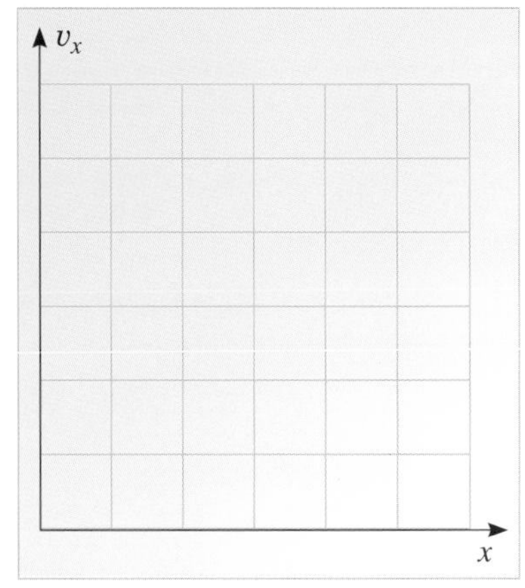

Figure 2.22 Phase cells illustrated schematically in two dimensions.

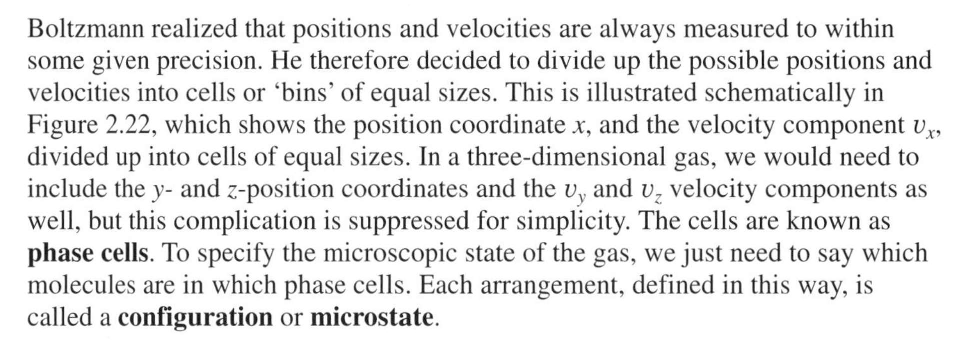

Boltzmann realized that positions and velocities are always measured to within some given precision. He therefore decided to divide up the possible positions and velocities into cells or 'bins' of equal sizes. This is illustrated schematically in Figure 2.22, which shows the position coordinate x, and the velocity component v_x, divided up into cells of equal sizes. In a three-dimensional gas, we would need to include the y- and z-position coordinates and the v_y and v_z velocity components as well, but this complication is suppressed for simplicity. The cells are known as **phase cells**. To specify the microscopic state of the gas, we just need to say which molecules are in which phase cells. Each arrangement, defined in this way, is called a **configuration** or **microstate**.

Configurations

Any arrangement of molecules into phase cells is said to define a **configuration** of the gas. Note that it matters *which* molecule is in *which* phase cell, so the two configurations shown in Figure 2.23 are different. There is no restriction on the number of molecules that can occupy a given phase cell.

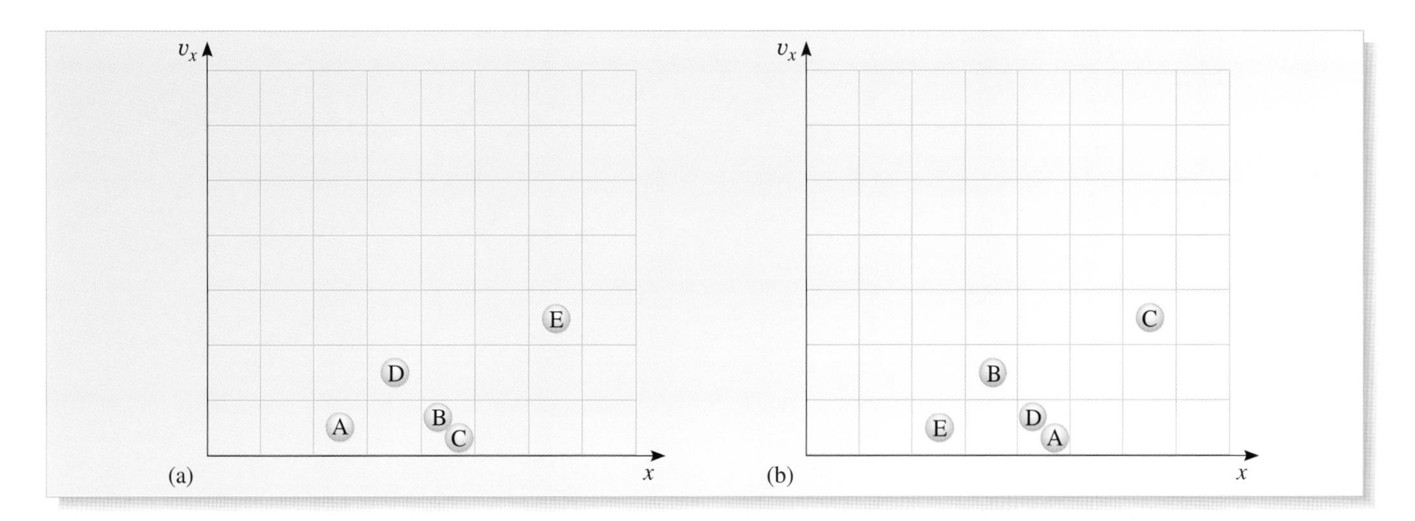

Figure 2.23 Two different configurations in a gas.

You may think that this definition is vague because we have not said how large a phase cell ought to be. Certainly, a phase cell should be small compared to any histogram we want to plot. Later in the course you will see that quantum mechanics imposes fundamental limits on our ability to make precise measurements, and therefore on the minimum sensible size for a phase cell. For the moment, we just note that the phase cells can be taken to be very small indeed, and that the precise choice will make no difference to our conclusions.

Because each phase cell specifies a particular position and a particular velocity (to within a certain precision), and because the molecules have definite masses, each phase cell also corresponds to a particular molecular energy (to within a certain precision). To avoid cumbersome sentences, we will talk of the 'energy of a phase cell'. This is a shorthand for the energy of a molecule of given mass whose position and velocity lie within that phase cell. The energy is only defined to within a certain precision, but this will not cause any problems. If we know the phase cell occupied by a given molecule, we essentially know the energy of that molecule. But the converse is not true; in general, many different phase cells have the same energy. For example, reversing the sign of v_x leads to a different phase cell but does not change the molecular energy.

5.3 Assigning probabilities to configurations

Having found a way of identifying distinct configurations, we must now decide on a way of assigning probabilities to them. This is something that cannot be deduced from Newtonian mechanics, or any other basic physical theory. What is needed is an inspired guess, that can be elevated to the status of a new physical principle. Boltzmann was prepared to make such a guess. He proposed that a gas in a container, that is isolated from the rest of the world and has settled down to a state of equilibrium, will obey the following two principles:

Boltzmann's principles of statistical mechanics

1 The conservation of energy

The energy of the gas does not depend on time, but has a constant value, E. The only possible configurations of the gas are those with the fixed energy, E.

2 Equal probabilities of allowed configurations

Each of these allowed configurations is equally likely.

The first principle is uncontroversial. Any isolated system will obey the law of conservation of energy. It is the assignment of *equal probabilities* to the various allowed configurations that is the bold and crucial step. This assumption is not obvious, but it has the great virtue of simplicity. Any other suggestion would, I think, be more artificial or more complicated. When a coin is tossed, heads and tails are generally taken to be equally likely. We have no reason to suspect that one configuration will be more likely than another, so why not make a similar assumption about configurations in a gas?

The role of equilibrium is important here. We could imagine deliberately setting the velocities of the gas molecules so that they are all more or less the same. This would produce a beam of molecules heading in a particular direction. In that case, we would have engineered things so that different allowed configurations were not equally likely. But this would not be an equilibrium situation. If we allow the molecules to collide with one another, and with the walls of their container, their velocities will change, and become randomized. Eventually, the gas will settle down to a macroscopic state of equilibrium, and it is in this situation that we can regard all configurations with the appropriate total energy as being equally likely. There is a direct analogy with a pack of playing cards. The pack could initially be rigged so that you are bound to receive aces, which is not a random situation. Shuffling the pack is analogous to allowing the molecules to collide; only when this randomizing process has been carried out sufficiently is it reasonable to use the normal rules of chance to estimate the probability of being dealt a winning hand.

5.4 The Boltzmann distribution law

The consequences of restricting the configurations to those with a fixed total energy, and taking all these allowed configurations to be equally likely, can now be explored. The key point to notice is that there is a very close analogy between our present discussion of molecules in a gas, and the discussion of a game of dice given at the end of Section 2. Table 2.2 compares the backgrounds of these two situations. They are clearly very similar, so we might expect that similar conclusions would be reached in each case.

Table 2.2 A close analogy can be established between a game played with many dice (Section 2) and the statistical behaviour of molecules in a gas.

Game of dice described in Section 2	Boltzmann's description of a gas
die	molecule
collection of dice	gas of molecules
score on die	phase cell occupied by molecule, which has an associated molecular energy
specification of scores on all the labelled dice (referred to as a *configuration*)	specification of phase cells occupied by all the labelled molecules (referred to as a *configuration*)
allowed configurations have total score restricted to a fixed value	allowed configurations have total energy restricted to a fixed value
all allowed configurations are equally likely	all allowed configurations are equally likely

Table 2.3 takes this analogy to its logical conclusion. The left-hand column reminds you of the conclusions reached for the game of dice. The right-hand column gives analogous conclusions for a gas. If you understood why the conclusions in the left column apply to our game of dice, you will understand why the conclusions in the right column apply to a gas. The reasons are the same.

Table 2.3 Conclusions following from the analogy outlined in Table 2.2.

Game of dice described in Section 2	Boltzmann's description of a gas
more configurations correspond to die A having a low score than having a high score	more configurations correspond to molecule A occupying a given phase cell of low energy than a given phase cell of high energy
the probability of die A having a low score is greater than the probability of die A having a high score	the probability of molecule A occupying a given phase cell of low energy is greater than the probability of molecule A occupying a given phase cell of high energy
with a large number of dice, the probability of a given die scoring *n* decreases exponentially as *n* increases	with a large number of molecules, the probability of a given molecule occupying a given phase cell decreases exponentially as the energy of that phase cell increases

At the risk of labouring the point, let me go through the argument again. The total energy of the gas is fixed, and each distinct way of distributing that energy amongst the molecules of the gas is taken to be equally likely. What do we mean by a distinct way of distributing the energy? We mean a definite configuration, based on the allocation of molecules to phase cells as shown in Figure 2.23.

Now imagine listing the enormous number of configurations that have the required total energy (just as we listed the configurations of dice that had a given total score). We concentrate on a particular molecule (molecule A, say) and compare the number of allowed configurations (N_{low}) in which this molecule is in a given phase cell of low energy with the number of allowed configurations (N_{high}) in which the molecule is in a given phase cell of high energy. When this is done, N_{low} is found to be greater than N_{high}. The greater number of configurations associated with the low-energy phase cell, coupled with the assumption that all configurations are equally likely, means that we are more likely to find molecule A in the low-energy phase cell than in the high-energy phase cell.

Why is N_{low} greater than N_{high}? One way of understanding this is to note that, when molecule A has less energy, the remaining molecules have more energy, and the extra freedom associated with allocating this energy amongst the remaining molecules allows more configurations to be constructed. If this does not convince you, go back to the dice analogy. If you can understand why many configurations in Table 2.1 correspond to a low score on die A, while few configurations correspond to a high score, you will have grasped the point.

A minor technicality is that the game of dice described in Section 2 was based on rather low total scores. You might wonder whether this leads to any misleading conclusions. It does not. Dice are unlike molecules in one important respect: each die has a *maximum* score of '6', whereas molecules have no maximum translational energy. It turns out that the existence of a maximum score for a die reduces the number of configurations that have high total scores. Thus, by avoiding high total scores, we avoided introducing effects that have no counterpart for molecules. The conclusions reached with dice using relatively low total scores can be extended to molecules, no matter how great the total energy.

The most important conclusion is contained in the last row of Table 2.3. By explicitly counting configurations for a gas in equilibrium, Boltzmann showed that the probability of finding a given molecule in a given phase cell of energy E is proportional to an exponential factor $e^{-\beta E}$. This is directly analogous to the formula used to fit the data of Figures 2.3 and 2.4. The physical meaning of β is not immediately obvious from Boltzmann's counting of configurations, but he was able to show that his theory would agree with standard equations of physics (such as the equation of state of an ideal gas) *provided that*

$$\beta = \frac{1}{kT} \tag{2.27}$$

where T is the absolute temperature and $k = 1.381 \times 10^{-23}\,\mathrm{J\,K^{-1}}$ is Boltzmann's constant. Bearing this in mind, we can now state Boltzmann's great discovery:

Boltzmann's distribution law

Consider a gas in equilibrium at temperature T. Then the probability of finding a given molecule in a given phase cell of energy E is

$$p = A\mathrm{e}^{-E/kT} \tag{2.28}$$

where A has the same value for all phase cells, no matter what their energy. The proportionality constant A is called the **normalization factor**, and the exponential term $e^{-E/kT}$ is called the **Boltzmann factor**.

This law may seem rather abstract, far removed from experimental results (let alone everyday conversation!), so it is worth pausing for a moment to consider its significance.

Remember that a phase cell defines a position and a velocity (to within a chosen precision). Because Boltzmann's distribution law tells us the probability of finding a molecule in any given phase cell, it gives us a detailed statistical description of the likely whereabouts and motions of the molecules. This is the 'holy grail' of statistical mechanics, from which much else can be deduced.

You may be tempted to forget that Boltzmann's law refers to phase cells and paraphrase it along the lines 'the probability that a molecule has a certain energy is proportional to the Boltzmann factor.' This certainly sounds straightforward, and has even been used in some semi-popular books, but is actually WRONG. In classical statistical physics, we *must* use phase cells to describe the state of a molecule. Since different energies can correspond to different numbers of phase cells, it is important to stick to Boltzmann's precise and correct formulation, as given by Equation 2.28.

If you are still uncomfortable with the concept of probability, remember that it tells us the fraction of times something is expected to happen. As a gas molecule moves and collides with others, it passes through a succession of phase cells. The probability of finding the molecule in a given phase cell is *the fraction of time* spent by the molecule in that phase cell. If many phase cells are sampled by the molecule, this probability will be very small. The normalization factor A is chosen to make sure that the sum of the probabilities for all the phase cells is equal to 1 in accordance with the normalization rule for probabilities. What really counts, though, is the *relative* probability of different phase cells.

For example, consider a gas in equilibrium at temperature T. If we take two different phase cells, labelled 1 and 2, with energies E_1 and E_2, the probabilities of finding a gas molecule in these phase cells are

$$p_1 = A\mathrm{e}^{-E_1/kT}$$

$$p_2 = A\mathrm{e}^{-E_2/kT}$$

So $$\frac{p_2}{p_1} = \frac{e^{-E_2/kT}}{e^{-E_1/kT}}.$$

Using some standard properties of the exponential function, this can also be written as

$$\frac{p_2}{p_1} = e^{-E_2/kT} \times e^{+E_1/kT}$$

$$\frac{p_2}{p_1} = e^{-(E_2-E_1)/kT}. \qquad (2.29)$$

This equation is plotted as a graph in Figure 2.24, where we have taken E_2 to be greater than E_1. Notice that the relative likelihoods of the two phase cells are determined by the size of their energy difference in comparison to kT. If the difference in their energies is much smaller than kT, the two phase cells are almost equally likely. If the difference in their energies is much larger than kT, the high-energy phase cell is very much less likely than the low-energy phase cell. If, for example, $E_2 - E_1 = 10\,kT$, the high-energy phase cell is about 0.000 05 times as likely as the low-energy phase cell.

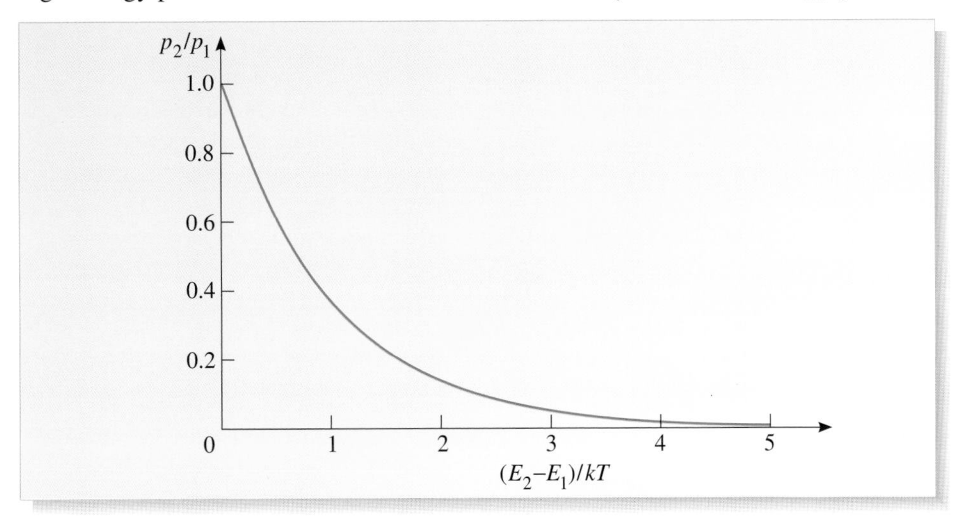

Figure 2.24 The relative probability of finding a given molecule in two different phase cells.

This discussion gives us a glimpse of the fundamental meaning of temperature. At low temperatures, a molecule is almost certain to be found in a low-energy phase cell: the odds are heavily stacked against finding it in a high-energy phase cell. Although the molecule is always more likely to be found in a given low-energy phase cell than in a given high-energy phase cell, the odds become more evenly balanced as the temperature rises (corresponding to moving from right to left across Figure 2.24). You can think of temperature as a parameter that sets the odds of finding a molecule in phase cells of different energies. At low temperatures, phase cells of low energy are much more likely but, as the temperature rises, the probabilities of phase cells of different energies become more and more equal. A temperature of 0 K corresponds to all the molecules being in phase cells of the lowest possible energy. An infinite temperature corresponds to all phase cells being equally likely.

It is worth noting that this interpretation of temperature only applies to systems that are in equilibrium, or at least close to equilibrium. Strictly speaking, a system that is far from equilibrium does not have a temperature at all! You might think that the average translational energy could always be used to define a temperature via Equation 2.13, but this is not good enough: a definite temperature implies that the Boltzmann distribution law is satisfied in detail, and this is only true for systems that are in equilibrium.

Finally, let me remind you of the assumptions that lie behind Boltzmann's distribution law. We assume that a configuration of a gas can be defined by declaring which molecules are in which phase cells. The conservation of energy requires all allowed configurations to have the same total energy, and all such allowed configurations are taken to be equally likely. This does not mean that all phase cells are equally likely — far from it, as the appearance of an exponential factor, favouring low-energy phase cells, shows. The bias towards low-energy phase cells arises because, at any fixed total energy, the *number* of configurations in which a given molecule occupies a given low-energy phase cell is greater than the number of configurations in which that molecule occupies a given high-energy phase cell. It is as simple (or as hard) as that. Notice that the law of conservation of energy plays a key role, by restricting the configurations to those with a fixed energy. If all configurations were allowed, there would be no bias towards low-energy phase cells and no Boltzmann factor (see Box 2.2).

Box 2.2 A memorable seminar

The role of the law of conservation of energy in justifying the Boltzmann factor was brought home to me in a dramatic incident. A visiting biology professor was giving a seminar in the physics department where I was a postgraduate student. The seminar was part of a regular series, each scheduled to last for about an hour. Twenty minutes into his talk, the professor defined some biological variables and argued that one of these played a role 'similar to energy', and another played a role 'similar to temperature' so it would be natural to introduce an exponential factor into his equations, which would play a role similar to the Boltzmann factor. The rest of the talk was designed to explore the biological consequences of this idea. Abruptly, one of the audience (who had given every appearance of being asleep) asked: 'where is your conservation law?'. The professor wanted more details on the thinking behind this question, and was presented with a devastating argument: there was no reason to believe that the biological variable (described as being 'similar to energy') would be conserved, so there was no reason to expect that anything like a Boltzmann factor would be appropriate. The seminar ended there and then, as all agreed there would be little point in continuing.

Question 2.10 A nitrogen molecule in a gas in equilibrium at 300 K spends, on average, a fraction 1.0×10^{-10} of its time in a tiny phase cell (1) centred on $x = 0\,\text{m}$, $y = 0\,\text{m}$, $z = 0\,\text{m}$, $v_x = 100\,\text{m s}^{-1}$, $v_y = 100\,\text{m s}^{-1}$, $v_z = 100\,\text{m s}^{-1}$. What fraction of its time is spent, on average, in a similar phase cell (2) centred on $x = 0\,\text{m}$, $y = 0\,\text{m}$, $z = 0\,\text{m}$, $v_x = 200\,\text{m s}^{-1}$, $v_y = 200\,\text{m s}^{-1}$, $v_z = 200\,\text{m s}^{-1}$? The mass of a nitrogen molecule is $4.65 \times 10^{-26}\,\text{kg}$. ■

5.5 The distribution of molecular speed

In practice, we don't need all the detail that is given by Boltzmann's distribution law. We scarcely ever need to know the probability of finding a molecule in a single phase cell; what is generally needed is the probability of finding a molecule in some special collection of phase cells. For example, consider the distribution of molecular speeds in a gas. In this case, you have seen that interest focuses on the speed distribution function, $f(v)$, where for a small speed range, Δv,

$$f(v)\,\Delta v = \text{fraction of molecules with speeds between } v \text{ and } v + \Delta v.$$

The right-hand side of this equation can be expressed in various ways: it is also the fraction of time that a given molecule has speeds between v and $v + \Delta v$, and it is the probability of finding a given molecule with a speed in that range.

We will now assume that the gas is composed of structureless molecules of mass m which only have translational energy

$$E = \tfrac{1}{2}mv^2 \tag{2.30}$$

and then show how the speed distribution function $f(v)$ can be derived from Boltzmann's distribution law.

In deriving the speed distribution function, we are interested in molecules that have speeds in a given small range, from v to $v + \Delta v$, so we must concentrate on phase cells that are characterized by speeds in this range. For brevity, we describe these as the *relevant* phase cells. According to the Boltzmann distribution law, the probability of a phase cell is proportional to the Boltzmann factor:

$$p \propto \mathrm{e}^{-E/kT} = \mathrm{e}^{-mv^2/2kT}. \tag{2.31}$$

Notice that this depends on speed, but not on position or direction of motion. So, the relevant phase cells, which all lie within a small range of speeds, can all be taken to have the *same* probability.

The only remaining question is: how many relevant phase cells are there? This can be answered using a geometric argument, based on Figure 2.25. In this diagram, the spherical shell contains the velocities that correspond to the given speed range, from v to $v + \Delta v$. Each of the tiny cubes within this shell represents the velocity range associated with a relevant phase cell. The number of cubes within this shell, and hence the number of relevant phase cells, is proportional to the volume of the spherical shell. The volume of a spherical shell of radius v and thickness Δv is proportional to $v^2 \Delta v$ (the square of its radius times its thickness), so we conclude that

$$\begin{pmatrix}\text{number of phase cells associated with}\\ \text{speed range from } v \text{ to } v + \Delta v\end{pmatrix} \propto v^2\,\Delta v. \tag{2.32}$$

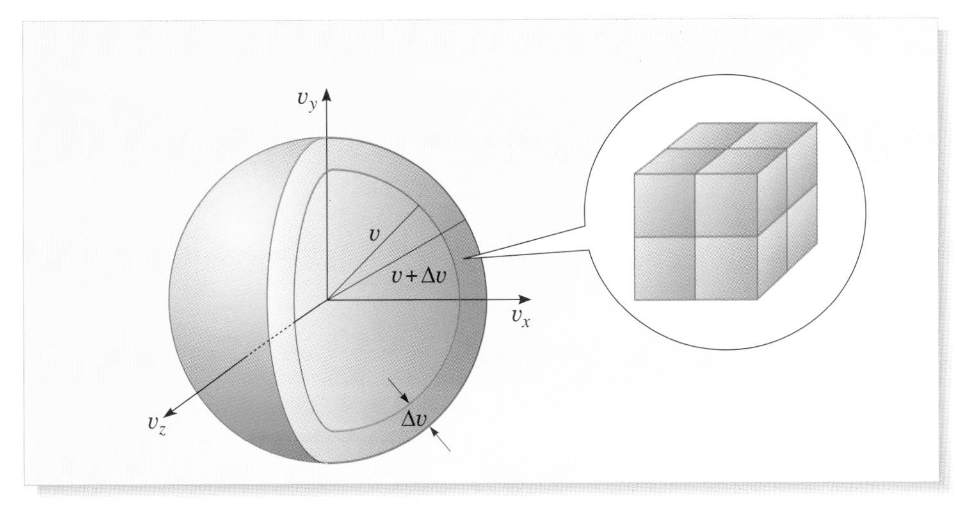

Figure 2.25 The number of phase cells associated with the speed range between v and $v + \Delta v$ is proportional to the volume of the spherical shell.

All of these thoughts can be brought together by noting that

$$\begin{pmatrix}\text{probability of finding a}\\ \text{given molecule in the speed}\\ \text{range between } v \text{ and } v + \Delta v\end{pmatrix} = \begin{pmatrix}\text{no. of phase}\\ \text{cells in the}\\ \text{speed range}\end{pmatrix} \times \begin{pmatrix}\text{probability of finding the}\\ \text{molecule in a given phase}\\ \text{cell in the speed range}\end{pmatrix}$$

The left-hand side of this equation is $f(v)\,\Delta v$, and the two terms on the right-hand side are proportional to $v^2\,\Delta v$ (Equation 2.32) and the Boltzmann factor $e^{-mv^2/2kT}$ (Equation 2.31). We therefore have

$$f(v)\,\Delta v \propto (v^2\,\Delta v) \times e^{-mv^2/2kT}$$

so we conclude that

$$f(v) = Bv^2 e^{-mv^2/2kT} \tag{2.33}$$

where B is a proportionality constant, which is independent of v.

Because $f(v)$ is a distribution function, the entire area under the graph of $f(v)$ against v must be equal to 1 or, expressed in terms of integrals,

$$\int_0^\infty f(v)\,dv = 1. \qquad \text{(Eqn 2.24b)}$$

The value of B must be chosen to ensure that this is true. The messy details lie beyond the scope of this course, but they are only a technicality. For the record, it turns out that

$$B = 4\pi\left(\frac{m}{2\pi kT}\right)^{3/2}. \tag{2.34}$$

This value of B ensures that any probability calculated from Equation 2.33 is scaled in the usual way (i.e. normalized), with 1 corresponding to certainty.

Equation 2.33, with B defined by Equation 2.34, was first written down by Maxwell in 1860, and is known as the **Maxwell speed distribution**. Maxwell gave an incomplete proof: his derivation rested on some assumptions which at first seemed plausible, but which later turned out to be highly contentious. After various attempts, the first full and rigorous proof was given by Boltzmann in 1896, using arguments similar to those given here. So, Equation 2.33 is also known as the **Maxwell–Boltzmann speed distribution**. Figure 2.26 plots this distribution for

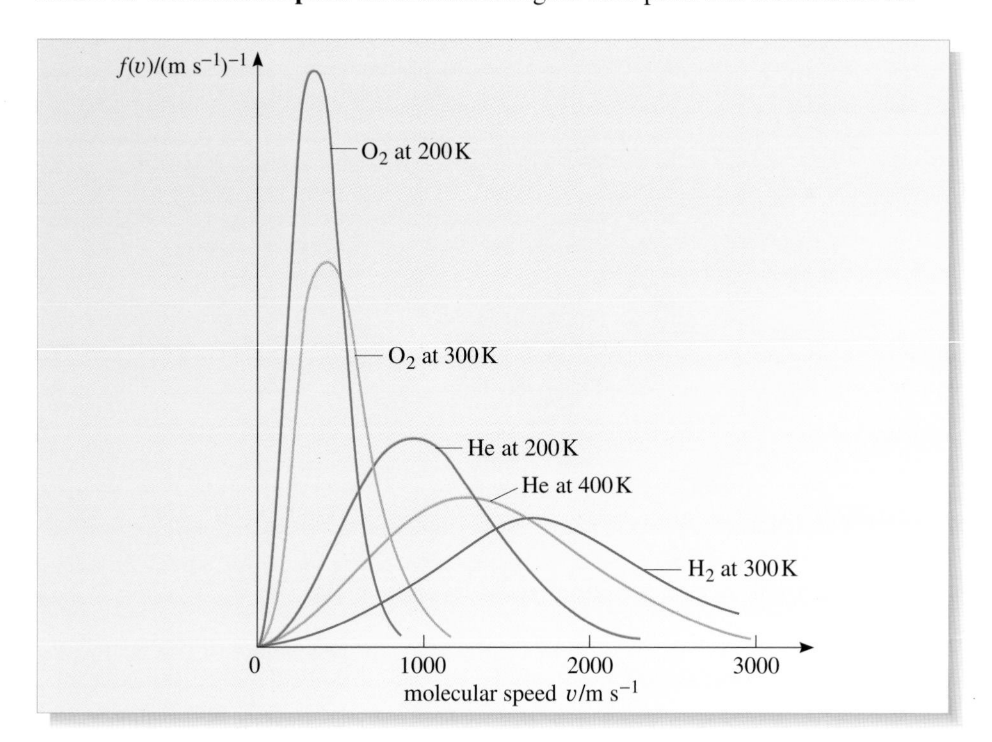

Figure 2.26 Graphs of the Maxwell speed distribution for oxygen, hydrogen and helium at particular temperatures. The masses of the molecules are such that $m_{O_2} > m_{He} > m_{H_2}$.

some typical cases. The fact that these curves agree with experimental results, and with computer simulations (e.g. Figure A1), is a triumphant vindication of the methods of statistical mechanics.

Early in this course, we encouraged you to get beneath the squiggles and symbols of equations, to try to understand their meaning. The complexity of the Maxwell speed distribution is rather off-putting, so let me come clean. I have never tried to memorize Equation 2.34 — whenever I need it, I look up the details. Equation 2.33, on the other hand, has a much more transparent interpretation. It shows that the speed distribution function is the product of three factors:

- an overall constant of proportionality (B);
- a term proportional to the number of relevant phase cells (v^2);
- a term proportional to the probability of finding a molecule in one of the relevant phase cells (the Boltzmann factor, $e^{-mv^2/2kT}$).

At very low speeds, v^2 is tiny. At very high speeds, the Boltzmann factor is tiny. The peak of the distribution lies between these two extremes, where there is a reasonable number of phase cells, each with a reasonable chance of being occupied.

One other feature can be read directly from the above equations. The speed distribution function $f(v)$ depends on the absolute temperature T and on the molecular mass m, but in a very special way: Equations 2.33 and 2.34 show that only the *ratio* m/T is significant. Thus, if one gas has molecular mass m_1 and temperature T_1, and another gas has molecular mass m_2 and temperature T_2, the two gases will have exactly the same speed distribution provided that $m_1/T_1 = m_2/T_2$. For example, helium molecules have twice the mass of hydrogen molecules, so helium at 600 K will behave just like hydrogen at 300 K.

Now that an expression for the speed distribution function has been found, it can be used to derive explicit expressions for the most probable speed, v_{mp}, average speed $\langle v \rangle$ and the root mean square speed, v_{rms}. The derivations are beyond the scope of this chapter, but here are the final results:

$$v_{mp} = \sqrt{\frac{2kT}{m}} \tag{2.35}$$

$$\langle v \rangle = \sqrt{\frac{8kT}{\pi m}} \tag{2.36}$$

$$v_{rms} = \sqrt{\frac{3kT}{m}}\,. \tag{2.37}$$

Any of these speeds may be taken to be a 'typical' speed for a molecule in a gas at temperature T, but they are not quite the same:

$$v_{mp} < \langle v \rangle < v_{rms}.$$

Because the whole shape of the Maxwell speed distribution depends only on the *ratio* m/T, it is hardly surprising that these speeds also depend only on m/T.

5.6 The distribution of molecular translational energy

The translational energy of a molecule is

$$E = \tfrac{1}{2}mv^2 \qquad \text{(Eqn 2.30)}$$

so the molecules with the highest translational energies are those with the highest speeds. If we could identify the molecules that have translational energies between E and $E + \Delta E$, these molecules would have speeds in some range, from v to $v + \Delta v$. Counting up such molecules, first in terms of their translational energy, and then in terms of their speed, gives

$$g(E)\,\Delta E = f(v)\,\Delta v \tag{2.38}$$

where $g(E)$ is the translational energy distribution function and $f(v)$ is the speed distribution function. We can therefore use the Maxwell speed distribution to obtain an explicit formula for the translational energy distribution function.

Using Equations 2.30 and 2.33, we can express $f(v)$ in terms of translational energy:

$$f(v) \propto E\mathrm{e}^{-E/kT}\,. \tag{2.39}$$

The relationship between the speed range, Δv, and the energy range, ΔE, is also needed. It can be found by noting that

$$\frac{\Delta E}{\Delta v} \approx \frac{\mathrm{d}E}{\mathrm{d}v}\,.$$

Differentiating Equation 2.30 with respect to v gives

$$\frac{\mathrm{d}E}{\mathrm{d}v} = mv\,.$$

Since the mass of the molecule is constant and the translational energy is proportional to the *square* of the speed, we have

$$\frac{\Delta E}{\Delta v} \propto \sqrt{E}$$

so

$$\Delta v \propto \frac{1}{\sqrt{E}}\,\Delta E\,. \tag{2.40}$$

Combining Equations 2.38, 2.39 and 2.40, we conclude that

$$g(E)\,\Delta E \propto E\,\mathrm{e}^{-E/kT} \times \frac{1}{\sqrt{E}}\,\Delta E\,.$$

So, our final expression for the energy distribution function is

$$g(E) = C\sqrt{E}\mathrm{e}^{-E/kT} \tag{2.41}$$

where C is a proportionality constant, which is independent of E.

Just as for the speed distribution, we must choose the value of the constant C so that the area under the graph of $g(E)$, from zero to infinity, is equal to 1. This makes sure that any probability calculated from Equation 2.41 is scaled in the conventional way (i.e. normalized), with 1 corresponding to certainty. Again, we will not carry out the explicit integration needed to evaluate C, but for the record here is the value that is obtained:

$$C = \frac{2}{\sqrt{\pi}}\left(\frac{1}{kT}\right)^{3/2}\,. \tag{2.42}$$

Equations 2.41 and 2.42 are known as the **Maxwell–Boltzmann energy distribution**. (In fact, Maxwell may never have written them down, but they are sufficiently close to the Maxwell–Boltzmann speed distribution that this name is appropriate, and it is useful to have a name that is clearly different from the Boltzmann distribution law given earlier.)

The Maxwell–Boltzmann energy distribution performs the same role for translational energy as the Maxwell–Boltzmann speed distribution did for speed. Figure 2.27 plots the Maxwell–Boltzmann energy distribution for a variety of cases. The fact that there is excellent agreement with experiment, and with computer simulations (e.g. Figure A2), provides further convincing support for statistical mechanics.

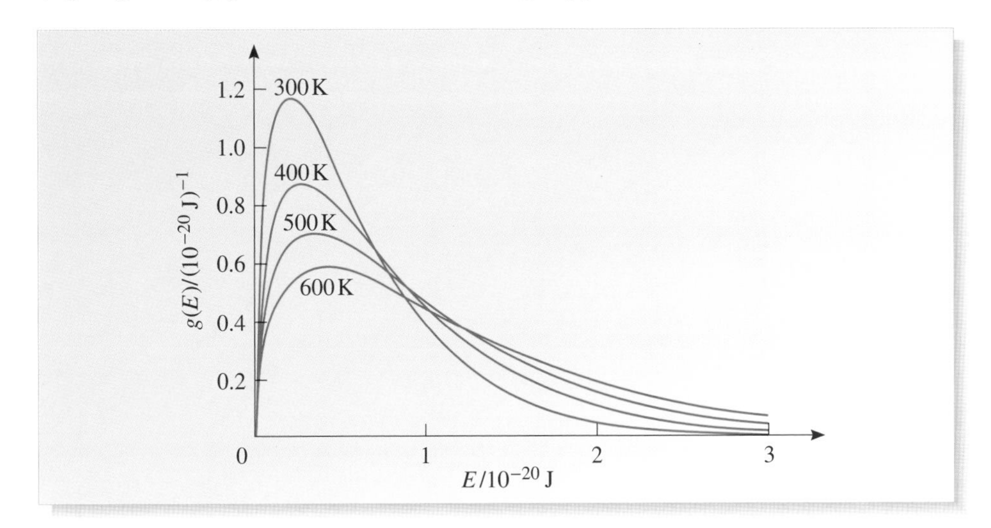

Figure 2.27 Graphs of the Maxwell–Boltzmann energy distribution at different temperatures.

Perhaps the most striking feature of the energy distribution in comparison with the speed distribution is that it *only* depends on temperature. At a given temperature, different gases, with molecules of different masses, have the same distribution of translational energies. Translational energy is proportional to the product of the mass and the square of the speed. So lighter molecules, with their lower mass, must be moving faster on average than heavier molecules — their more rapid motion compensates for their lower mass and ensures that all types of molecule have the same distribution of translational energy.

Air is a mixture of different gases: nitrogen, oxygen, carbon dioxide, water vapour, argon etc. Each type of molecule has the *same* distribution of translational energy but carbon dioxide molecules, with a relative molecular mass of 44, tend to be significantly more sluggish than water molecules with a relative molecular mass of 18. There is almost no hydrogen in the Earth's atmosphere; there was once, but hydrogen molecules are the lightest of all, with a relative molecular mass of 2. They move so rapidly that they have long since escaped the pull of Earth's gravity and drifted off into outer space. In the near future, ideas like this might be applied to oxygen molecules in planets beyond the Solar System to help discover whether they have conditions suitable for the development and sustenance of life.

The precise form of the Maxwell–Boltzmann energy distribution (Equations 2.41 and 2.42) can be used to derive explicit expressions for the most probable translational energy, E_{mp}, and the average translational energy $\langle E \rangle$:

$$E_{\text{mp}} = \tfrac{1}{2}kT \qquad (2.43)$$

$$\langle E \rangle = \tfrac{3}{2}kT. \qquad (2.44)$$

We will not carry out the detailed mathematics needed to establish these formulae, but you should be in no doubt that they are consequences of the Maxwell–Boltzmann distribution. Equation 2.44, for example, is obtained from the integral:

$$\langle E \rangle = \int_0^\infty E g(E)\,\mathrm{d}E$$

where $g(E)$ is given by Equations 2.41 and 2.42. Depending on your mathematical background you may find this daunting or intriguing. The important point, though, is that evaluation of the average translational energy has been reduced to a purely mathematical question, which can be handed over to a mathematician, or a computer algebra program such as *Physica*.

The final results are in full agreement with those obtained earlier in this chapter. For example, Equations 2.43 and 2.44 show that the average translational energy is three times the most probable translational energy, in agreement with Equation 2.20; and Equations 2.7 and 2.44 show that:

$$\tfrac{1}{2}mv_{\text{rms}}^2 = \tfrac{3}{2}kT$$

so that $$v_{\text{rms}} = \sqrt{\frac{3kT}{m}},$$

in agreement with Equation 2.37.

Given that Equation 2.44 follows from the Maxwell–Boltzmann energy distribution, we can now review our progress in linking the microscopic and macroscopic descriptions of gases — the problem set out in the introduction to this chapter.

In Section 3, we used the simple gas model to establish that

$$PV = \tfrac{2}{3}N\langle E_{\text{trans}}\rangle. \qquad \text{(Eqn 2.11)}$$

You have now seen that statistical mechanics leads to the conclusion that

$$\langle E_{\text{trans}}\rangle = \tfrac{3}{2}kT.$$

We can now put these results together to obtain

$$PV = NkT.$$

In other words, we have succeeded in using microscopic models and theories to derive the ideal gas equation of state. After a lengthy argument involving many new ideas, we have finally achieved the goal of uniting the microscopic and macroscopic descriptions of a gas.

Example 2.1

Figure 2.28 shows the initial distributions of speed and translational energy for molecules in a sample of monatomic helium gas. This is not an equilibrium distribution, but molecular collisions ensure that it soon comes to equilibrium, with no change in the total translational energy of the gas. What are the temperature, average translational energy and average speed of molecules in the final equilibrium distribution? The mass of a helium atom is 6.64×10^{-27} kg.

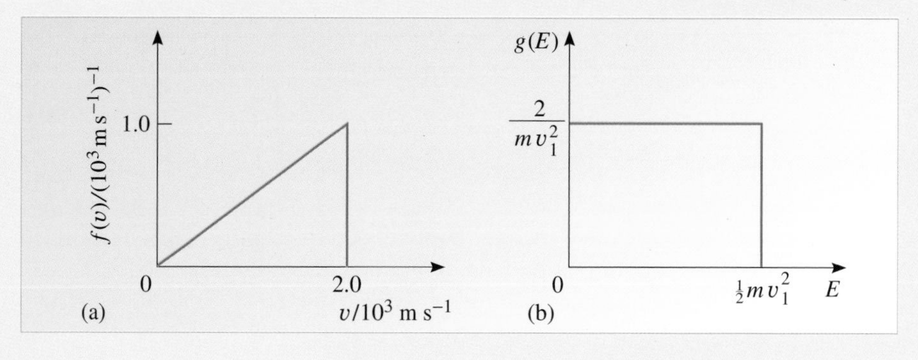

Figure 2.28 A non-equilibrium distribution of (a) speed and (b) translational energy. We use the notation m for the mass of a molecule and v_1 for the maximum speed of a molecule in the initial distribution (i.e. $2.0 \times 10^3\,\text{m s}^{-1}$)

Solution

This question can be answered by examining the translational energy distribution function. Because the initial distribution of translational energies is symmetrical, the average translational energy corresponds to the midpoint of the distribution. Thus:

$$\langle E \rangle = \frac{\frac{1}{2}mv_1^2}{2} = \tfrac{1}{4} \times 6.64 \times 10^{-27}\,\text{kg} \times (2.00 \times 10^3\,\text{m}\,\text{s}^{-1})^2$$
$$= 6.64 \times 10^{-21}\,\text{J}.$$

The final distribution has the same energy as the initial distribution, so we again have

$$\langle E \rangle = 6.64 \times 10^{-21}\,\text{J}.$$

When the system finally reaches equilibrium, it has a definite temperature, T, with

$$\langle E \rangle = \tfrac{3}{2}kT$$

so the temperature is

$$T = \frac{2\langle E \rangle}{3k} = \frac{2 \times 6.64 \times 10^{-21}\,\text{J}}{3 \times 1.38 \times 10^{-23}\,\text{J}\,\text{K}^{-1}} = 321\,\text{K}.$$

Corresponding to this temperature, we can find the final average speed of the molecules from Equation 2.36:

$$\langle v \rangle = \sqrt{\frac{8kT}{\pi m}} = \sqrt{\frac{8 \times 1.38 \times 10^{-23}\,\text{J}\,\text{K}^{-1} \times 321\,\text{K}}{\pi \times 6.64 \times 10^{-27}\,\text{kg}}}$$
$$= 1.30 \times 10^3\,\text{m}\,\text{s}^{-1}.$$

It is interesting to compare these results with the initial distributions of speed and energy. You have seen that the average translational energy is the same in the initial and final distributions. The same cannot be said for the average speed. In fact, the average speed in the initial distribution is $1.33 \times 10^3\,\text{m}\,\text{s}^{-1}$. It may surprise you that the molecules have slowed down slightly, on average. You might ask where the missing speed has gone to, but this would not be a good question. There should be no expectation that the average speed will remain the same because there is no law of conservation of speed. By contrast, the average translational energy does remain constant because there is a law of conservation of energy and the collisions are taken to be elastic. Note, too, that we cannot compare temperatures in the initial and final distributions: the initial distribution has no temperature associated with it because the gas is not in equilibrium.

Question 2.11 The quantity

$$\Delta v_{\text{sd}} = \sqrt{\langle v^2 \rangle - \langle v \rangle^2}$$

is known as the *standard deviation* of the molecular speed. It gives a measure of the extent by which molecular speeds are spread on either side of the average value. Use the Maxwell speed distribution and the Maxwell–Boltzmann energy distribution to show that the standard deviation of the molecular speed in a gas is proportional to $\sqrt{T/m}$, where T is the absolute temperature and m is the molecular mass.

Question 2.12 The peak of the energy distribution function occurs at a translational energy of $kT/2$. Use this fact to show that the peak value of the energy distribution function is inversely proportional to the absolute temperature. ■

6 The equipartition of energy

The ideas of statistical mechanics, presented in the previous section for a gas of structureless particles, can be extended to more complex systems. We will not give any detailed derivations, but will leap ahead to state a very general conclusion, known as the equipartition of energy theorem, which has many important applications.

Before this 'theorem' can be stated, we need to analyse the types of energy a molecule can have (Figure 2.29).

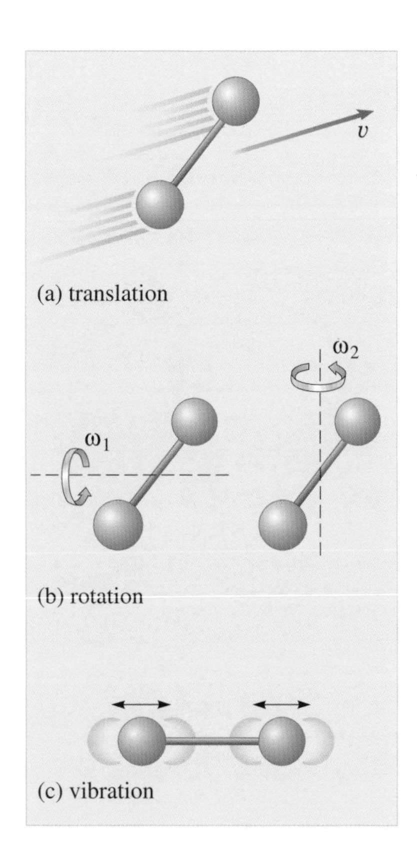

Figure 2.29 Types of energy in a molecule.

1 **Translational energy** appears when the molecule moves, as a whole, through space (Figure 2.29a). The expression for translational energy can be written out in full as

$$E_{\text{trans}} = \tfrac{1}{2}MV_x^2 + \tfrac{1}{2}MV_y^2 + \tfrac{1}{2}MV_z^2 \tag{2.45}$$

where M is the total mass of the molecule and V_x, V_y and V_z are the velocity components of the molecule's centre of mass.

2 **Rotational energy** appears when the molecule rotates about an axis (Figure 2.29b). We assume that no rotation is possible for a single atom. This may not fit with the classical image of an atom as a tiny ball, but it turns out to be a valid assumption, as you will see later in the course. If the molecule is diatomic, we also assume that the atoms cannot rotate about the line that joins their centres, but there are two independent axes, perpendicular to the line that joins the atoms, about which rotation can take place. The expression for the rotational energy of a diatomic molecule can therefore be written as

$$E_{\text{rot}} = \tfrac{1}{2}I\omega_1^2 + \tfrac{1}{2}I\omega_2^2 \tag{2.46a}$$

where I is the moment of inertia of the diatomic molecule about an axis through its centre of mass, perpendicular to the line joining the two atoms, and ω_1 and ω_2 are the angular speeds about the two axes of rotation. For a non-linear but symmetrical molecule, rotation can take place about three different axes, and the rotational energy can be written as

$$E_{\text{rot}} = \tfrac{1}{2}I_1\omega_1^2 + \tfrac{1}{2}I_2\omega_2^2 + \tfrac{1}{2}I_3\omega_3^2. \tag{2.46b}$$

3 **Vibrational energy** appears when the relative positions of individual atoms oscillate with simple harmonic motion. We assume that no vibration is possible for a single atom, but a diatomic molecule can vibrate with its atoms moving to and fro along the line that joins them. You saw in Chapter 1 that a vibrating diatomic molecule behaves like a one-dimensional simple harmonic oscillator, so its vibrational energy can be written as

$$E_{\text{vib}} = \tfrac{1}{2}mv^2 + \tfrac{1}{2}k_sx^2. \tag{2.47}$$

I shall not bother to relate m, v and x to other variables (such as the masses of the atoms in the molecule, or their separation) since the only thing that matters here is the general shape of Equation 2.47. For larger molecules, complicated modes of vibration are possible, but each mode is described by an equation similar to Equation 2.47.

4 **Intermolecular potential energy** appears when a molecule interacts with its neighbours. This is a considerable complication, but is one that can be safely ignored in the context of ideal gases. In an ideal gas, the intermolecular forces and potential energies are taken to be negligible.

Finally, the total energy of a molecule in a gas can be taken to be the sum of contributions due to translation, rotation and vibration:

$$E_{\text{tot}} = E_{\text{trans}} + E_{\text{rot}} + E_{\text{vib}}\,. \qquad (2.48)$$

The conclusion from all this is that the total energy of a molecule in a gas can generally be written as a sum of independent terms, each of which is proportional to either the *square* of a displacement coordinate (e.g. $\frac{1}{2}kx^2$), or the square of a velocity component (e.g. $\frac{1}{2}Mv_x^2$) or the square of an angular velocity component (e.g. $\frac{1}{2}I\omega_1^2$).

We can now state the central result of this section.

The equipartition of energy theorem

Suppose that the total energy of a molecule can be written as the sum of f independent terms, *each* of which is proportional to either the *square* of a displacement coordinate, or the square of a velocity component, or the square of an angular velocity component. The molecule is then said to have f **degrees of freedom**.

In equilibrium at temperature T, each degree of freedom contributes $kT/2$ to the average energy per molecule. If the molecule has f degrees of freedom, its average energy is therefore

$$\langle E_{\text{tot}} \rangle = \frac{f}{2}kT\,. \qquad (2.49)$$

John Waterston (1811–1883)

One of the first to suggest the equipartition of energy theorem was John Waterston (Figure 2.30). Waterston achieved extraordinary things, but was particularly unfortunate in his attempts to make them known. In 1843, he published a book entitled *Thoughts on the Mental Functions*, which undertook the (perhaps over-ambitious) task of explaining human behaviour in terms of mathematics. Hidden away at the end of this book was a full and accurate account of the kinetic theory of gases. Not surprisingly, this did not receive much attention, so Waterston decided to write up his ideas for publication by the prestigious Royal Society of London. His paper, written in 1845, included many original ideas, including a limited version of the equipartition of energy theorem. Unfortunately, it was rejected out of hand as being 'nothing but nonsense'. Not only did the Royal Society refuse to publish Waterston's paper — they refused to let him have it back! Waterston eventually moved on to other projects. Around 1883, he mysteriously disappeared, and is believed to have fallen into the sea. In 1891, Lord Rayleigh, one the great physicists of the nineteenth century, discovered Waterston's paper in the Royal Society archives. The paper had lost its impact, having become overtaken by the work of others, but Rayleigh insisted that it should be published as a belated recognition of the quality of Waterston's work.

Figure 2.30 John Waterston.

6.1 Application to gases

Consider again an ideal monatomic gas. In this case, only translational motion is possible, so there are *three* degrees of freedom (Equation 2.45). The equipartition of energy theorem then shows immediately that

$$\langle E_{\text{tot}} \rangle = \tfrac{3}{2}kT \tag{2.50}$$

which is just what we deduced from the Maxwell–Boltzmann energy distribution.

For a diatomic gas we would expect three translational degrees of freedom (Equation 2.45), two rotational degrees of freedom (Equation 2.46a) and two vibrational degrees of freedom (Equation 2.47), giving seven degrees of freedom in all. The equipartition of energy theorem therefore predicts that, for a diatomic gas,

$$\langle E_{\text{tot}} \rangle = \tfrac{7}{2}kT. \tag{2.51}$$

Similar calculations can be carried out for more complicated gas molecules. The average energy per particle is again predicted to be proportional to the absolute temperature but, because there will be more degrees of freedom, the constant of proportionality will be larger.

Corresponding to these average values, we can obtain expressions for the molar internal energy. This is the sum of all the energies of the molecules. Since one mole contains Avogadro's number of molecules,

$$U_{\text{m}} = N_{\text{m}}\langle E_{\text{tot}} \rangle$$

and the molar internal energies of monatomic and diatomic gases are predicted to be $3RT/2$ and $7RT/2$, respectively, where $R = N_{\text{m}}k$ is the universal gas constant. No one actually measures the molar internal energy, but it is easy to measure the *rate of change* of molar internal energy with temperature. This quantity is called the constant volume molar heat capacity, and is given by:

$$C_{V,\text{m}} = \frac{\text{d}U_{\text{m}}}{\text{d}T}.$$

Differentiating the molar internal energy with respect to temperature, we see that the equipartition of energy theorem predicts that $C_{V,\text{m}}$ is a constant, independent of temperature. All monatomic gases are predicted to have the same value, $3R/2$. All diatomic gases are predicted to have the same value $7R/2$. More complicated gases are predicted to have higher constant values, which can be calculated on an individual basis. How well do these predictions stand up to the test of experiment?

There are many successes. Around room temperature, most gases do have almost constant heat capacities, and the value obtained depends on the complexity of the molecule, with simple molecules having lower heat capacities than more complicated molecules. For monatomic gases, the prediction $C_{V,\text{m}} = 3R/2$ is very accurate. But, for diatomic molecules, there is an unexpected hitch. Around room temperature, most diatomic gases have $C_{V,\text{m}} = 5R/2$ instead of the $7R/2$ predicted from theory.

One way of looking at this unwelcome discovery is to say that, at room temperature, a typical diatomic gas has only five *effective* degrees of freedom. It is as if some types of motion were forbidden — that diatomic molecules translate and rotate, but do not vibrate. This interpretation is given further support by Figure 2.31, which shows the constant volume molar heat capacities of a selection of gases over a wide range of temperatures. (Each curve is limited in extent because the gases liquefy at low temperatures, and their molecules disintegrate into individual atoms at high temperatures.) Most diatomic gases behave like hydrogen, with a room temperature constant molar heat capacity of around $5R/2$. Well above room temperature, the heat capacity rises, and, at very high temperatures, the $7R/2$ prediction of the equipartition of energy theorem becomes fairly accurate. (In a few exceptional cases, such as fluorine or iodine, the $7R/2$ prediction is close to being satisfied, even at room temperature.)

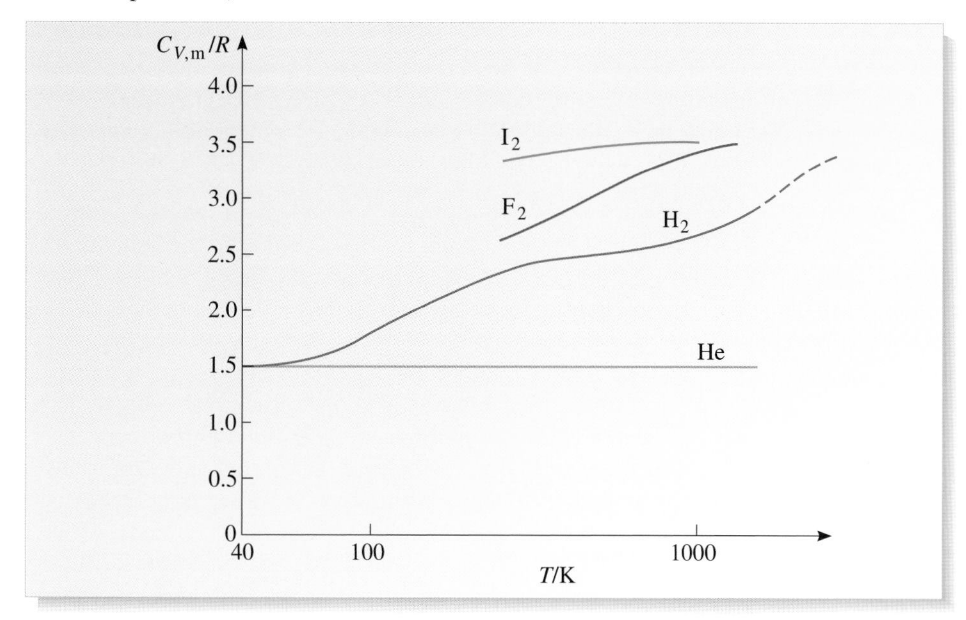

Figure 2.31 Constant volume molar heat capacities for various gases over a wide range of temperatures. Helium (He) is monatomic, while hydrogen (H_2), fluorine (F_2) and iodine (I_2) are diatomic.

Finally, if we cool a diatomic gas to well below room temperature, and the gas does not liquefy, the constant volume molar heat capacity drops towards $3R/2$. This behaviour would make sense if translational, rotational and vibrational degrees of freedom were active at the highest temperatures, if just translational and rotational degrees of freedom were active at room temperature, and if only translational degrees of freedom were active at the lowest temperatures. In the jargon of statistical mechanics, we say that certain types of degrees of freedom appear to be *frozen out* as the temperature drops.

● It is common to summarize the above results by writing

$$\langle E_{\text{tot}} \rangle = \frac{f}{2} kT$$

where f is the *effective* number of degrees of freedom. What value of f is appropriate for (a) a monatomic gas and (b) a typical diatomic gas at room temperature?

❍ (a) $f = 3$ (a monatomic gas has three translational degrees of freedom).

(b) $f = 5$ (a typical diatomic gas at room temperature has three translational degrees of freedom and two rotational degrees of freedom; the two vibrational degrees of freedom are frozen out at room temperature.) ■

Let me stress that this interpretation is an artificial one, arrived at in order to gain a semblance of agreement with experiment. You might think that it is reasonable for the vibrations to be suppressed at low temperatures. One would certainly expect the amplitude of vibration to be small at low temperatures, and this is confirmed by the fact that the light emitted by molecules, at low temperatures, contains almost no contribution from the frequencies associated with molecular vibrations. However, that is not the point. The fact that vibrations are possible is all that is needed for the equipartition of energy theorem to make its prediction. Because two degrees of freedom are associated with vibrations, the average vibrational energy of a molecule is predicted to be kT at any temperature. This expected contribution to the total energy appears to be missing — except at the highest temperatures.

The discovery of this discrepancy alarmed many scientists; and the closer they looked, the worse it got. When it became clear that atoms contain electrons, it followed that there would be degrees of freedom associated with the motion of these electrons. The equipartition of energy theorem would then predict many more contributions to the internal energy, and to the heat capacity, which are not detected in experiments.

Naturally, physicists looked hard at the assumptions that lie behind the equipartition of energy theorem to see if there were any loopholes that could save the situation, but they found it hard to fix the problem, whilst preserving the undoubted successes of the rest of statistical mechanics. I suspect this is why they called it a *theorem*. The word implies a certain authority, and reflects the fact that they saw no way of disproving it, much as they would have liked to have done so. This was certainly a setback for classical statistical mechanics, but was also one of its great achievements because the theory was, at least, advanced enough to show that a real problem existed. While this may seem a curious sort of achievement, the revelation of such points of crisis is a major aim of science, and is ultimately how progress is made. Eventually, the problem was resolved by the creation of quantum mechanics, which showed that a certain minimum 'quantum' of energy was needed to initiate each type of motion. It was not the statistical ideas that were to blame, but the whole world-view of Newtonian mechanics.

Question 2.13 One mole of nitrogen gas is held in a container of constant volume $1.0 \times 10^{-2}\,\text{m}^3$. Figure 2.32 shows how the distribution of molecular translational energies changes when the temperature is increased from T_1 to T_2. Calculate the energy that must be put into the gas to achieve this. What is the pressure of the gas at the higher temperature? (*Hint*: The simplest feature to measure on an energy distribution curve is the position of the peak. You may assume that, in the relevant temperature range, a nitrogen molecule has five effective degrees of freedom.) ■

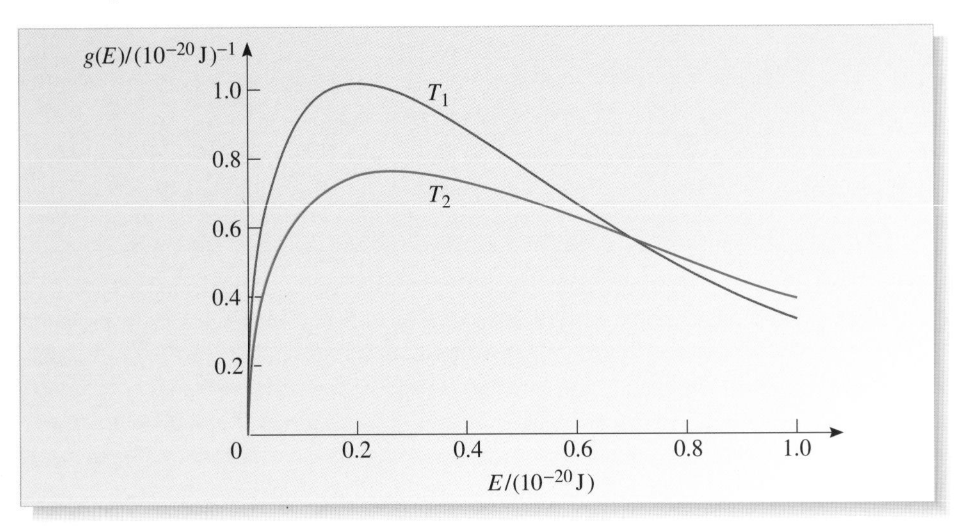

Figure 2.32 Energy distribution functions, for use in Question 2.13.

6.2 Application to solids

This chapter has concentrated on the behaviour of gases, since they are the simplest phase of matter to understand. Nevertheless, it is important to realize that the methods of statistical mechanics are completely general. The final topic in this chapter looks at what can be learnt by applying the equipartition of energy theorem to solids. We again find it convenient to introduce a model based on simplifying assumptions:

The simple solid model

A simple solid is one in which the atoms oscillate in three dimensions about fixed equilibrium positions (Figure 2.33). The atoms are treated as independent particles.

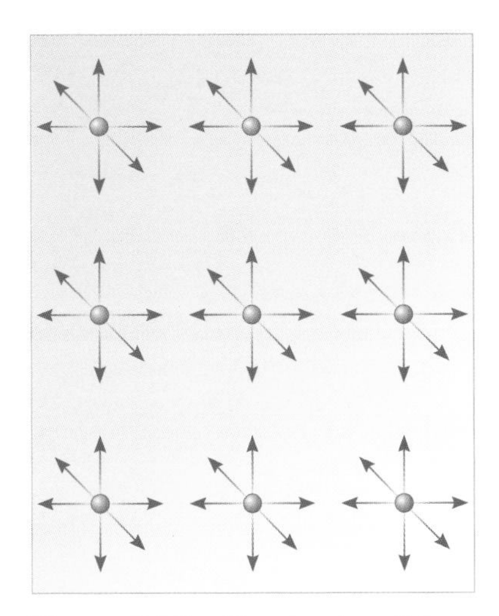

Figure 2.33 The simple solid model.

In a real solid, the atoms are not really independent of one another, and waves of atomic displacement can propagate through the solid in a highly ordered way. Even so, the simple solid retains the essential idea that the atoms vibrate in three dimensions about fixed equilibrium positions, and this is what matters for present purposes. Because each atom behaves as a three-dimensional simple harmonic oscillator, its energy can be written as

$$E_{\text{tot}} = \tfrac{1}{2}m(v_x^2 + v_y^2 + v_z^2) + \tfrac{1}{2}k_s(x^2 + y^2 + z^2) \tag{2.52}$$

where m is the atom's mass, v_x, v_y and v_z are components of the atom's velocity, k_s is a constant indicating how strongly the atom is attracted toward its equilibrium position and x, y and z are components of the atom's displacement from its equilibrium position. The details scarcely matter: the important point is that there are *six* independent squared terms in the expression for the atom's energy, so there are *six* degrees of freedom. The equipartition of energy theorem then gives

$$\langle E_{\text{tot}} \rangle = 3kT$$

so

$$C_{V,\text{m}} = 3R. \tag{2.53}$$

When Boltzmann obtained this result in 1876, he was providing a solution to a long-standing puzzle. As early as 1819, Pierre Dulong and Alexis Petit had measured the heat capacities of several solid elements at room temperature (Table 2.4). In modern terms, they observed that all the molar heat capacities were about $25\,\text{J}\,\text{K}^{-1}\,\text{mol}^{-1}$. This is known as the **Dulong–Petit law**. Inserting $R = 8.314\,\text{J}\,\text{K}^{-1}\,\text{mol}^{-1}$ in Equation 2.53 finally explained this pattern.

Later, exceptions to the Dulong–Petit law were found. At room temperature, the molar heat capacity of diamond is only $6.3\,\text{J}\,\text{K}^{-1}\,\text{mol}^{-1}$ and, at very low temperatures, the molar heat capacities of all substances start to fall. These exceptions are further examples of the problem noted earlier for diatomic gases. Again, quantum mechanics is needed to resolve this problem, as you will see in the last book of this course, *Quantum physics of matter*.

Question 2.14 When a solid is heated, the atoms vibrate more and more vigorously until the solid 'shakes itself to pieces' and forms a liquid. Frederick Lindemann (1886–1957) studied this phenomenon using the simple solid model. He proposed that solid elements melt when

Lindemann later became Lord Cherwell and was Scientific Adviser to Churchill during World War II.

$$x_{\text{rms}} = \sqrt{\langle x^2 \rangle} = \frac{d}{10}$$

where x appears in Equation 2.52 and d is the equilibrium interatomic spacing. Use the equipartition of energy theorem to predict the melting temperature of tungsten, which has $d = 2.5 \times 10^{-10}\,\text{m}$, and force constant $k_s = 100\,\text{N}\,\text{m}^{-1}$. ■

Table 2.4 Dulong and Petit's original data, converted to SI units. These measurements were taken at constant pressure but, for solids, there is little difference between constant pressure and constant volume heat capacities.

Element	Specific heat capacity / $J\,K^{-1}\,kg^{-1}$	Molar heat capacity / $J\,K^{-1}\,mol^{-1}$
bismuth	120.4	25.77
lead	122.5	25.48
gold	124.6	24.92
platinum	131.3	23.63
tin	214.8	25.35
silver	232.8	25.38
zinc	387.5	25.15
tellurium	381.2	24.74
copper	395.4	25.19
nickel	432.6	25.70
iron	459.9	25.06
cobalt	626.2	24.80
sulfur	785.8	25.50

7 Closing items

7.1 Chapter summary

1. The probability of an outcome is the fraction of times that outcome is expected to happen in the long run. A probability of 1 corresponds to certainty. The probability of two independent outcomes occurring together is the product of their probabilities; the probability of *either* of two alternative outcomes happening is the sum of their probabilities. The sum of probabilities for all the alternative outcomes is equal to 1. ('Alternative' means 'mutually exclusive' in this context.)
2. The average value of a quantity x is defined by multiplying each possible value x_i by its probability p_i and then adding together the results:

$$\langle x \rangle = \sum_{i=1}^{N} p_i x_i \,. \qquad \text{(Eqn 2.2b)}$$

3. If our best estimate is that an outcome will occur N times, we should not be surprised if it occurred between $N-\sqrt{N}$ and $N+\sqrt{N}$ times but would be *astonished* if the outcome occurred less than $N-10\sqrt{N}$ times or more than $N+10\sqrt{N}$ times.
4. The vast number of molecules present in any normal sample of gas means that the macroscopic properties of the gas can be deduced from the average behaviour of the molecules. Random fluctuations can be neglected.
5. The pressure of a gas, detected on the walls of a container, is due to the ceaseless random bombardment by gas molecules.

6 The simple gas model treats molecules as structureless particles in random motion. The molecules collide with one another and with the walls of their container, subject to the laws of Newtonian mechanics. Gravity and intermolecular forces between collisions are neglected and all collisions are taken to be elastic.

7 The simple gas model predicts that the pressure exerted by a gas depends on the average translational energy of the gas molecules:

$$PV = \tfrac{2}{3}N\langle E_{\text{trans}}\rangle. \qquad \text{(Eqn 2.11)}$$

8 The distribution of molecular speeds in a gas can be represented by a histogram in which the height of each bar is the fractional frequency of molecules with speeds in the range of the width of the bar. The sum of the heights of all the histogram bars is equal to 1. The distribution of molecular translational energies can likewise be represented by a histogram.

9 For any small range of speeds, Δv, the speed distribution function $f(v)$ is defined so that

$$f(v)\,\Delta v = \text{fraction of molecules with speeds between } v \text{ and } v + \Delta v. \qquad \text{(Eqn 2.21)}$$

The fraction of molecules with speeds between v_1 and v_2 is equal to the area under the speed distribution curve between v_1 and v_2. The total area under the speed distribution curve is equal to 1. The average speed is

$$\langle v\rangle = \int_0^\infty v\,f(v)\,\mathrm{d}v. \qquad \text{(Eqn 2.25)}$$

10 For any small range of translational energies, ΔE, the energy distribution function $g(E)$ is defined so that

$$g(E)\,\Delta E = \text{fraction of molecules with translational energies between } E \text{ and } E + \Delta E. \qquad \text{(Eqn 2.26)}$$

The fraction of molecules with translational energies between E_1 and E_2 is equal to the area under the energy distribution curve between E_1 and E_2. The total area under the energy distribution curve is equal to 1. The average translational energy is

$$\langle E\rangle = \int_0^\infty E g(E)\,\mathrm{d}E.$$

11 The speed distribution is slightly asymmetric, with a tail extending to high speeds. As the temperature increases, the most likely speed (corresponding to the peak of the distribution) and the average speed both increase. The peak becomes less prominent and the distribution becomes broader. Similar effects are observed when the mass of the molecule is reduced at a fixed temperature.

12 The energy distribution is strongly asymmetric, with a long tail extending to high translational energies. As the temperature increases, the most likely translational energy (corresponding to the peak of the distribution) and the average translational energy both increase. The peak becomes less prominent and the distribution becomes broader. The energy distribution does not depend on the mass of the molecule: all gases at the same temperature have the same distribution of translational energy.

13 A phase cell specifies the position and velocity of a particle to within a certain precision. The energy of a molecule occupying a given phase cell is loosely termed the energy of the phase cell. A configuration of a gas is defined by stating which molecules are in which phase cells.

14 Statistical mechanics of gases is based on two principles:

- The energy of the gas does not depend on time, but has a constant value, E. The only allowed configurations of the gas are those with the fixed energy, E.
- Each of the allowed configurations is equally likely.

15 There are more allowed configurations with a given molecule in a low-energy phase cell than in a high-energy phase cell. This is quantified by Boltzmann's distribution law which states that, for a gas in equilibrium at temperature T, the probability p of finding a given molecule in a given phase cell of energy E is

$$p = A\mathrm{e}^{-E/kT} \qquad \text{(Eqn 2.28)}$$

where the normalization factor A has the same value for all phase cells, no matter what their energy. The exponential term $\mathrm{e}^{-E/kT}$ is called the Boltzmann factor.

16 Temperature can be interpreted as a parameter that sets the relative chances of finding a molecule in different phase cells. At low temperatures, the odds strongly favour low-energy phase cells, but as the temperature rises, the odds become more and more equal for phase cells of different energies. This interpretation only applies to systems that are in equilibrium, or close to equilibrium.

17 Boltzmann's distribution law can be used to derive the Maxwell speed distribution

$$f(v) = Bv^2\mathrm{e}^{-mv^2/2kT} \qquad \text{(Eqn 2.33)}$$

where $B = 4\pi\left(\dfrac{m}{2\pi kT}\right)^{3/2}$ (Eqn 2.34)

and the Maxwell–Boltzmann energy distribution

$$g(E) = C\sqrt{E}\mathrm{e}^{-E/kT} \qquad \text{(Eqn 2.41)}$$

where $C = \dfrac{2}{\sqrt{\pi}}\left(\dfrac{1}{kT}\right)^{3/2}$. (Eqn 2.42)

18 The Maxwell–Boltzmann energy distribution has an average translational energy

$$\langle E_{\text{trans}} \rangle = \tfrac{3}{2}kT.$$

Combined with the expression for the pressure of a gas, this provides a microscopic derivation of the ideal gas equation of state,

$$PV = NkT.$$

19 If the total energy of a molecule can be written as the sum of f independent terms, each of which is proportional to either the *square* of a displacement coordinate, or the square of a velocity component or the square of an angular velocity component, the molecule is said to have f degrees of freedom.

20 The equipartition of energy theorem states that, in equilibrium at temperature T, each degree of freedom contributes $kT/2$ to the average energy per molecule. If the molecule has f degrees of freedom, its average energy is $fkT/2$.

21 Monatomic gases ($f = 3$) obey the equipartition of energy theorem. Diatomic gases ($f = 7$) obey the equipartition of energy theorem at very high temperatures, but at room temperature most behave as if they had only five *effective* degrees of freedom. At very low temperatures, diatomic gases behave as if they had only three effective degrees of freedom.

22 The Dulong–Petit law states that all solid elements have molar heat capacities of about $3R$. It is valid for most elements at room temperature, but is not true for diamond, or for other solids at temperatures well below room temperature.

23 A simple solid is one in which independent atoms oscillate in three dimensions about fixed equilibrium positions. Each atom has six degrees of freedom. The Dulong–Petit law is accounted for by applying the equipartition of energy theorem to the simple solid model.

24 Difficulties associated with applying the equipartition of energy theorem to diatomic gases, diamond or low-temperature solids will be resolved by using quantum physics rather than Newtonian physics, but the concepts of statistical mechanics will be maintained.

7.2 Achievements

Now that you have completed this chapter, you should be able to:

A1 Understand the meaning of all the newly defined (emboldened) terms introduced in this chapter.

A2 Perform simple calculations using probabilities and average values, and judge whether a given fluctuation is surprising or not.

A3 State the assumptions of the simple gas model and discuss why they are reasonable.

A4 Define what is meant by a distribution function in the context of molecular speeds or translational energies.

A5 Sketch the speed distribution and energy distribution for a variety of temperatures and molecular masses and describe the main features of these distributions.

A6 Use the area under a distribution curve to estimate the fraction of molecules that have speeds or translational energies in a given range.

A7 State Boltzmann's principles of statistical mechanics and explain qualitatively how they lead to the Boltzmann distribution law.

A8 Use the Boltzmann factor to estimate the relative probabilities of finding a molecule in different phase cells.

A9 Interpret temperature as a parameter that determines the relative chances of finding a given molecule in phase cells of different energies.

A10 Recall how the Maxwell speed distribution depends on speed and how the Maxwell–Boltzmann energy distribution depends on translational energy.

A11 Use the Maxwell speed distribution and the Maxwell–Boltzmann energy distribution in simple calculations (not involving differentiation or integration).

A12 Define the simple solid model.

A13 State the equipartition of energy theorem and apply it in simple cases. Give examples of this theorem agreeing and disagreeing with experiment.

7.3 End-of-chapter questions

Question 2.15 A horizontal tube contains N identical non-interacting molecules of a simple gas. Each molecule is equally likely to be found in the left-hand and right-hand halves of the tube. Calculate the probability of finding all the molecules in the right-hand half of the tube for the following values of N: (a) 1; (b) 2; (c) 8; (d) 100.

Question 2.16 (a) A stationary car contains $6.6\,m^3$ of air at atmospheric pressure and a temperature of 300 K. What is the total translational kinetic energy of the gas molecules in the air? (b) The car accelerates and reaches a steady speed of $30\,m\,s^{-1}$.

By what fraction has the total translational kinetic energy of the gas molecules increased? (Take the mass of a typical molecule in air to be 4.65×10^{-26} kg.)

Question 2.17 At room temperature, nitrogen molecules have an average speed of about $500\,\mathrm{m\,s^{-1}}$. Oxygen molecules have similar speeds. Approximately how many gas molecules per second will strike a square millimetre of a surface exposed to air at room temperature and atmospheric pressure? *Hint*: For a rough estimate, it is sufficient to use the Joule classification, with one-sixth of the molecules heading directly towards the surface at $500\,\mathrm{m\,s^{-1}}$, and the remaining molecules travelling away from the surface or tangential to it.

Question 2.18 In Question 2.5, we found that $v_{\mathrm{mp}} = 296\,\mathrm{m\,s^{-1}}$ for the speed at which particles are most commonly seen to move in the beam of bismuth used for the experiment described in Section 4.1. Calculate the temperature of the furnace holding the bismuth, assuming that the bismuth particles in the beam are all single atoms. (Bismuth has a relative atomic mass of 209.)

Question 2.19 A molar sample of an ideal gas is in equilibrium at a given temperature. Figure 2.34 shows the distribution function for translational energy of molecules in this sample. Use this graph to estimate the number of molecules with translational energies between E_1 and $E_1 + \Delta E$, where $E_1 = 1.00 \times 10^{-20}$ J and $\Delta E = 1.00 \times 10^{-25}$ J. Would it be surprising if this number varied by one part in 10^8?

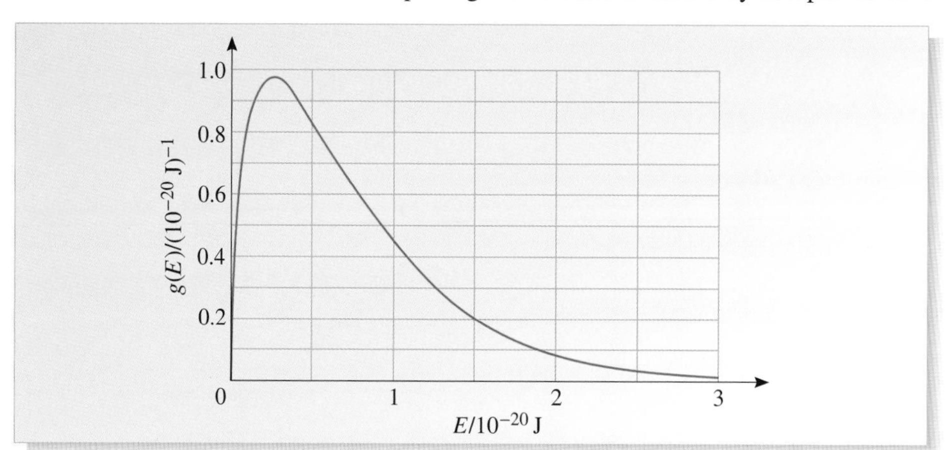

Figure 2.34 An energy distribution function for use with Question 2.19.

Question 2.20 The outer layer of the Earth's atmosphere is heated by the absorption of solar radiation and by collisions with high-energy cosmic and solar particles. It has a temperature of about 1500 K and the gas there consists mainly of single atoms. These atoms can only escape from the atmosphere if they are moving away from the Earth at speeds in excess of the escape speed ($11.2\,\mathrm{km\,s^{-1}}$). Assuming that the gas in the outer layer of the Earth's atmosphere is in equilibrium at 1500 K, estimate the fraction of oxygen molecules that have speeds greater than the escape speed. Is there much danger of the Earth's oxygen leaking away into outer space?

Useful mathematical fact: For values of E_1 that are many times kT, the following approximation is valid:

$$\int_{E_1}^{\infty} \sqrt{E}\,\mathrm{e}^{-E/kT}\,\mathrm{d}E \approx kT\sqrt{E_1}\,\mathrm{e}^{-E_1/kT}.$$

(Take the mass of an oxygen atom to be 2.66×10^{-26} kg.) ■

Appendix to Chapter 2 – Computer simulations

Computer simulations can be carried out using a small number of molecules confined to a tiny cube. The molecules are taken to be hard spheres which collide elastically with one another and with the walls of the container. The molecules have translational energy, but no rotational, vibrational or potential energy. To ensure that collisions between molecules are frequent, the molecules are taken to be larger than real molecules. This does not affect the equilibrium speed and energy distributions, but accelerates the approach to equilibrium.

Speed and energy histograms can be calculated at regular time intervals. Because computer simulations are carried out with a relatively small number of molecules, the histograms are likely to fluctuate significantly from one instant to the next. Cumulative histograms are used to overcome this. In a cumulative histogram, the results are collected over an extended period of time. The data are allowed to accumulate until the histograms have settled down and are no longer changing or fluctuating. The cumulative histograms then have similar shapes to the equilibrium histograms that apply, instant by instant, in a real gas which contains a vastly greater number of molecules. Computer simulations can therefore be used to explore the speed and energy distributions in detail.

Fluctuations

Fluctuations in the equilibrium histograms can be investigated by varying the number of molecules and the sampling time used in the simulation. The fluctuations are observed to decrease when the number of molecules is increased and the sampling time is increased. (In a real gas, there are vastly more molecules than in the computer simulation, leading to fluctuations that are negligible even if the sampling time is extremely small.)

The approach to equilibrium

It is possible to choose a variety of starting conditions for the molecules in the computer simulations. The starting conditions can be very far from equilibrium — for example, all the molecules can start with the same energy. Once the simulation is started, and the molecules collide with one another, the initial distribution of energy and speed changes. No matter what the starting distribution, the histograms of energy and speed settle down into well-defined equilibrium patterns. For a given gas, and a given initial average energy, the final equilibrium distributions are independent of the details of the starting distribution.

Factors that affect the equilibrium distributions

Simulations can be used to explore how the equilibrium distributions depend on various factors: the molecular mass, the temperature and the volume of the gas (for a given number of molecules). The main findings are as follows:

- The volume of the gas has no influence on the speed or the energy distributions.
- As the temperature rises, the equilibrium speed distribution becomes broader and shifts to higher speeds and the peak of the distribution becomes less pronounced (compare Figure A1a and Figure A1b). Similarly, the equilibrium energy distribution becomes broader and shifts to higher energies, and the peak of the distribution becomes less pronounced (compare Figure A2a and A2b).
- In more detail, the average speed is proportional to the square root of the absolute temperature and the average energy is proportional to the absolute temperature.

- As the molecular mass decreases, the equilibrium speed distribution becomes broader and shifts to higher speeds, and the peak of the distribution becomes less pronounced (Figure A1a and A1c). However, the equilibrium energy distribution is independent of molecular mass: at a fixed temperature, all gases have the same equilibrium distribution of translational energy (Figure A2a and A2c).
- In more detail, the average speed is inversely proportional to the square root of the molecular mass.

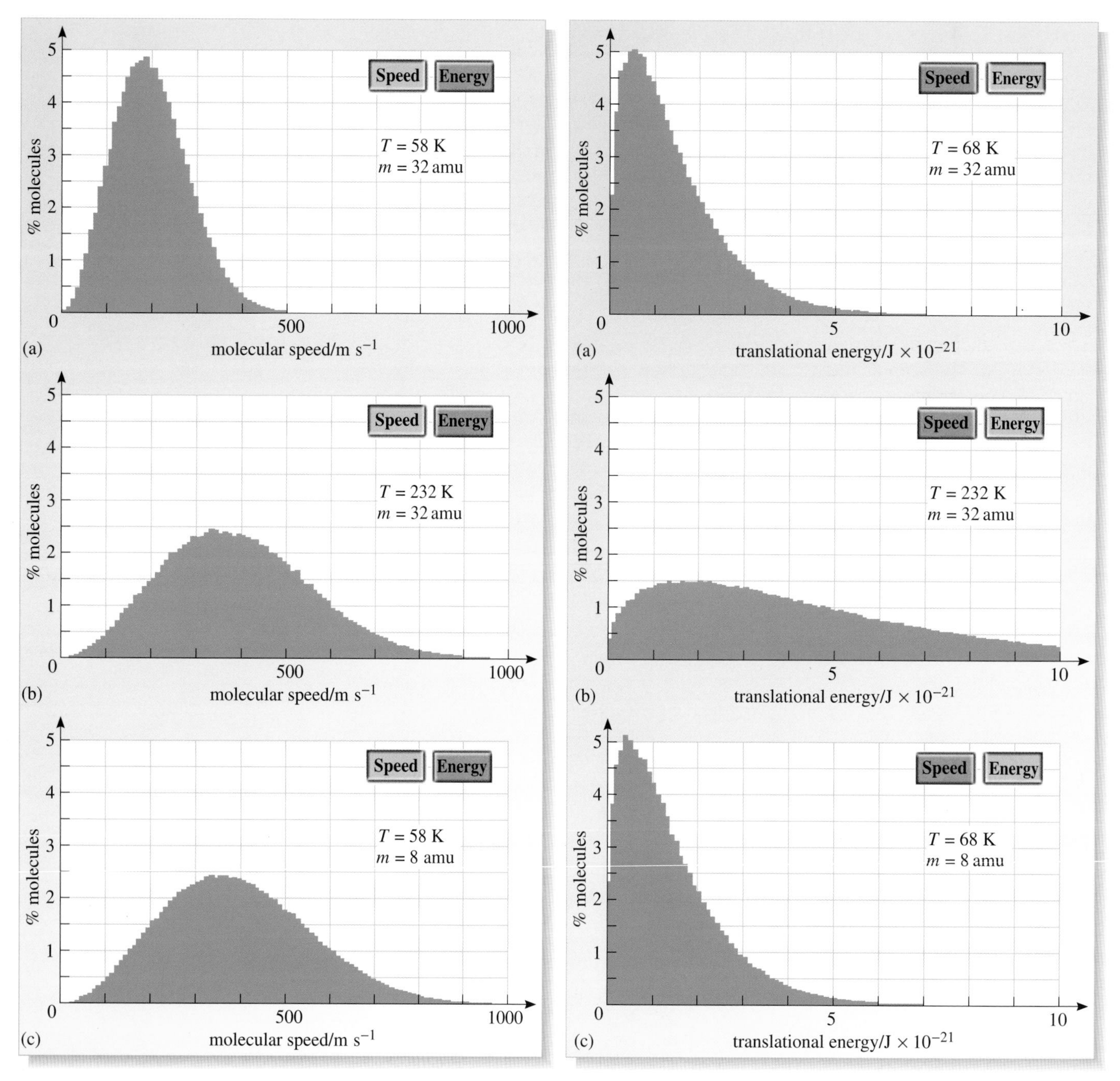

Figure A1 Speed histograms taken from computer simulations.

Figure A2 Energy histograms taken from computer simulations.

Chapter 3 Thermodynamics and entropy

1 Engine efficiency — an effect of thermodynamics

Imagine you are sitting in a car and about to start the engine (Figure 3.1). By turning the ignition key, you connect the electrically powered starter motor to the battery. A current produced by a chemical reaction in the battery flows through the starter motor causing your vehicle's petrol engine to 'turn over', thus giving it some rotational kinetic energy. Petrol is drawn into the cylinders and ignited. This heats the gas in each cylinder, raising its temperature and causing it to expand. The expansion pushes down a piston, doing work, which, via a mechanical linkage, further increases the rotational kinetic energy of the engine. Once the engine is running and up to speed, you engage gear, release the brakes, let out the clutch and drive away. As your vehicle moves off, rotational energy in the engine is converted into translational energy of the car.

Figure 3.1 A petrol-driven internal combustion engine. The engine burns petrol and uses the energy released to produce rotational kinetic energy that may be stored in a flywheel.

The basic function of a car engine is clear; it burns petrol in order to release useful energy, needed to keep the car moving, with all its systems active. Given this function, you can measure the *efficiency* of an engine by determining how much of the energy released by the burning petrol is actually converted into useful energy (Figure 3.2). The balance sheet is pretty much the same for all modern petrol engines: under typical conditions, they are about 30% efficient. Only about one-third of the energy released from the petrol is available to move your car. This means that more than half the money you spend on petrol is in effect wasted, no matter where you go.

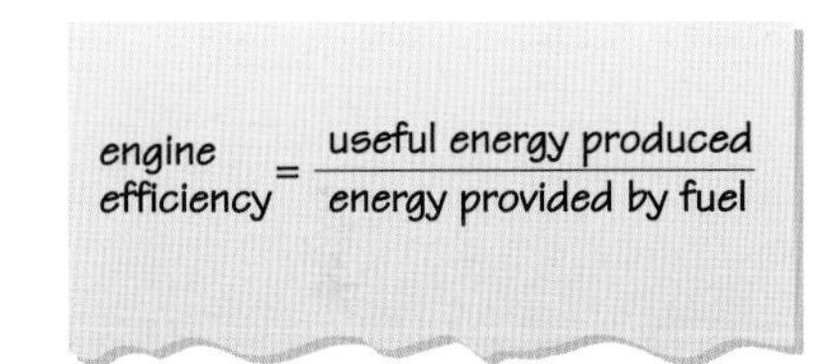

$$\text{engine efficiency} = \frac{\text{useful energy produced}}{\text{energy provided by fuel}}$$

Figure 3.2 A measure of engine efficiency.

Clearly, one way of saving money would be to improve the efficiency of the engine. Why cannot 60%, or even 90%, of the energy released by the petrol be converted into usable kinetic energy? Being realistic, you will realize that 100% efficiency is unattainable; dissipative effects, such as friction and viscosity, will inevitably lead to some wastage of energy, but surely not as much as two-thirds of the energy supplied? Why don't motor manufacturers build more efficient engines, so that we can all benefit?

The fact is that modern petrol engines are almost as efficient as they can be. Apart from the unavoidable effects of friction, most of the wasted energy is transferred to the surrounding environment as heat. The cooling system of your car is designed to facilitate this transfer of energy, particularly through the radiator. The hot gases expelled through the exhaust also carry energy away from the engine. Minimizing unnecessary energy losses is an important feature of engine design, but the laws of physics show that a considerable amount of heat loss is an unavoidable side effect of producing useful work. A different type of engine, such as a diesel engine, might be more efficient, but the conventional petrol engine is just about as efficient as it is ever likely to be. If you drive a car, you had better resign yourself to buying fuel to heat the environment. Even if your car caused no other pollution, its 'heat pollution' is an essential part of its operation and you will keep paying for it as long as you continue to drive.

The part of classical physics that explains the limits on engine efficiency is **thermodynamics** — the study of heat and its relationship to energy in general. It is

Figure 3.3 Three founders of thermodynamics:
(a) Sadi Carnot (1796–1832);
(b) William Thomson (Lord Kelvin) (1824–1907);
(c) Rudolf Clausius (1822–1888).

a well-developed subject that is based on a number of laws. The first of these laws is a general statement of energy conservation that takes account of heat. The second law, which gives thermodynamics much of its particular character, is intimately related to the important concept of *entropy*. Entropy is similar to energy in some respects (though it is measured in units of joules per kelvin rather than joules), but unlike energy it is *not* conserved. In fact, in isolated systems, entropy always tends to increase — it never decreases. It is this requirement that imposes the limits on engine efficiency, as you will see towards the end of this chapter. The first law of thermodynamics (energy conservation) prevents your engine from being more than 100% efficient, but it is the second law that prevents your engine from even getting close to that limit.

This chapter will introduce you to the first and second laws of thermodynamics, and to entropy, before returning to more practical issues that affect the design of efficient engines and effective refrigerators. By the end of the chapter, you should have a better understanding of why some processes occur and others do not, even when they obey all the known conservation laws. You will also learn why some processes are common in nature and occur spontaneously, while others are rare and only occur with intelligent intervention. (The breaking of eggs is common in nature, but the making of an omelette is always the work of a cook.)

The birth of thermodynamics

One of the originators of thermodynamics was the French scientist Nicolas Léonard Sadi Carnot (Figure 3.3a). Carnot started out as a military engineer, and as a young man became convinced that part of the reason for Napoleon's downfall was the superiority of British industrial technology. This was one of the factors that prompted Carnot to undertake a detailed study of steam engines. In a relatively short life, he produced just one influential work: *Reflexions sur la Puissance Motrice du Feu* (*Reflections on the Motive Power of Heat*) in which he discussed the factors that would limit the efficiency of a frictionless ideal steam engine. Carnot concluded that the efficiency of an engine depended on the difference in temperature between its hottest and coldest parts. This result is now expressed in the equation

$$\text{efficiency} = \frac{T_\text{h} - T_\text{c}}{T_\text{h}}$$

where T_h represents the absolute temperature of the hottest part (the boiler of a steam engine) and T_c represents the absolute temperature of the coldest part (the water-cooled condenser of a steam engine).

Carnot's book, published in 1824, was to some extent ahead of its time and failed to produce much immediate impact. However, his ideas were resurrected ten years later by another Frenchman, Emile Clapeyron (1799–1864), who presented them in a more convincing mathematical form that caught the attention of several scientists and engineers. The work was further developed by others, notably the British physicist William Thomson, who later became Lord Kelvin (Figure 3.3b). Thomson confirmed Carnot's basic result in his *Account of Carnot's Theorem* (1849), and then went on to use it in his development of the second law of thermodynamics. The second law was also developed independently, at about the same time, by the German physicist Rudolf Clausius (Figure 3.3c). The term 'thermodynamics' was coined by Kelvin in 1854.

The laws of thermodynamics are very broad in their coverage and therefore exceptionally powerful. So it is at the risk of sounding frivolous that I mention one other mystery that they can help to explain — the untidiness of my desk. Despite what seems like incessant tidying, things are always in a bit of a mess. As you will see later, there are deep links between entropy and *disorder*. It would not be correct to claim that the untidiness of my desk is ordained by nature (though it often seems that way), but physics *can* explain why tidiness is not a condition that is likely to arise spontaneously.

2 The first law of thermodynamics

The first law of thermodynamics is essentially a general statement of the conservation of energy. As such, you will already have a fairly good idea of what it says. Even so, we shall not write down the precise statement of the first law until Section 2.3. The reason for the delay is that thermodynamics is an area of study in which it is particularly important to work with precisely defined quantities. We therefore precede the statement of the first law by defining some of the terms we shall be using later.

2.1 Systems, states and processes

In physics, it is traditional to divide the world into two parts:

1 the **system** — the part of the world we are interested in at the moment, and

2 the **environment** — the rest of the world, which may interact with the system but is not of immediate interest.

The combination of a system and its environment is sometimes rather grandly referred to as the **Universe**, meaning the totality of all that physically exists (Figure 3.4).

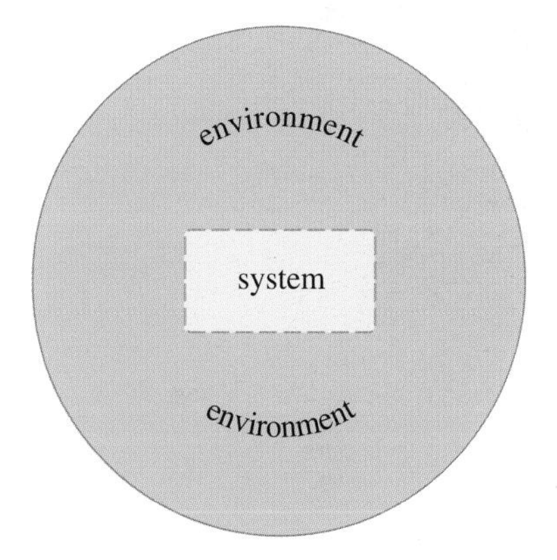

Figure 3.4 A system and its environment. Together, they constitute the Universe.

Thermodynamics is a part of classical physics, so it is concerned with systems that can be described in terms of classical physics, as opposed to quantum physics. It emerged in the nineteenth century from the tradition that tried to avoid making specific assumptions about the microscopic structure of matter, so thermodynamic systems are usually described without any reference to their detailed atomic composition. This is both a weakness and a strength. On the one hand, thermodynamics sheds little light on processes that occur at an atomic level; on the other hand, its predictions are very robust and general, precisely because they rely on so few assumptions. Provided it is macroscopic in scale, a thermodynamic system can be almost anything you care to make it — some gas in a cylinder, a car battery, the Earth's atmosphere, a single raindrop, or a supercluster of galaxies.

The environment of a thermodynamic system can usually supply energy to the system or accept energy from it; in some cases, it may also supply or remove matter as well. For instance, if the system is a lake, water may flow into it from rivers, and the Sun will heat the lake and cause evaporation. An **isolated system** is one that exchanges no energy or matter with its environment, and a **closed system** is one that exchanges no matter with its environment. So, a lake is neither isolated nor closed.

Thermodynamics is mainly concerned with systems that are in (or at least close to) **thermodynamic equilibrium**. As you saw in Chapter 1, equilibrium is especially simple to describe. In thermodynamic equilibrium, the system is in a settled, unchanging state. Its macroscopic properties (pressure, temperature, volume, internal energy, and so on) have definite values that do not change or fluctuate with time. Some special simplifications apply in equilibrium. First, quantities like pressure and temperature become uniform throughout the system so we can, for example, use a *single* value to characterize the temperature of the whole system. Secondly, there are definite equilibrium relationships between various macroscopic variables. This means that any particular **state** of a system, in thermodynamic equilibrium, can be specified by giving the values of just a few macroscopic variables: once these values are known, the values of all the other macroscopic variables can be deduced from the relationships.

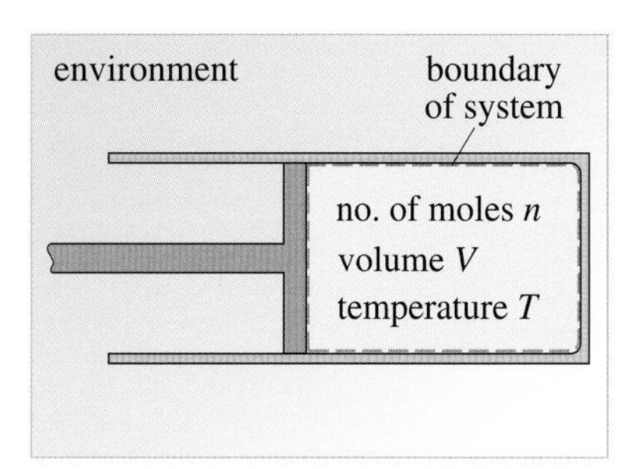

Figure 3.5 A typical thermodynamic system together with part of its environment. In this case, the system is the gas in the cylinder. The equipment that monitors the pressure, volume and temperature of the gas, and the cylinder that contains the gas, are all part of the environment.

For example, an equilibrium state of an ideal monatomic gas can be specified by giving the number of moles, the volume and the temperature (Figure 3.5) — the pressure and internal energy of the gas then follow automatically from the equation of state and the internal energy equation of an ideal monatomic gas. Of course, not all systems can be treated as simply as an ideal gas. Specifying the state of a battery, say, would involve various chemical and electrical quantities as well as pressure, volume and temperature, but the principle remains the same. For a typical thermodynamic system in equilibrium, a state can be specified by strikingly few variables; three or four are often sufficient even though the system might contain 10^{18} atoms, or more. The reason for this degree of descriptive economy was explained in Chapter 2: temperature, pressure and the like are statistical averages of microscopic quantities such as molecular translational kinetic energy and molecular momentum. Nonetheless, it is amazing that the laws of physics allow so much detailed information to be meaningfully summarized so concisely. This economy lies at the heart of thermodynamics.

The macroscopic variables that are used to characterize a state are called **state variables**, or **functions of state**. Any macroscopic quantity can be chosen as a state variable, provided its value is completely determined by the state of the system, and is independent of the detailed past history that the system has gone through. Certainly, pressure, temperature, volume and internal energy are suitable state variables.

Despite its emphasis on equilibrium, thermodynamics is also concerned with **processes** — the changes that occur when a system undergoes a transition from one state to another. Thermodynamics manages to combine an emphasis on equilibrium states with a concern about processes by concentrating on **quasi-static processes**. A quasi-static process is one that happens sufficiently slowly that, at each stage, it is a good approximation to regard the system as being in an equilibrium state. Thus at each stage in a quasi-static process the state of the system can be specified in terms of macroscopic variables which have definite values, and obey the same relationships as would be found in equilibrium.

Quasi-static processes can be visualized using the ***PVT* surface** of a substance, which was introduced in Chapter 1 — an example of such a surface is shown in Figure 3.6. The *PVT* surface, you will recall, provides a graphical representation of the **equation(s) of state** of a substance; it shows at a glance all the equilibrium states of a fixed amount of the substance. If you choose two states on the surface, and draw a continuous pathway on the surface from one state to the other, then that pathway represents a possible quasi-static process that would lead from one of the chosen states to the other. It is a quasi-static process precisely because it is depicted by a path *on* the *PVT* surface and is therefore represented by an equilibrium state at every stage.

● Write down the equation of state of an ideal gas. Specify three pairs of macroscopic variables, each of which could be used to specify a particular equilibrium state of a given quantity of such a gas.

❍ The equation of state of an ideal gas can be written as

$$PV = nRT \qquad (3.1)$$

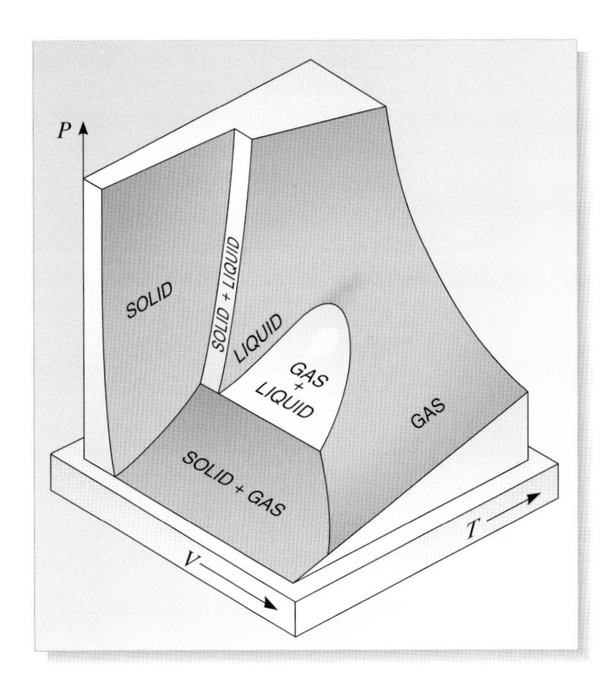

Figure 3.6 A *PVT* surface illustrating the equilibrium states of a pure substance that expands on melting. The *PVT* surfaces of other substances may differ markedly from this. For instance, water contracts on melting, and an ideal gas remains a gas under all conditions of temperature and pressure.

where n is the number of moles of the gas and $R = 8.314\,\mathrm{J\,K^{-1}\,mol^{-1}}$ is the molar gas constant. For a given quantity of gas, the value of n is fixed. Moreover, the equation of state provides a relationship between P, V and T (for a given value of n), so the values of any two of these macroscopic variables will determine the third. Hence, in order to specify a particular equilibrium state of the given sample, it would be sufficient to quote the values of P and V, *or* of V and T, *or* of P and T. ■

2.2 Heat and work

Heat and work were both briefly discussed in Chapter 1. Both terms are used to describe quantities of energy that may be transferred to or from a system. Consequently, both heat and work can be measured in joules (J). What distinguishes heat from work is the method by which the transfer of energy is accomplished.

Heat is the term used to refer to any quantity of energy transferred between a system and its environment as a result of temperature difference. We will use the symbol Q to denote a heat transfer *into* a system from the environment. A negative value of Q indicates energy flow in the opposite direction, from the system into the environment.

'Heat' is a word, and a concept, that is used a lot in everyday life, and everybody has some idea what it means. But beware! Everyday speech tends to draw little distinction between heat (energy transferred by virtue of a temperature difference) and temperature (a parameter that determines the direction in which heat will flow, from hot to cold). For instance, when we say that one body is hotter than another, what we generally mean is that it has a higher temperature, not that it has gained or lost more heat. Also note that it makes no sense to speak of the heat of a system, only the heat gained or lost by a system. A system may have a temperature, but it does not have a 'heat'.

Everyday experience shows that there are a number of different ways by which heat transfer can occur. To give one example, if you put a cold metal spoon into some hot

(a) conduction

(b) radiation

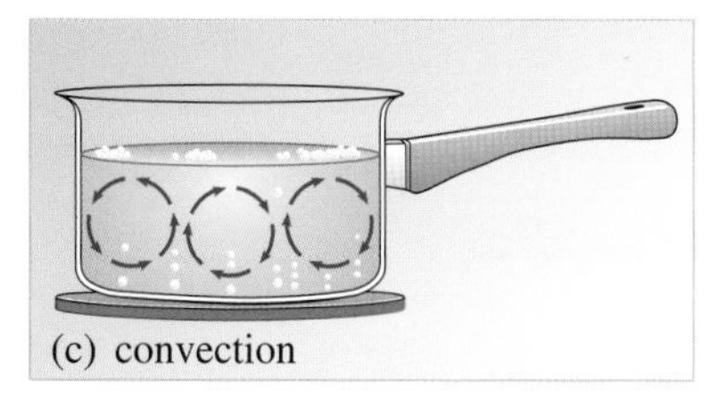

(c) convection

Figure 3.7 Three mechanisms of heat transfer: (a) conduction; (b) radiation; (c) convection.

soup, the handle of the spoon will soon become hot (Figure 3.7a). This is because energy is transferred from the soup to the bowl of the spoon and thence to the handle by **conduction**. In this case, electrons in a relatively hot part of the spoon will have, on average, more kinetic energy than electrons in a relatively cool part. Collisions will then tend to even out the distribution of energy, transferring energy from the hotter to the cooler region, causing the cooler region to warm up and the hotter region to cool down.

An entirely different way of transferring energy between objects at different temperatures is by means of **radiation**. All objects emit radiation. The Sun obviously emits visible light, and your body is a source of infrared radiation, along with all other objects at or near room temperature. When you lie in direct sunlight, your body absorbs some of the Sun's radiation and acquires the energy being carried by that radiation (Figure 3.7b). (The Sun will also receive some of the radiation emitted by your body, but because the surface of the Sun is at a much higher temperature than the surface of your body, the Sun has a much greater effect on you than you have on it.)

A third method of heat transfer is **convection**. This mechanism occurs when a hot body is in contact with a fluid such as air or water. The warmth of the body heats the fluid in its immediate environment, causing the fluid to expand and reduce its density. As a result, the heated fluid becomes more buoyant than the surrounding cooler fluid and it rises, allowing cooler fluid to replace it (Figure 3.7c). Under the right circumstances, these rising 'convection currents' provide a very effective means of transporting energy away from the body. This mechanism helps keep you cool in summer. Conversely, inhabitants of cold countries wear hats during the winter, largely in order to reduce convection currents flowing away from their heads. And the fact that the roof of a greenhouse prevents convection currents from rising into the atmosphere is one of the main reasons why greenhouses stay warm inside — it's all good physics!

Not all systems exchange heat with their environment. A vacuum flask keeps hot drinks warm and cold ones cool by inhibiting the transfer of heat. A system that cannot exchange heat with its environment is said to be in a state of **thermal isolation** or to be **adiabatic**. Perfect thermal isolation is an idealization; real systems are always to some extent in **thermal contact** with their environment, implying that they can exchange heat with it. (Even coffee in a vacuum flask will eventually reach room temperature if you leave it long enough.) Note though that thermal contact does not imply direct physical contact in the sense of touching. A sunbather is in thermal contact with the Sun via its radiation, but the sunbather certainly cannot reach out and touch the Sun.

Calories and joules

In the nineteenth century and the first half of the twentieth century, it was common to measure heat not in joules, but in **calories**. The calorie was defined as the amount of energy necessary to raise the temperature of 1 g of water by 1°C. It is related to the joule by the equation

$$1\ \text{calorie} = 4.186\ \text{J}.$$

The calorie is now obsolete in most sciences, but is still used by dieticians and nutritionists. Unfortunately, they usually use the kilocalorie, but refer to *this* as the Calorie (spelt with a capital C); thus 1 Calorie = 1000 calories = 4186 J! Grown-ups who 'would rather not grow any more' are generally advised to maintain an average daily food intake with an energy equivalent of 2200 Cal (women) or 2700 Cal (men), i.e. 9.2 MJ (women) and 11.3 MJ (men).

In thermodynamics, all ways of transferring energy to or from a system are classified as being either heat or work. We now turn to the other way of transferring energy to or from a system — work.

Work is the term used to refer to any quantity of energy transferred between a system and its environment that is *not* classified as heat. Thus, work is energy transferred between a system and its environment, irrespective of any temperature differences.

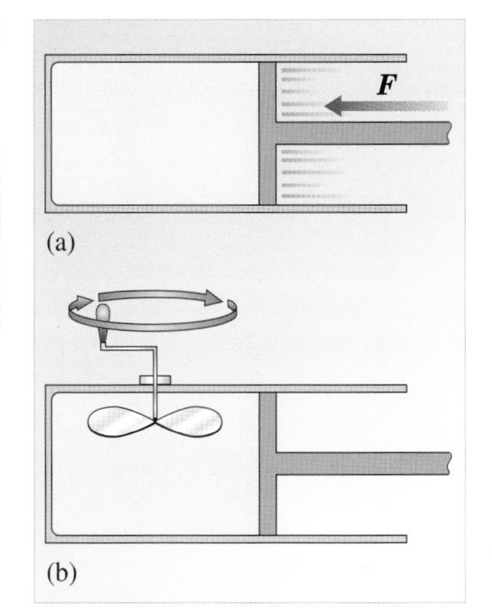

Figure 3.8 Two ways of doing work on a gas confined in a cylinder. (a) The gas can be compressed by exerting a force $\boldsymbol{F}$ that moves the piston. (b) The gas can be stirred by applying a torque that turns the paddle wheel while the piston remains fixed.

Work was discussed extensively in the context of mechanics in *Predicting motion*. That discussion showed, for example, that the work done by a constant force F_x, when its point of application moves through a displacement Δx, is

$$W = F_x \Delta x\,. \tag{3.2}$$

Similarly, if a constant torque Γ_z, directed along the z-axis, acts over an angular displacement $\Delta\theta$ about the z-axis, then the work done by the torque is

$$W = \Gamma_z \Delta\theta\,. \tag{3.3}$$

So, if the environment exerts unbalanced forces or torques on a system, work is done on the system and energy is transferred to it.

Two ways of doing work on a simple system are shown in Figure 3.8. The system consists of a fixed quantity of gas confined in a cylinder by a piston. Work may be done on the system by exerting a force on the piston and compressing the gas, or by applying a torque to a paddle wheel and stirring the gas. In both cases, let's assume there is no friction in the moving parts, so all of the work done by the force or the torque results in energy being transferred to the gas. These are both examples of work because neither relies on any difference in temperature to transfer energy from the environment to the system.

The compression process shown in Figure 3.8a features in many thermodynamic discussions and therefore merits further consideration. Figure 3.9 provides a more detailed view of the process in a case where the piston moves through a small displacement Δx. By restricting our attention to the case where Δx is *small*, we can assume that the pressure P of the confined gas remains constant throughout the process. Now, if the cylinder has cross-sectional area A, the force that the gas exerts on the piston will have magnitude PA. This is the force that the piston must overcome so, in a quasi-static compression, the force exerted by the piston on the system is

If there is friction between the piston and the cylinder, the magnitude of the applied force may have to be greater than PA. However, the extra work done by this greater force is performed on the cylinder, not on the gas, so it is irrelevant to this discussion.

$$F_x = -PA\,. \tag{3.4}$$

The minus sign arises because the external force acts from right to left in Figure 3.9, opposite to our choice of x-axis. The work done on the gas by the external force is then

$$W = F_x \Delta x = -PA\,\Delta x\,.$$

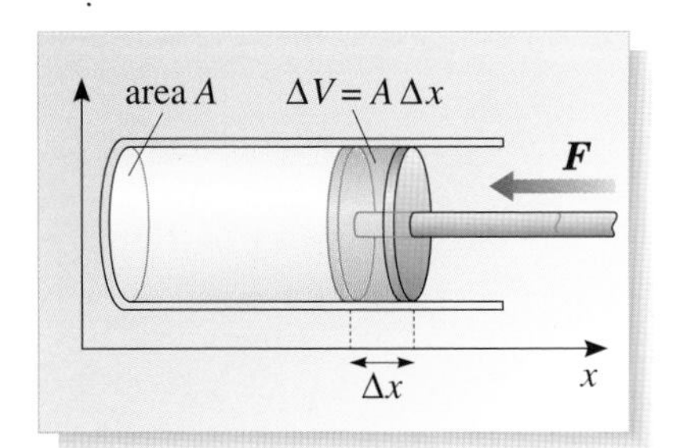

Figure 3.9 Gas at pressure P is confined in a cylinder of cross-sectional area A by a piston that can move in the x-direction. A displacement Δx causes the volume of the gas to change by $\Delta V = A\,\Delta x$. For a compression, both Δx and ΔV are negative.

If we introduce the symbol ΔV to represent the change in the volume of the gas, then

$$\Delta V = A\,\Delta x\,.$$

So, for a small change in volume, ΔV, in which the pressure has a practically constant value, P, the work done on the gas is

Note the minus sign in this equation for the work done *on* a gas.

$$W = -P\,\Delta V. \tag{3.5}$$

It is important to realize that W is the work done *on* the gas, corresponding to an energy flow *into* the gas and that ΔV is the change in volume of the gas, which is negative when the gas is compressed and positive when it is expanded. When a gas is compressed, ΔV is negative, and Equation 3.5 shows that positive work is done on the gas and energy flows from the environment into the gas. By contrast, when a gas is expanded, ΔV is positive and negative work is done on the gas. This means that the gas does work on its surroundings, so energy flows from the gas into the environment. This agrees with common experience; in order to compress the air in a bicycle pump, you (as part of the environment) have to do work *on* the air. If you allow gas to expand, it can do work *on* you.

Equation 3.5 is just one among many expressions for the work done on a given thermodynamic system in a specified process. Other expressions can be written down to describe, for example, the work done by the electric current that powers a device such as a fan, or the work done in blowing up a balloon. These too are processes that involve energy transfer, but the transfer is not caused by any temperature difference, so the transferred energy must be an example of work rather than heat (it has to be one or the other).

An alternative approach to thermodynamics uses such formulae (including Equation 3.5 and many others) to define exactly what is meant by work, and then defines *heat* as any energy transfer that is not accounted for by these formulae. This approach is slightly more rigorous than that used here, but requires encyclopaedic knowledge of many aspects of physics before a complete catalogue of the contributions to work can be assembled. In the end, though, our starting point makes little difference. The important point is that all energy transfers are classified as being either heat or work and, in practice, it is easy enough to distinguish one from the other.

Question 3.1 A fixed quantity of gas, maintained at a constant pressure $P = 1.06 \times 10^5$ Pa, is allowed to increase its volume by 3.50×10^{-3} m^3. How much work is done *on* the gas as a result of the expansion? How much work is done *by* the gas? ■

2.3 The first law of thermodynamics

Having discussed systems, states, functions of state, processes, heat and work, we are finally in a position to write down the first law of thermodynamics (often known as the 'first law'). This may be stated as follows.

The first law of thermodynamics

When a system undergoes a change from one equilibrium state to another, the sum of the heat transferred *to* the system and the work done *on* the system depends on the initial and final equilibrium states of the system, but not on the process by which the change is brought about.

This is a precisely worded statement which needs to be carefully examined if it is to be properly understood. The first thing to note is that although this law concerns processes that start and end with equilibrium states, it is not restricted to quasi-static processes. The process causing the change can be quite general, provided it starts and ends with an equilibrium state.

The next thing to notice is that the law is mainly concerned with the sum of the heat transferred to the system, Q say, and the work done on the system, W. This sum, $Q + W$, is, of course, the total energy transferred *to* the system. So the first law tells us that the total energy transferred to a system is determined by the initial and final states of the system, and not by the particular process that led from one state to the other.

There are a number of consequences that follow from this.

1 The first law justifies the introduction of an **internal energy** U for any thermodynamic system. Transferring an amount of energy, $Q + W$, to a system increases the system's internal energy from U to $U + \Delta U$, where

$$\Delta U = Q + W. \tag{3.6}$$

This equation is usually referred to as the 'mathematical form' of the first law. Note that Q and W may each be positive or negative, so their sum $Q + W$ may also be positive or negative. This means that although we speak of energy being transferred *to* the system, Equation 3.6 also applies to cases in which ΔU is negative, when energy is transferred *from* the system *to* its environment.

2 The first law implies that the internal energy is determined by the state of the system. Every equilibrium state corresponds to some particular value of internal energy so, in the language introduced earlier, the internal energy is said to be a function of state. The significance of this fact may become clearer if you realize that neither heat nor work is a function of state — we cannot say that either of these quantities has a definite value in an equilibrium state. In passing from one equilibrium state to another, various different combinations of heat and work may be supplied to the system, but the change in internal energy is definite, and fixed.

3 By recognizing that systems have an internal energy that remains constant unless the state of the system changes, and by accepting that heat as well as work can bring about such changes, the first law justifies the very general **principle of energy conservation**. The internal energy of a thermodynamic system might consist of the kinetic and potential energy of its constituents, but it might also involve other forms of energy, such as the energy associated with electric and magnetic fields, or radiation. The fact that changes in the internal energy of a system involve the transfer of energy to or from the environment implies that the total energy of a system and its environment (i.e. the energy of the Universe) is constant. Or, to quote Chapter 2 of *Predicting motion*, 'The total amount of energy in any isolated system is always constant.'

Now, you should already be familiar with the idea of internal energy from earlier chapters. In particular, in Chapter 1 you saw that a given sample of ideal gas has an internal energy that is determined by its absolute temperature T. In the case of a monatomic ideal gas, for example, the internal energy is

$$U = \tfrac{3}{2}nRT \tag{3.7}$$

where n is the number of moles of gas. Similarly, for a typical diatomic ideal gas, under moderate conditions, the internal energy is given by

$$U = \tfrac{5}{2}nRT. \tag{3.8}$$

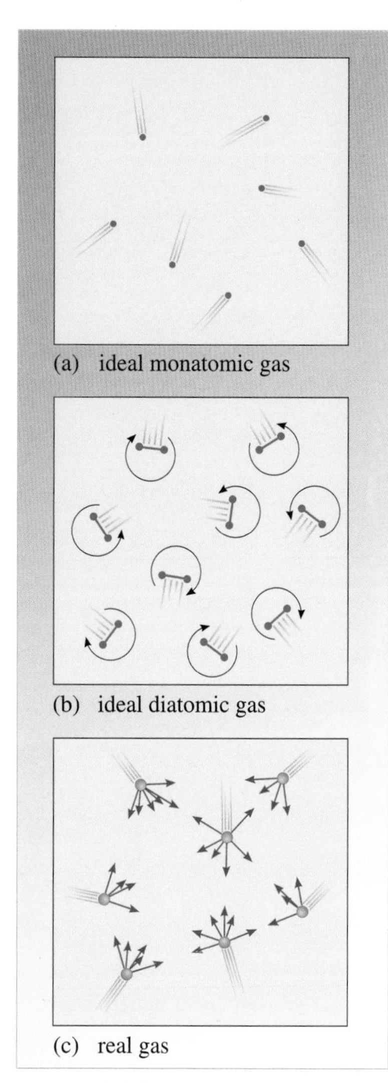

Figure 3.10 The internal energy of a gas arises from the motion of its constituents and (in non-ideal gases) from their mutual interaction.

These examples bear out the general claims made by the first law. For both systems the internal energy can be easily defined and interpreted — it is just the total kinetic energy of the molecules that make up the gas, as indicated in Figure 3.10a and b. It is also clear that the internal energy is a function of state, that depends on the precise nature of the gas (whether it is monatomic or diatomic) and on the state of the gas, specified by number of moles, n, and the absolute temperature, T. Other cases may be treated similarly, though most will not be as simple as an ideal gas. In a dense gas, for example, the molecules have potential energy due to their mutual interactions (Figure 3.10c) as well as kinetic energy, and the internal energy will depend on volume, as well as on n and T, but the first law still guarantees that there will be an internal energy, and that it will be a function of state.

The fact that the internal energy of a system is a function of state means that it is always possible to plot a graph or a surface which shows how the internal energy varies from state to state. Just such a surface, for the case of a monatomic ideal gas, is shown in Figure 3.11. Note that we have chosen to characterize the equilibrium state of the gas in terms of the variables V and T, but we could equally well have chosen P and T, or even P and V.

A similar plot, again showing U as a function of V and T, but for a more realistic monatomic gas, is shown in Figure 3.12. When the volume is large, so that molecular interactions are relatively unimportant, the real gas behaves in a way that is similar to an ideal gas. However, when it is strongly compressed, deviations from the ideal gas behaviour become evident.

As a third example of internal energy, and as a reminder of the generality of thermodynamic concepts, consider a system consisting entirely of electromagnetic radiation, contained in a box of volume V with walls that are maintained at a constant temperature T. Due to their temperature, each of the walls will constantly emit radiation that will be absorbed by the other walls. If the walls are maintained at

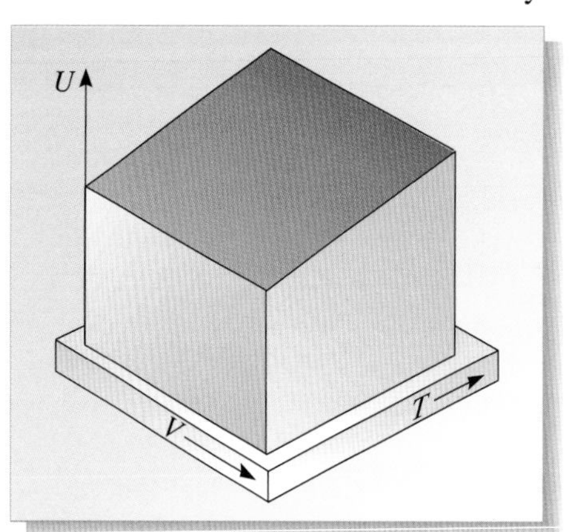

Figure 3.11 The internal energy U of a fixed quantity of ideal gas plotted against the variables V and T that determine the state of the gas. The fact that a single value of U corresponds to each pair of values for V and T indicates that U is a function of state, as the first law implies. In this case, the internal energy happens to be independent of V.

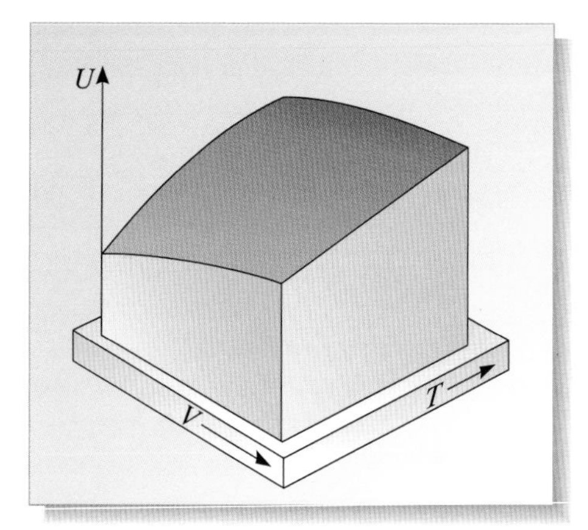

Figure 3.12 The internal energy U of a fixed quantity of realistic monatomic gas, plotted against V and T.

room temperature, the radiation will be mainly infrared radiation similar to that emitted by your body. If the wall temperature was somewhat higher, about 6000 K say, like the surface of the Sun, then the radiation would contain a significant amount of visible light; and if the temperature was more like 10^6 K, the radiation would be predominantly X-rays. In any case, whatever the temperature of the walls, provided the system is in equilibrium, the internal energy of the confined radiation is given by

$$U = (7.5 \times 10^{-16}\,\mathrm{J\,m^{-3}\,K^{-4}})\,VT^4. \qquad (3.9)$$

The UVT surface for this system is shown in Figure 3.13. In case the idea of a system consisting entirely of radiation at a temperature T sounds a bit odd, you should remember that there is good evidence that the Universe is filled with microwave radiation at a temperature of about 3 K, left over from the Big Bang. So this sort of system is more representative of the Universe in general than any sort of atomic gas.

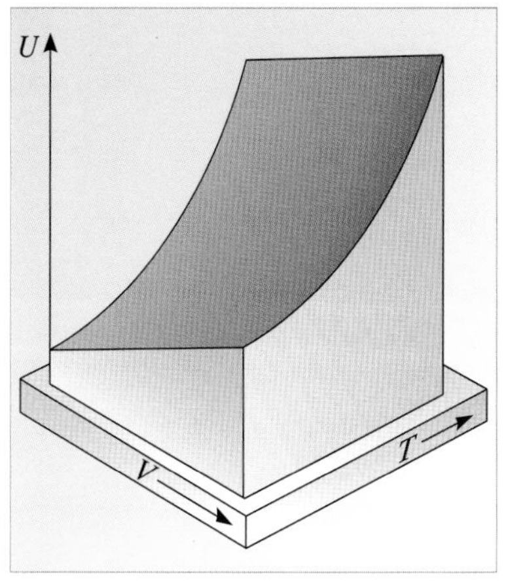

Figure 3.13 The internal energy U of a system consisting of radiation in equilibrium with the walls of a container at temperature T. The internal energy is plotted against the volume V and temperature T of the radiation.

Having explored the first law and its meaning, we now turn to its applications, starting with a worked example.

Example 3.1

A specimen of copper is simultaneously subjected to two forms of energy transfer: 20 J are transferred *from* it as heat, while 30 J are supplied *to* it as work. What is the resulting change in the copper's internal energy?

Solution

Preparation Clearly, this problem involves the mathematical form of the first law:

$$\Delta U = Q + W. \qquad \text{(Eqn 3.6)}$$

It should be straightforward, but care is always needed in such calculations over the signs of Q, W and ΔU. We must remember that Q is the heat transferred *to* the system, W is the work done *on* the system and ΔU is the *increase* in the internal energy of the system, but any of these quantities may be negative.

In this case, 20 J are transferred *from* the system as heat, so $Q = -20$ J. In addition, 30 J are transferred *to* the system as work, so $W = +30$ J.

Working Using the first law of thermodynamics,

$$\Delta U = Q + W = -20\,\mathrm{J} + 30\,\mathrm{J}$$

so $$\Delta U = 10\,\mathrm{J}.$$

Checking The answer is in joules, as it should be. The calculation is very straightforward in this case, but it would be a good idea to re-read the question to make sure that the correct signs have been used. The positive result is consistent with the fact that more energy was transferred to the system than was taken from it.

One of the main uses of the first law in practical problem-solving is in working out the heat that has been transferred to a system under specified circumstances. The following questions will give you an opportunity to practise this but, as always, you will need to pay careful attention to the signs of the quantities involved.

Question 3.2 One mole of monatomic ideal gas is in a cylinder like that shown in Figure 3.9. Initially it is at a temperature of 300 K and a pressure of 1.00×10^5 Pa, and it occupies a volume of $2.49 \times 10^{-2}\,m^3$. Suppose the temperature is raised to 350 K. Explain why, strictly speaking, the information you have just been given is insufficient to allow you to calculate the heat transferred to the gas.

Question 3.3 Now suppose that in addition to the information given in Question 3.2 you also know that, throughout the process, either (a) the volume is kept constant, or (b) the pressure is kept constant. In each of these cases calculate the heat transferred to the gas, and the work done on the gas. ■

Box 3.1 The discovery of the first law of thermodynamics

The discovery of the first law of thermodynamics (and with it the conservation of energy) is often presented as an example of 'simultaneous discovery'. Many scientists realized the need for such a law at about the same time, though some had a clearer view than others and none would have used exactly the words we used in presenting it.

One of the first to suspect the existence of such a general principle was Julius Robert Mayer (1814–1878), a German physicist who originally trained as a physician. Around 1840, while serving as a ship's doctor, Mayer noticed that the blood in human veins was redder in hot climates than in cool ones. He correctly interpreted this as an indicating that maintaining body temperature in hot climates consumes less of the oxygen carried by the blood. Further thought led him to realize that the food humans consume can eventually lead to the transfer of either heat or work, and that these two quantities must therefore be of the same fundamental nature. Mayer did not formulate this principle with particular clarity, even though his 1845 pamphlet, *Organic Motion Related to Digestion*, tried to put it on a quantitative footing. As a result, his work was overlooked for many years, and had little direct impact on the development of science.

The same cannot be said of the work of the British brewer and amateur scientist James Prescott Joule (1818–1889). Joule was a talented experimentalist who concentrated on making accurate measurements. As early as 1840, he was already engaged in measuring the heating effect of an electric current and by 1847 had demonstrated to his own satisfaction that work and heat could be measured in the same units. Joule was a quiet man and his discovery initially got a rather unsympathetic reception. However, he had the good fortune to meet William Thomson (1824–1907), later Lord Kelvin, one of the most gifted scientists of the nineteenth century. Thomson, who had recently been appointed to a professorial position in Glasgow (a job he held for 53 years), became Joule's champion and was instrumental in arranging Joule's celebrated presentation of a paper, *On the Mechanical Equivalent of Heat*, to the Royal Society of London in 1849. Joule, with the aid of Thomson, certainly deserves a good deal of the credit for demonstrating the conservation of energy, but the first clear, quantitative formulation of that principle belongs to another German.

Hermann von Helmholtz (1821–1894), like Mayer, had been a student of medicine. During a long and distinguished career, he carried out detailed studies of the eye and ear, and was the first to measure the speed of a nerve impulse. He taught physiology as well as physics, but his greatest achievement was the publication, in 1847, of a detailed account of energy conservation. This work clearly enunciated the equivalence of work and heat, and even quoted some of Joule's early results. It discussed at length the energy of humans and animals, though it was written independently of Mayer's work, and it also covered other matters such as the energy associated with electrical systems and electromagnetic radiation.

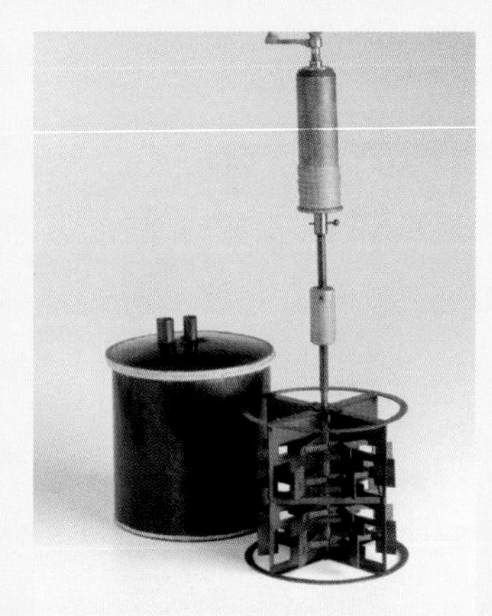

Figure 3.14 Part of the equipment used by Joule to demonstrate that a certain quantity of work, in this case provided by the stirring of a fluid, has the same effect as a given amount of heat. Both raise the temperature of the fluid by the same amount.

2.4 The first law and heat capacities

The concept of a heat capacity emerges from the observation that some systems can have a considerable amount of heat transferred to them without displaying much increase in temperature: such systems are said to have a high heat capacity. The heat capacity of a system depends on the conditions under which heat transfer takes place — the simplest case to consider is where the system is at constant volume. We define the **heat capacity at constant volume,** C_V, as the heat input, Q, needed to warm a body, divided by the corresponding temperature rise ΔT, where the temperature rise is small, *and the heating is carried out at constant volume*:

$$C_V = \frac{Q}{\Delta T} \quad \text{(at constant volume)}. \tag{3.10}$$

The subscript V serves to remind us that the volume is being held constant. You may have been tempted to think of this as the rate of change of heat with temperature, but I hope you will now see why this should not be done: heat is *not* a function of state, so we should *not* ascribe a certain 'heat' to the body, nor think of the 'heat of the body' as changing. What is changing is the body's internal energy. However, if we restrict ourselves to simple situations, such as that shown in Figure 3.8a, where compression is the only way of doing work, the fact that the volume remains constant means that no work can be done on the body so,

$$Q = \Delta U$$

and we can then write the heat capacity at constant volume as the rate of change of internal energy with temperature. Taking the limit of a very small change in temperature,

$$C_V = \frac{\mathrm{d}U}{\mathrm{d}T}. \tag{3.11}$$

Often, we will consider one mole of a substance, and we are then interested in the **molar heat capacity at constant volume**. This is the change in molar internal energy with temperature:

$$C_{V,\mathrm{m}} = \frac{\mathrm{d}U_\mathrm{m}}{\mathrm{d}T}. \tag{3.12}$$

You saw in Chapter 2 that this quantity is important from a theoretical point of view, especially when we try to explain measured values in terms of an underlying microscopic model (details that lie beyond the limits of thermodynamics). From a practical point of view, however, and for many situations observed in nature, it is usually unrealistic to assume that the volume of the system remains constant. When a substance is heated, it generally expands unless special precautions are taken (e.g. enclosing a gas in a rigid container). For solids and liquids, the forces required to prevent thermal expansion are so great that it is very hard to guarantee that the volume remains constant. If we just let the expansion happen, with no restraint, the volume will change but the pressure exerted on the system by the environment will remain fixed. We are therefore led to introduce the **heat capacity at constant pressure**, C_P, as the heat input, Q, needed to warm a body, divided by the corresponding temperature rise, ΔT, where the temperature rise is small, *and the heating is carried out at constant pressure*:

$$C_P = \frac{Q}{\Delta T} \quad \text{(at constant pressure)}. \tag{3.13}$$

The subscript P serves to remind us that the pressure is being held constant. The symbol $C_{P,\mathrm{m}}$ is used for the corresponding **molar heat capacity at constant pressure**.

The relationship between the heat capacity at constant volume and the heat capacity at constant pressure — the former of theoretical importance, and the latter of practical importance — will now be discussed, using the first law of thermodynamics. We begin with the case of an ideal gas. You saw in Chapter 2 that the internal energy of an ideal gas can generally be written as

$$U = \frac{f}{2}nRT \tag{3.14}$$

where f is the number of 'effective' degrees of freedom per molecule (3 for a monatomic gas, 5 for a typical diatomic gas at room temperature, and so on) and n is the number of moles of gas. Differentiating this expression with respect to temperature immediately gives

$$C_V = \frac{\mathrm{d}U}{\mathrm{d}T} = \frac{f}{2}nR. \tag{3.15}$$

So, for an ideal monatomic gas,

$$C_V = \tfrac{3}{2}nR. \tag{3.16}$$

and for a typical diatomic gas at room temperature

$$C_V = \tfrac{5}{2}nR. \tag{3.17}$$

One extra step is needed to calculate the corresponding heat capacities at constant pressure: we must take account of the fact that the gas expands. Fortunately, we can calculate how much expansion takes place. Provided the heating takes place quasi-statically, the gas will, at each stage, obey the ideal gas equation of state, $PV = nRT$, so

$$V = \frac{nRT}{P}.$$

Because n, R and P are all constant, the expansion that accompanies an increase ΔT in temperature is

$$\Delta V = \frac{nR}{P}\,\Delta T.$$

This expansion means that the gas does work on its environment, so some of the energy supplied to the gas will be transferred into the environment, and the gas will not warm up as much as if it were held at constant volume. Combining our general expression for work done with the expansion ΔV calculated above, we see that

$$W = -P\,\Delta V = -nR\,\Delta T.$$

(Here, the minus sign makes good sense: it shows that negative work is done *on* the gas, which is sensible because the gas actually does work on the environment as it expands.)

To find the heat capacity at constant pressure we need to form the ratio, $Q/\Delta T$, of the heat transferred to the gas to the corresponding temperature rise, under conditions of constant pressure. The first law of thermodynamics provides the key for this. The first law shows that the amount of energy transferred to the gas *as heat* is

$$Q = \Delta U - W = \Delta U + nR\,\Delta T$$

and the heat capacity at constant pressure is then given by

$$C_P = \frac{Q}{\Delta T} = \frac{\Delta U}{\Delta T} + nR.$$

The quantity $\Delta U/\Delta T$ is the rate of change of internal energy with temperature, and we have seen that is the heat capacity at constant volume. We therefore conclude that, for an ideal gas, the heat capacity at constant pressure exceeds the heat capacity at constant volume by an amount nR:

$$C_P = C_V + nR\,. \tag{3.18}$$

In particular, the molar heat capacities are related by

$$C_{P,\mathrm{m}} = C_{V,\mathrm{m}} + R\,. \tag{3.19}$$

So, for example, the constant pressure molar heat capacity of a monatomic gas is $5R/2$, while that of a typical diatomic gas is $7R/2$. Remember, heat capacities at constant pressure are larger than those at constant volume because some of the heat supplied to the gas is converted into work that the gas does on the environment as it expands, and so is not available for increasing the temperature.

As you will see later in this chapter, one quantity that has an important role in explaining the behaviour of gases is the **ratio of heat capacities**, C_P/C_V. In forming this ratio, any fixed quantity of gas can be chosen, such as a mole or a kilogram. Not only is this ratio easy to measure in the laboratory, it is also directly related to the speed of sound in the gas. We therefore give it a special symbol, γ, and define

γ is the lower case Greek letter gamma.

$$\gamma = \frac{C_P}{C_V}. \tag{3.20}$$

For an ideal gas with f effective degrees of freedom, we have seen that the constant volume heat capacity is

$$C_V = \frac{f}{2}nR. \qquad \text{(Eqn 3.15)}$$

Using Equation 3.18, the constant pressure heat capacity is then

$$C_P = \left(\frac{f}{2} + 1\right)nR \tag{3.21}$$

so

$$\gamma = \frac{f+2}{f} = 1 + \frac{2}{f}. \tag{3.22}$$

Thus, under moderate conditions, the value of γ will be 1.67 for a monatomic gas, 1.40 for a diatomic gas and approach 1 for gases with large and complicated molecules having many effective degrees of freedom. Experiments with real gases support these predictions, as indicated in Table 3.1.

Table 3.1 The ratio of heat capacities for a range of gases under moderate conditions of temperature and pressure.

Gas	$\gamma = C_P/C_V$
helium (He)	1.67
argon (Ar)	1.67
hydrogen (H_2)	1.41
nitrogen (N_2)	1.40
oxygen (O_2)	1.40
carbon dioxide (CO_2)	1.30

The detailed arguments given above are specific to ideal gases. This is revealed by the fact that we have used the ideal gas equation of state and the internal energy equation for an ideal gas to arrive at Equation 3.18. Solids and liquids do not obey Equation 3.18, although we would generally expect the heat capacity at constant pressure to be greater than that at constant volume, to allow for the fact that expansion at constant pressure leads to a transfer of energy out of the system via work done on the environment.

In fact, it is possible to use thermodynamics to develop a much more general relationship than Equation 3.18, namely

$$C_P - C_V = TV\frac{\alpha^2}{\beta} \qquad (3.23)$$

where α is the isobaric expansivity and β is the isothermal compressibility — quantities introduced in Chapter 1. We shall not derive this result here, nor even ask you to use it. It is included just to give you a glimpse of the truly universal nature of thermodynamics, and its power to derive unexpected relationships between macroscopic variables that are valid for all systems, no matter what their equation of state or internal energy equation may be.

The Dulong–Petit law was introduced in Chapter 2, Section 6.2.

According to classical physics, if a solid is composed of individual atoms rather than molecules, its constant volume molar heat capacity should obey the **Dulong–Petit law**, which asserts that

$$C_{V,\mathrm{m}} = 3R. \qquad (3.24)$$

Equation 3.23 allows us to take experimentally measured quantities $C_{P,\mathrm{m}}$, α and β, and calculate the corresponding value of $C_{V,\mathrm{m}}$ (which is very difficult to measure). In this way, experiment can be compared with theory. The examples in Table 3.2 show that at room temperature the Dulong–Petit law is well-obeyed by some substances but not by others, such as carbon. Carbon in fact does fall into line with the law, but only at much higher temperatures. Explaining this behaviour is beyond the scope of thermodynamics. The first good explanation was provided by Albert Einstein, and was one of the early triumphs of quantum physics.

Table 3.2 Molar heat capacities of solids at room temperature.

Solid	$C_{P,\mathrm{m}}/R$	$C_{V,\mathrm{m}}/R$
aluminium	2.96	2.83
copper	2.94	2.85
iron	3.02	2.98
sodium	3.40	3.15
carbon (graphite)	1.03	1.03

Incidentally, when dealing with practical problems, it is common to discuss the heat capacity per kilogram rather than the heat capacity per mole. The heat capacity per kilogram of any substance is called the **specific heat capacity** (usually shortened to 'specific heat'). The units of specific heat are $\mathrm{J\,K^{-1}\,kg^{-1}}$. If there is no mention of whether the specific heat is at constant pressure or at constant volume, you should assume that the constant pressure value is supplied. The definition of specific heat means that:

heat transferred = mass of system × specific heat × rise in temperature.

This fact is often used in problems.

- The specific heat of copper is $385\,\mathrm{J\,K^{-1}\,kg^{-1}}$. How much energy, transferred as heat, is required to raise the temperature of 100 g of copper from the freezing point to the boiling point of water?

○ The mass of copper is 0.100 kg, and the temperature difference between the starting and finishing values is 100 K, so the heat required is

$$Q = (385\,\mathrm{J\,K^{-1}\,kg^{-1}}) \times (0.100\,\mathrm{kg}) \times (100\,\mathrm{K}) = 3850\,\mathrm{J}.$$

● How much energy is needed to heat 1.00 litre ($10^{-3}\,\mathrm{m^3}$) of water from room temperature (20 °C) to boiling point? The specific heat of water is $4.19\,\mathrm{kJ\,K^{-1}\,kg^{-1}}$. If the water is in a 2.50 kW electric kettle, how long would the process take? (You may take the mass of 1.00 litre of water to be 1.00 kg, and ignore any energy associated with warming the kettle itself, or its surroundings.)

○ Energy required = mass × specific heat × temperature rise
$= (1.00\,\mathrm{kg}) \times (4.19 \times 10^3\,\mathrm{J\,K^{-1}\,kg^{-1}}) \times (80\,\mathrm{K}) = 3.35 \times 10^5\,\mathrm{J}$.
Time taken $= (3.35 \times 10^5\,\mathrm{J})/(2.50 \times 10^3\,\mathrm{J\,s^{-1}}) = 134\,\mathrm{s}$. ■

The conservation of energy, implied by the first law of thermodynamics, ensures that when the energy of a system decreases, the energy lost by the system must be transferred to its environment. This is the guiding principle behind the following example.

Example 3.2

A brass weight of mass 500 g and specific heat $370\,\mathrm{J\,K^{-1}\,kg^{-1}}$ is heated to a temperature of 535 K and then dropped into 1.00 litre of water at 300 K. What is the final common temperature of the brass and water, and how much heat has been transferred to the water when this temperature is attained? (You may ignore any evaporation of the water.)

Solution

This is an example of a common problem, which is solved by noting that the amount of heat transferred from the hotter body must be the energy gained by the cooler one. Suppose that the final temperature is T, and, in order to avoid mixing numbers and algebraic symbols, let the temperature of the hot brass be $T_\mathrm{h} = 535\,\mathrm{K}$, and the temperature of the cold water $T_\mathrm{c} = 300\,\mathrm{K}$. Let the mass and specific heat of the brass be m_B and C_B and the mass and specific heat of the water be m_W and C_W. Then

$$\text{energy lost by brass} = m_\mathrm{B}C_\mathrm{B}(T_\mathrm{h} - T);$$

$$\text{energy gained by water} = m_\mathrm{W}C_\mathrm{W}(T - T_\mathrm{c}).$$

Equating these two gives

$$m_\mathrm{B}C_\mathrm{B}(T_\mathrm{h} - T) = m_\mathrm{W}C_\mathrm{W}(T - T_\mathrm{c})$$

so $$m_\mathrm{B}C_\mathrm{B}T_\mathrm{h} + m_\mathrm{W}C_\mathrm{W}T_\mathrm{c} = T(m_\mathrm{W}C_\mathrm{W} + m_\mathrm{B}C_\mathrm{B})$$

hence $$T = (m_\mathrm{B}C_\mathrm{B}T_\mathrm{h} + m_\mathrm{W}C_\mathrm{W}T_\mathrm{c}) / (m_\mathrm{W}C_\mathrm{W} + m_\mathrm{B}C_\mathrm{B}).$$

Substituting the relevant values (including the units):

$$T = \frac{(0.50\,\mathrm{kg} \times 370\,\mathrm{J\,K^{-1}\,kg^{-1}} \times 535\,\mathrm{K} + 1.00\,\mathrm{kg} \times 4.19 \times 10^3\,\mathrm{J\,K^{-1}\,kg^{-1}} \times 300\,\mathrm{K})}{(1.00\,\mathrm{kg} \times 4.19 \times 10^3\,\mathrm{J\,K^{-1}\,kg^{-1}} + 0.50\,\mathrm{kg} \times 370\,\mathrm{J\,K^{-1}\,kg^{-1}})}$$

i.e. $$T = \frac{(9.90 \times 10^4\,\mathrm{J} + 1.26 \times 10^6\,\mathrm{J})}{(4.19 \times 10^3\,\mathrm{J\,K^{-1}} + 185\,\mathrm{J\,K^{-1}})} = \frac{(1.36 \times 10^6\,\mathrm{J})}{(4.38 \times 10^3\,\mathrm{J\,K^{-1}})}$$

$$= 311\,\mathrm{K}.$$

The water has been heated by 11 K, so the amount of heat transferred to it from the brass is

$$(1.00\,\mathrm{kg}) \times (4.19 \times 10^3\,\mathrm{J\,K^{-1}\,kg^{-1}}) \times (11\,\mathrm{K}) = 4.6 \times 10^4\,\mathrm{J}.$$

At first sight it may seem that the rise in temperature is small, considering how hot the brass was originally; the reason is that the specific heat of the water is very large. In fact, water has by far the highest specific heat of any common substance. This has a significant effect on the behaviour of the world around us, particularly on climate. The oceans need large amounts of energy to heat them up or cool them down. As a result, their temperature varies very much less than does the temperature of adjacent land masses. As residents of an island, we feel the benefit of this; our climate is much milder than that of places at the same latitude further east — the latitudes of Glasgow and Moscow are much the same, yet their mean winter temperatures are very different (Figure 3.15).

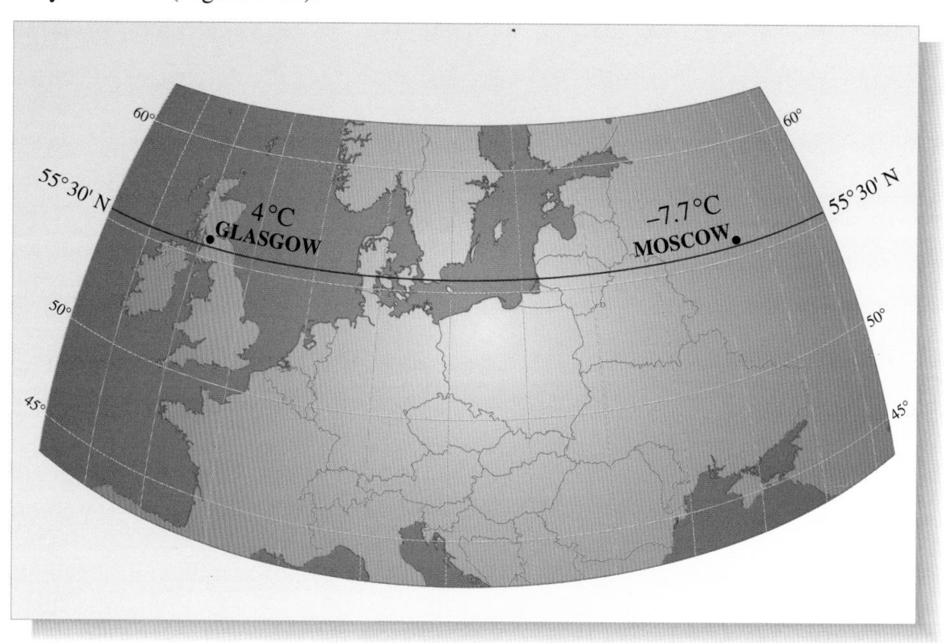

Figure 3.15 Points at the same latitude may have very different climates, depending on their location relative to large bodies of water.

Question 3.4 An aluminium kettle of mass 0.90 kg contains 1.3 kg of water at 12 °C. Assuming that energy losses to the surroundings are negligible, how much energy must be supplied to bring the water to the boil? (The specific heat of aluminium is $913\,\mathrm{J\,K^{-1}\,kg^{-1}}$, and that of water is $4.19 \times 10^3\,\mathrm{J\,K^{-1}\,kg^{-1}}$.) ■

2.5 Isothermal and adiabatic processes

In the previous section, we were concerned with processes that occur while either the pressure or the volume of a system is held constant. In this section, we shall consider two other kinds of process — *isothermal* and *adiabatic*. In each case we shall demand that the process is quasi-static (so that the system can always be regarded as being in equilibrium) and the only system we shall consider is a fixed quantity of an ideal gas, confined in a cylinder by a moveable piston. This means that any work done will involve moving the piston and changing the volume of the gas — we will rule out, for example, the possibility of stirring the gas. The reason for considering such a limited system and for concentrating on isothermal and adiabatic processes within it, will become clear in the next section.

We start by considering isothermal processes. Here is the general definition of such a process:

An **isothermal process** is one in which there is no change of temperature.

One way of ensuring that a process is isothermal is to place the system in good thermal contact with an environment that can supply heat to the system or accept heat from the system without itself undergoing any significant change of temperature. Provided the process is quasi-static, the heat flows can take place rapidly enough for the system to maintain a constant temperature. (If this sounds paradoxical, remember that the temperature of the system will rise when work is done on it, and heat flowing out of the system can compensate for this — as usual in thermodynamics, precise definitions are important!) An environment that stabilizes the temperature of a system is called a **heat bath** or **thermal reservoir**. A typical example is a bath containing a large quantity of water at a carefully regulated temperature (Figure 3.16).

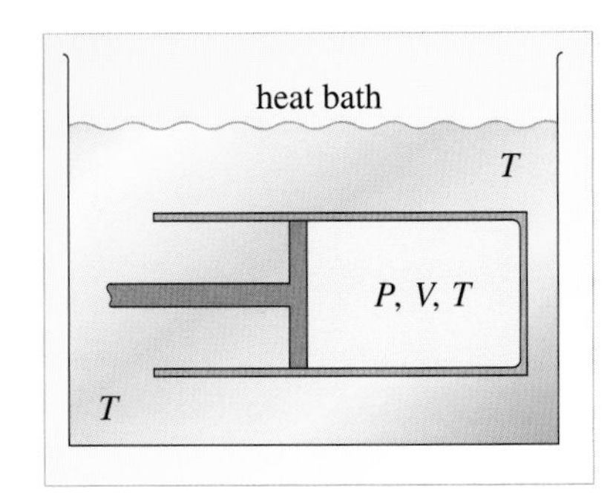

Figure 3.16 A fixed amount of ideal gas in thermal contact with a heat bath at constant temperature T. Changes in the volume or pressure of the gas will cause heat to flow to or from the gas, but the temperature will remain constant.

The ideal gas in the cylinder must, of course, obey the ideal gas equation of state. So, no matter what quasi-static process is considered, we must have

$$PV = nRT. \qquad \text{(Eqn 3.1)}$$

In addition, since the temperature is constant in an isothermal process, the right-hand side of this equation is constant, so the gas must obey a condition of the form

$$PV = I \qquad (3.25)$$

where $I = nRT$ is a parameter that has a constant value for any particular isothermal process, but which has different values for isothermal processes that take place at different temperatures. Equation 3.25 is known as the **isothermal condition**.

It is important to understand the distinction between Equations 3.1 and 3.25. For a given quantity of gas, Equation 3.1 determines a two-dimensional surface in a three-dimensional space with axes P, V and T. Equation 3.25 (for a given value of I) selects a particular pathway on that surface, corresponding to a particular isothermal process. Such a pathway is known as an **isotherm**. A number of isotherms, corresponding to different values of T, are shown in Figure 3.17.

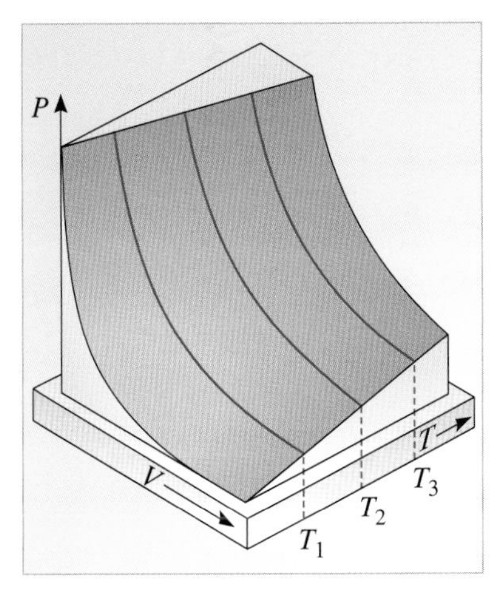

Figure 3.17 Some isotherms on the PVT surface of an ideal gas. Each isotherm corresponds to a different temperature T and therefore to a different value of PV.

We now turn to adiabatic processes. Here is the general definition:

An **adiabatic process** is one in which no heat is transferred.

We can ensure that a process is adiabatic by insulating the system so that no heat can be transferred to or from it, as shown in Figure 3.18. Although no insulation is perfect, heat flow should be negligible whilst the process is taking place. Some very rapid processes can be treated as adiabatic, even without special insulation, simply because they take place too quickly to be affected by heat flow. By contrast, very slow processes may require considerable insulation. Adiabatic processes are of great importance in physics. They include processes occurring inside a vacuum flask, the passage of sound through air (which occurs too rapidly for heat transfers to take place) and many kinds of explosion. Even the Big Bang must have been an adiabatic process.

We are concerned here with quasi-static processes in an ideal gas. The ideal gas must, of course, obey the ideal gas equation of state. So, at all times

$$PV = nRT. \qquad \text{(Eqn 3.1)}$$

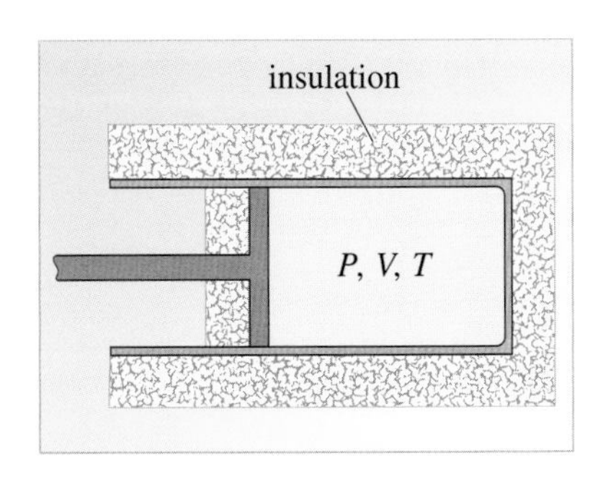

Figure 3.18 A fixed amount of ideal gas in an insulated cylinder. The gas is thermally isolated from the rest of its environment so that no heat can enter or leave the gas. Such a system is said to be adiabatic and the processes that occur in the system are adiabatic processes.

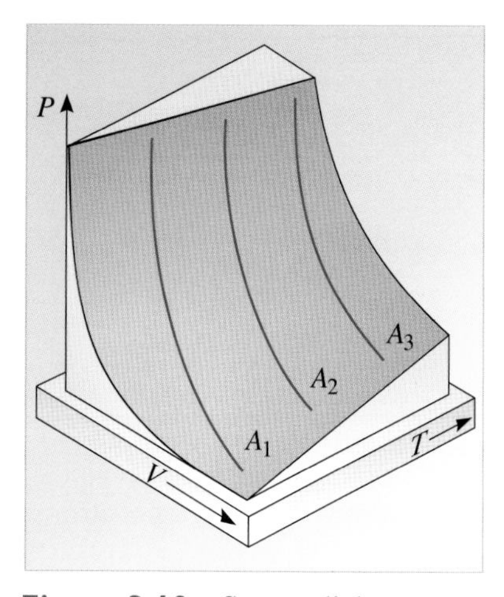

Figure 3.19 Some adiabats on the *PVT* surface of an ideal gas. Each adiabat corresponds to a different value of *A*, and therefore to a different value of PV^γ.

But what additional condition must the gas satisfy in an adiabatic process? What is the adiabatic analogue of Equation 3.25? The answer, as shown in Box 3.2, involves $\gamma = C_P/C_V$, the ratio of the heat capacities at constant pressure and constant volume. The condition itself, known as the **adiabatic condition**, is

$$PV^\gamma = A \tag{3.26}$$

where *A* is a parameter that has a constant value for any particular adiabatic process, but has different values for different adiabatic processes. For a given quantity of gas, a particular value of *A* corresponds to a particular path across the *PVT* surface. Such a path is called an **adiabat**. A number of adiabats, corresponding to different values of *A*, are shown in Figure 3.19.

Box 3.2 Establishing the adiabatic condition

The following derivation is optional reading. It is included for completeness, but you can skip it if you don't mind taking Equation 3.26 on trust.

The derivation of the adiabatic condition rests on four equations:

$$\Delta U = W$$

$$W = -P\,\Delta V$$

$$PV = nRT$$

$$U = \frac{f}{2}nRT.$$

The first equation applies because the process is *adiabatic*. Since no heat can enter or leave the gas, any change in its internal energy must be due to work. The second equation is an expression for the work done on the gas when its volume changes. (Remember, we are assuming that such volume changes provide the only means of doing work on the gas.) The third and fourth equations are the *equation of state* and the *internal energy equation* for an ideal gas with *f* effective degrees of freedom. These equilibrium equations apply so long as the process is *quasi-static*.

Now, we can combine the first three equations to obtain an expression for the change in internal energy of the gas:

$$\Delta U = W = -P\,\Delta V = -\frac{nRT}{V}\Delta V.$$

The fourth equation gives another expression for the change in internal energy of the gas:

$$\Delta U = \frac{f}{2}nR\Delta T.$$

Equating these two expressions for ΔU gives

$$\frac{f}{2}nR\Delta T = -\frac{nRT}{V}\Delta V$$

and further rearrangement gives

$$\frac{\Delta T}{\Delta V} = -\frac{2}{f}\frac{T}{V}.$$

In the limit as ΔV becomes vanishingly small, this leads to the differential equation

$$\frac{\mathrm{d}T}{\mathrm{d}V} = -\frac{2}{f}\frac{T}{V}.$$

The solution to this differential equation turns out to be

$$T = BV^{-2/f}$$

where B is an arbitrary constant. (You can check this, if you like, by differentiating both sides of the solution with respect to V and showing that the result is indeed equivalent to the original differential equation.)

Using the equation of state again, this time in the form $T = PV/nR$, we can eliminate T to obtain

$$\frac{PV}{nR} = BV^{-2/f}.$$

Introducing a new constant, $A = nRB$, we then have

$$PV^{1+2/f} = A.$$

Finally, Equation 3.22 shows that $1 + 2/f = \gamma$, so we can rewrite our last result as

$$PV^{\gamma} = A, \qquad \text{(Eqn 3.26)}$$

thus establishing the adiabatic condition for an ideal gas. In this context, the quantity γ (the ratio of heat capacities) is sometimes called the *adiabatic exponent*.

It is worth noting that Equation 3.26 does not apply to every adiabatic process in every system. As mentioned at the beginning of this section, we are concentrating on *quasi-static* processes in *ideal gases*, and are assuming that any work done involves a change in volume of the gas. Although these restrictions may seem severe, Equation 3.26 will be important later on, when we discuss the entropy of an ideal gas.

It is instructive to compare the isotherms and adiabats of an ideal gas. This is done in Figure 3.20 (overleaf), which shows paths on the PVT surface, and their corresponding projections onto the P–V plane. As the volume decreases, the adiabat rises more steeply than the corresponding isotherm. This is because adiabatic compression causes an increase in temperature, so that P increases more rapidly than in the isothermal case, where the temperature remains constant.

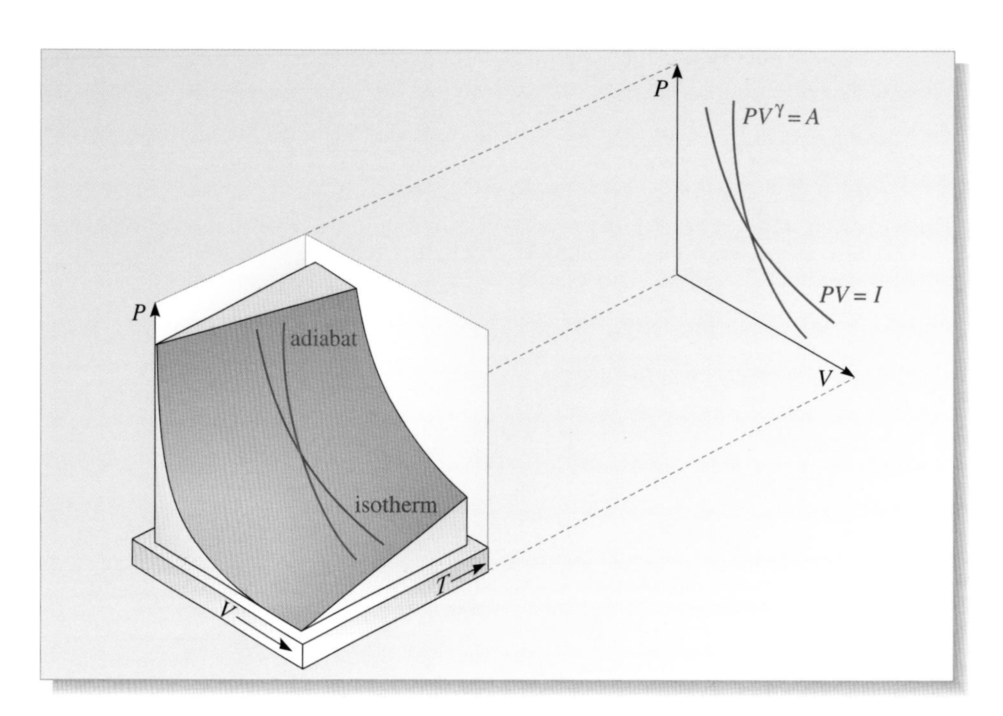

Figure 3.20 Comparison of isotherms and adiabats for an ideal gas on the *PVT* surface and on its projection onto the *P–V* plane.

The comparison is taken further in Table 3.3, which records the energy flows that take place in four different types of process. Two of the entries in the table (those for W and Q in row 3) are difficult to derive without using integrals, and are included only for completeness. The remaining entries can be derived in a fairly straightforward way, and form the basis of some of the following exercises. The general point to note, though, is that different types of process have totally different patterns of heat and work transfer.

Table 3.3 Four different types of quasi-static process in an ideal gas, from an initial state (P_1, V_1, T_1) to a final state (P_2, V_2, T_2). Each process has a distinctive balance of work done on the gas (W), heat transferred to the gas (Q) and change in internal energy (ΔU). It is assumed that the only way of doing work on the gas is by changing its volume. Throughout the table, C_V and C_P are the constant volume and constant pressure heat capacities, given by Equations 3.15 and 3.21. The effective number of degrees of freedom, f, is related to the ratio of heat capacities, γ, by $\gamma = 1 + 2/f$.

Process	W	Q	ΔU	Comments
constant pressure	$-P(V_2 - V_1)$	$C_P(T_2 - T_1)$	$C_V(T_2 - T_1)$	$P_1 = P_2$
constant volume	0	$C_V(T_2 - T_1)$	$C_V(T_2 - T_1)$	$V_1 = V_2$ $\Delta U = Q$
isothermal	$-nRT \log_e\left[\frac{V_2}{V_1}\right]$	$nRT \log_e\left[\frac{V_2}{V_1}\right]$	0	$T_1 = T_2$ $Q = -W$
adiabatic	$\frac{f}{2}(P_2V_2 - P_1V_1)$	0	$C_V(T_2 - T_1)$	$\Delta U = W$

Question 3.5 Justify the entries in the first row of Table 3.3 and show that they are consistent with the first law of thermodynamics.

Question 3.6 It is a common misconception to think that an isothermal process cannot involve any transfer of heat, but the third row of Table 3.3 shows that this is not so. Explain in your own words why the isothermal change described in the table should be expected to involve heat transfer.

Question 3.7 Equation 3.26 expresses the adiabatic condition in terms of P and V. Use the equation of state of an ideal gas to express the condition in terms of (a) P and T, and (b) T and V. ■

3 The second law of thermodynamics

Not all the processes we can conceive of actually occur. This is so obvious that most people scarcely think about it. Nonetheless, it is a basic fact about the physical world and the laws of physics should account for it. If you have two isolated bodies at different temperatures and put them in thermal contact, then they will in time come to the same temperature; the hotter one will cool, and the cooler one will warm up. However, if you have two bodies at the same temperature that are in thermal contact you never find that one warms up while the other cools down. If the heat gained by one equalled the heat lost by the other, this process would be consistent with the first law of thermodynamics, but it still never happens. Why not? If I drop a plate on the floor and it breaks, I am not surprised by the damage. If, however, the pieces were to reform and leap back into my hand, I would be astounded. Again, though, this process need not violate the first law, so what stops it from happening?

Clearly, some additional law of physics is needed that will, broadly speaking, prevent those processes which do not happen but allow those that do. This is the role of the second law of thermodynamics.

There are several equivalent ways of formulating the second law. Some emphasize its implications for practical devices such as engines or refrigerators, while others concentrate on the justification it provides for the introduction of *entropy* as an important property of the physical world. We shall adopt this second approach. In this section then, apart from providing a statement of the second law, we shall introduce the entropy of a system and explain how it can be determined. Its practical implications will be discussed in Section 4.

3.1 Reversible and irreversible processes

Given any two states of a system, A and B, it is generally possible to find a process that leads from A to B, and a process that leads from B to A, but it is often the case that one of these processes is more complicated than the other. For instance, breaking a bottle is a simple process, but reforming the bottle is quite complicated; it may well involve melting the glass and remoulding the bottle. Breaking the bottle leaves the system (i.e. the bottle) in a radically altered state, but has little effect on the environment. Reforming the bottle returns the system to its original state but almost certainly involves radical changes in the environment. Often, these changes mean that the environment is warmed up.

Despite the difficulty of reversing most processes without causing some change to the environment, there are some processes that can be reversed without leaving any sign that they ever occurred. An example would be the kind of adiabatic process discussed in Section 2.5. In that case we considered a gas that was thermally isolated, so that it could not exchange any heat with its environment, and we demanded that the only way in which work could be done on or by the gas was by means of a volume change. If the piston is frictionless and all changes are quasi-static, then any work that the environment does on the gas by compressing it can be recovered by allowing the gas to expand back to its original volume, pushing back the piston and doing work on the environment in the process. Thus, the adiabatic process we discussed in Section 2.5 was actually a *reversible* adiabatic process, where by reversible we mean the following:

A *reversible* process in thermodynamics has much the same status as a *frictionless* process in mechanics.

A thermodynamic process is said to be **reversible** if, following the process, both the system *and its environment* can be returned to the states they were in prior to the process. Any process that is not reversible is said to be **irreversible**.

Notice that simply being able to reverse a process does not make it reversible. You can reform a broken bottle, but you cannot do so without changing the environment. For a process to be reversible, both the system, *and its environment,* must be able to return to their original states.

Reversible processes are, in fact, something of an idealization. All real processes are to some extent irreversible. Certainly, any process that involves friction or any other dissipative phenomenon will be irreversible. That is why we took care to exclude the possibility of transferring energy to the gas by means of a stirrer when we discussed adiabatic processes in Section 2.5. Stirring the gas would not have conflicted with the requirement that $Q = 0$ in an adiabatic process, but it would have meant that those particular adiabatic processes would not have been reversible. You can increase the internal energy of a gas by stirring it, but you cannot 'unstir' the gas. The adiabatic condition that we wrote down in Section 2.5

$$PV^{\gamma} = A \qquad \text{(Eqn 3.26)}$$

is in fact, the condition that characterizes a *reversible* adiabatic process in an ideal gas. Irreversible adiabatic processes (such as those caused by stirring the gas in a thermally insulated container) do not satisfy this condition. This means Equation 3.26 is more accurately described as the **reversible adiabatic condition** and the paths in Figure 3.19 are more accurately described as **reversible adiabats**. For brevity, we will continue to use the terms 'adiabatic condition' and 'adiabats' as shorthands.

Isothermal processes can also be reversible. For the system we were considering in Section 2.5, we simply require that any work done in such a process is performed reversibly, and that the compensatory heat transfers (required to avoid any overall change in temperature) are always the result of infinitesimal differences in temperature between the system and its environment.

Question 3.8 Which of the following processes *can be reversed* and which are *reversible*? (a) Smashing an egg on the floor. (b) Pushing a chest of drawers across a room. (c) Transferring energy to a gas by means of a quasi-static increase in its volume carried out under ideal conditions free from friction and other dissipative effects. (d) Reading this book from cover to cover. ■

3.2 Adiabatic accessibility and the second law

This section takes a closer look at the distinction between reversible and irreversible processes, and states one version of the second law of thermodynamics (you will see that there are many equivalent ways of formulating this law). It also introduces a new quantity — the adiabatic accessibility index — which will lead us towards a definition of entropy. The logic of the argument is subtle, so it is worth surveying the steps that lie ahead. Have a brief look at Box 3.3, which provides a route map through the next few sections.

Box 3.3 The path to entropy

This section considers processes that occur in an *adiabatic* system. Both reversible and irreversible processes are considered, but the fact that no heat can enter or leave the system creates a severe constraint which prevents many things from happening. Starting from a given initial state, you will see that states can be divided into three groups:

(a) those that can be reached by a reversible adiabatic process;

(b) those that can be reached by an irreversible adiabatic process;

(c) those that cannot be reached by any adiabatic process.

To make the discussion less abstract, we consider the special case of an ideal gas. In this case, it is possible to find an explicit quantity, A, which underpins the above classification. States in group (a) have the same value of A as the initial state; those in group (b) have greater values of A than the initial state, and those in group (c) have smaller values of A than the initial state. For this reason, A will be called the *adiabatic accessibility index.*

In Section 3.3, we will remove the constraint that the system is adiabatic, and allow heat to be exchanged between the system and its environment. This transforms the situation. In particular, states that were in group (b) can now be accessed by a reversible process in which heat flows into the system, and states that were in group (c) can be accessed by a reversible process in which heat flows out of the system. The heat flows that are required for these reversible processes to take place allow us to provide a general definition of entropy.

For the special case of an ideal gas, we will show that the general definition of entropy is closely linked to the adiabatic accessibility index. At this point, we will have enough information to discuss the significance of entropy in detail. Section 3.4 will show how changes in entropy determine the direction of heat flow and Section 3.5 will finally interpret entropy as a measure of disorder.

If Box 3.3 is unclear now, read on, and return to it from time to time to check where you have got to in the argument.

The second law is a very general statement, which is believed to apply to thermodynamic systems of all kinds. However, our first steps towards this law are made by discussing a very simple system with which you should now be quite familiar. This system is shown in Figure 3.21: it consists of a fixed quantity (n moles) of ideal gas confined in a cylinder by a frictionless piston. The volume of the gas can be controlled by moving the piston. There is also a paddle wheel which can be used to stir the gas, thus performing work on it without changing the volume. The cylinder is clad with good thermal insulation so that no heat can flow into or out of it. We shall also suppose that all changes occur quasi-statically, so the ideal gas equation of state $PV = nRT$ applies at all times. Finally, the apparatus is supplemented by whatever other equipment is needed to monitor the pressure P, volume V and temperature T of the gas.

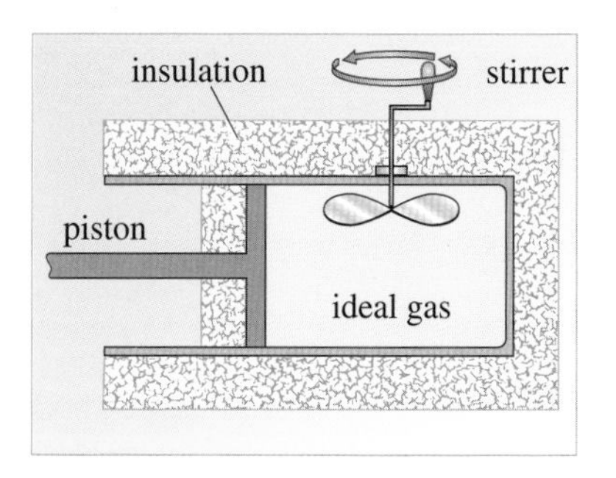

Figure 3.21 A thermally isolated system consisting of a fixed quantity of ideal gas, confined in a cylinder by a frictionless piston, with a stirrer.

Because the system is thermally insulated, no heat flows can occur, so any change in the state of the gas must be the result of work done in an adiabatic process. The work can be performed by sliding the piston, rotating the stirrer, or combining both these motions. Here is what would be observed in each case:

1 *Processes that involve moving the piston but not the stirrer.*
Depending on the direction of motion of the piston, these processes cause the state of the system to move back or forth along the adiabat that passes through the initial state of the gas, as indicated in Figure 3.22a. Since the initial state can be recovered, and the environment is left undisturbed (remember, there are no heat flows), these are *reversible* adiabatic processes.

2 *Processes that involve moving the stirrer but not the piston.*
Moving the stirrer does not involve any transfer of heat. However it does increase the internal energy of the gas and so increases its temperature. Since the gas always obeys the equation of state $PV = nRT$, any rise in temperature at constant volume must be accompanied by an increase in pressure. The effect of these changes is therefore to cause the state of the system to move vertically upwards on the P–V diagram from one adiabat to another, as indicated in Figure 3.22b. Note that it is not possible to undo this process and move vertically downwards back to the original adiabat (a stirred gas cannot be unstirred). So, these are *irreversible* adiabatic processes.

3 *Processes that involve a combination of movements of the piston and the stirrer.*
By choosing an appropriate combination of sliding the piston and rotating the stirrer, it is possible to move from an initial state to any other state that is *on or above* the adiabat that passes through the initial state. This is indicated in Figure 3.22c. These processes are reversible so long as no use is made of the stirrer. However, once the stirrer is brought into operation, as it must be to get away from the initial adiabat, the process immediately becomes *irreversible*.

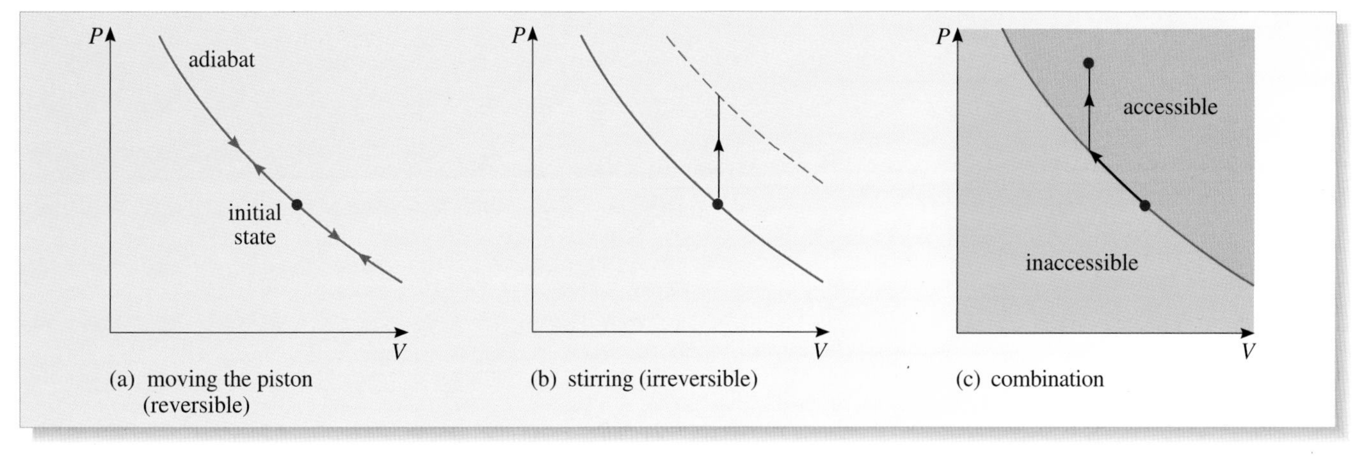

Figure 3.22 P–V diagrams showing state changes resulting from adiabatic processes taking place in the system of Figure 3.21. (a) The state of the gas may move back and forth along an adiabat as a result of a reversible movement of the piston. (b) The state may move to a 'higher' adiabat as a result of irreversible stirring without any change in volume. In such a case, the pressure (and the temperature) must increase. (c) The state may move to any point on or above the initial adiabat as a result of an appropriate combination of piston movements and stirring.

Our discussion of adiabatic processes in a gas can now be summarized in a slightly different way:

- Reversible adiabatic processes (quasi-statically sliding the piston) allow us to access any state on the same adiabat as the initial state.
- Irreversible adiabatic processes (rotating the stirrer, possibly in combination with sliding the piston) allow us to access any state on a higher adiabat than the initial state.

- There are no adiabatic processes that allow us to access a state on a lower adiabat than the initial state.

The phrases 'higher adiabat' and 'lower adiabat' may be clear enough in the context of Figure 3.22, but it would be useful to have a more precise formulation that did not rely on such a diagram. In the case of an ideal gas, this is easy to obtain. You already know that the adiabats of an ideal gas are described by the adiabatic condition of Equation 3.26, according to which

$$A = PV^{\gamma}$$

where γ is the ratio of heat capacities of the gas and A is a parameter that is constant for any given adiabat, but which takes different values on different adiabats. Now, suppose that the initial state of the gas is characterized by the values $P = P_0$ and $V = V_0$ and that we are interested in reaching each of the three states shown in Figure 3.23, characterized respectively by (P_1, V_1), (P_2, V_2) and (P_3, V_3).

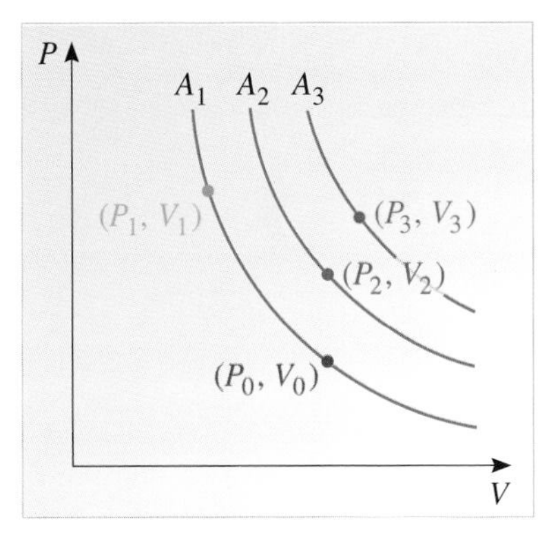

Figure 3.23 An initial state (P_0, V_0), and three other states that are accessible from it by adiabatic processes. Each of the three states is on a different adiabat, characterized by a different value of the parameter A. The relevant value of A for each adiabat can be determined by using the condition $A = PV^{\gamma}$.

As you can see, they are all on or above the adiabat that passes through the initial state, so all three should be accessible from the initial state by means of an appropriate adiabatic process.

The first of the three states is on the same adiabat as the initial state. We can identify that particular adiabat by saying that it corresponds to a particular value of A. If we call this value A_1, the adiabatic condition tells us that its value must be $A_1 = P_1 V_1^{\gamma}$. Since this is also the adiabat of the initial state, it will also be the case that $A_1 = P_0 V_0^{\gamma}$. The second of the three states is on an adiabat characterized by a different value A, let us call it A_2. In this case, $A_2 = P_2 V_2^{\gamma}$, but $V_2 = V_0$, while P_2 is greater than P_0, so A_2 must be greater than A_1. Similarly, the third state is on an adiabat characterized by $A = A_3 = P_3 V_3^{\gamma}$ where A_3 is greater than A_2.

- Formulate a convincing argument to show that A_3 is greater than A_2.

- All states on the adiabat characterized by the value $A = A_3$ obey the condition $PV^{\gamma} = A_3$. Among those states there will be one that has volume $V = V_2$, but its pressure P will be greater than P_2. Consequently, for that particular state, PV^{γ} (which is equal to A_3) will be greater than $P_2 V_2^{\gamma}$ (which is equal to A_2). It follows that A_3 is greater than A_2, as claimed. ■

Our earlier findings about which states are accessible by means of an adiabatic process from a given initial state can now be expressed in terms of the parameter A, which we shall henceforth refer to as the **adiabatic accessibility index**. Here are our conclusions:

Every state of an ideal gas corresponds to a particular value of the adiabatic accessibility index, given by $A = PV^{\gamma}$. The adiabatic accessibility index is therefore a function of state. Given a particular initial state:

- it is possible to reach any other state with the same value of A by means of a reversible adiabatic process;
- it is possible to reach any state with a higher value of A by means of an irreversible adiabatic process;
- it is impossible to reach any state with a lower value of A by any adiabatic process. Such states are therefore said to be **adiabatically inaccessible** from the initial state.

To take a definite case, suppose that a sample of helium gas (with $\gamma = 5/3$) is contained in a perfectly insulated vessel, with a pressure of 1 atmosphere and a volume of 1 m^3. The perfect insulation ensures that this is an adiabatic system so that only adiabatic processes are allowed. If you were asked to find an adiabatic process that would change the pressure of the gas to 2 atmospheres and its volume to 0.5 m^3, you could immediately say that no such adiabatic process exists, because the value of A in the final state is lower than the value of A in the initial state. The requested state is therefore adiabatically inaccessible from the initial state.

Before making such a sweeping statement as 'no such adiabatic process exists', it is advisable to be sure of the basis of your argument. So far, we have made a number of assertions that rely on experience and experimental observations, but we have not yet formulated a fundamental law that forbids certain processes from happening. It is now time to do so.

Although all our arguments in this section have involved ideal gases, the Greek mathematician Constantin Carathéodory (1873–1950) realized that the idea of adiabatic inaccessibility can be generalized to all thermodynamic systems. He recognized that the following principle is of fundamental importance:

Carathéodory's statement of the second law of thermodynamics

In the neighbourhood of any equilibrium state of a thermodynamic system, there are equilibrium states that are adiabatically inaccessible.

Given the above discussion, the truth of Carathéodory's statement is not so surprising, but the amazing thing is that this brief statement implies much more. In fact, Carathéodory's statement is one way of encapsulating the second law of thermodynamics, and so leads directly to the concept of entropy. Carathéodory used his principle to show that:

1. For any thermodynamic system there exists some quantity, similar to the adiabatic accessibility index, that determines which states are adiabatically accessible from a given initial state, and whether they are accessible by a reversible adiabatic process. This is the quantity called **entropy**.
2. The entropy of a system is a function of state. Thus, in any process that starts and finishes with an equilibrium state, the change in the entropy depends on the initial and final states but not on the details of the process.
3. In a reversible adiabatic process, there is no change in the entropy of a system; and in an irreversible adiabatic process, the entropy of a system increases. The entropy of an isolated system never decreases.

How the entropy of a system is evaluated will be dealt with in the next section.

3.3 Defining entropy

Our discussion so far has been rather abstract because we have not yet said what entropy is, except that its role is similar to the adiabatic accessibility index. You are no doubt wondering why we need to introduce the concept of entropy at all, and why it is different from the adiabatic accessibility index. There are two important reasons.

First, the adiabatic accessibility index applies to a very specific system — an ideal gas. The adiabatic condition, $PV^\gamma = A$, does not apply in other cases, and it would take a lot of effort to generalize it in an appropriate way for other systems, even if this could be done.

Secondly, the adiabatic accessibility index does not have the desirable property of being *additive*. Let me explain what this means. Suppose we have two different systems, with internal energies U_1 and U_2 and adiabatic accessibility indices A_1 and A_2. If these two systems are joined together and regarded as a single system, the internal energy of the whole system is $U_1 + U_2$. This fact is expressed by saying that internal energy is an additive quantity — the internal energy of the whole is the sum of the internal energy of its parts. The adiabatic accessibility index does not have this property. It is *not* true that the adiabatic accessibility index of the whole system is $A_1 + A_2$. The property of additivity turns out to be so useful that it is worth looking for a quantity that is related to the adiabatic accessibility index, but which *is* additive: this quantity is entropy.

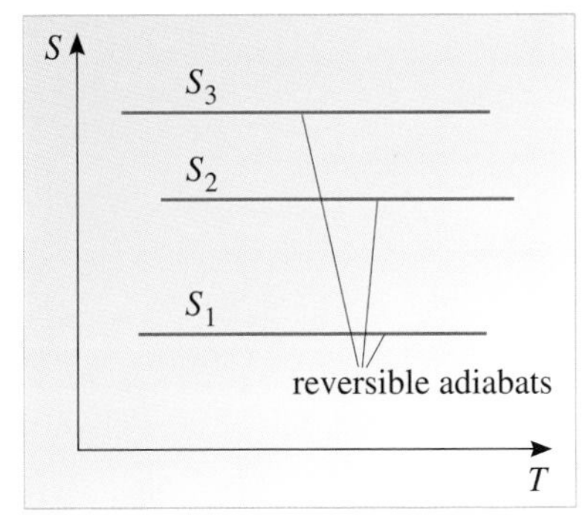

Figure 3.24 The entropy temperature diagram (S–T diagram) of a system can be thought of in much the same way as the P–V or V–T diagrams of an ideal gas. Lines in the S–T plane schematically represent processes linking different equilibrium states. Horizontal lines correspond to fixed values of S and hence represent reversible adiabatic processes.

How is entropy defined in general? Well, for any thermodynamic system, we can certainly make use of general thermodynamic concepts such as state, process, temperature, heat, work and internal energy. Since we want the entropy to 'play a similar role' to the adiabatic accessibility index, we shall insist that states which are mutually accessible by means of a reversible adiabatic process have the same entropy. In other words, all states that lie on the same (reversible) adiabat have the same entropy. This is indicated schematically in Figure 3.24, where, as is conventional, we use the symbol S to represent entropy of a system and show the reversible adiabats of the system as (horizontal) lines of constant entropy on a plot of entropy against temperature.

In order to distinguish states of different entropy, we need to identify some quantity that will enable us to label the different reversible adiabats of a general system (which is what A did for an ideal gas). The best way of finding such a quantity is to relax the constraint of requiring the system to be adiabatic. Once heat can flow into or out of the system, we will be able to move reversibly between one adiabat and another, and look for a suitable way of quantifying the change that has occurred.

One suggestion, for example, would be to measure the heat transfer that takes place as we move from one adiabat to another. Unfortunately, heat is a most unsatisfactory choice because it is not a function of state. Since the heat transferred depends on the precise process used to move between adiabats, it cannot be used to label adiabats. A suitable choice is suggested by the following result which applies to any thermodynamic system:

> Given any two states with the same absolute temperature T, and a reversible process that leads from one state to the other, then, if the heat supplied to the system during that process is Q_{rev}, the quantity Q_{rev}/T depends only on the states involved and not on the details of the process. Thus, Q_{rev}/T is a function of state.

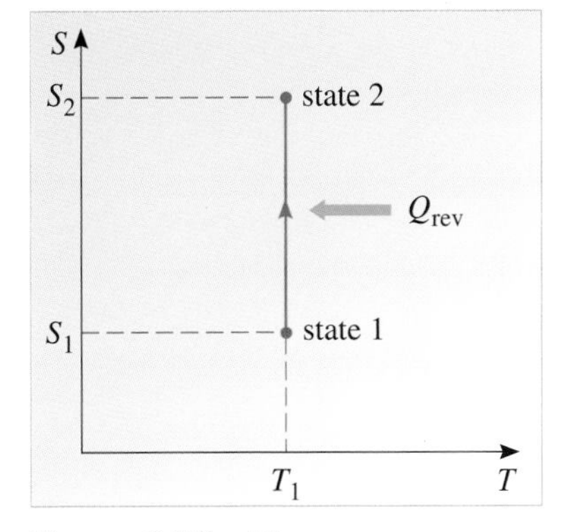

Figure 3.25 The entropy difference between two states with the same temperature T_1 is $\Delta S = S_2 - S_1 = Q_{\text{rev}}/T_1$, where Q_{rev} is the heat transferred to the system in any reversible process that leads from the first state to the second.

Using this result (which is not proved here), we can now give a definition for the entropy change that occurs when we move between two states at the same temperature. If we let the entropy of the first state be S_1 and the entropy of the second be S_2, then the entropy difference between the two states, $\Delta S = S_2 - S_1$, is defined to be

$$\Delta S = \frac{Q_{\text{rev}}}{T} \tag{3.27}$$

where Q_{rev} is the heat that is transferred to the system during any reversible process that leads from the first state to the second. This relationship is indicated schematically in Figure 3.25.

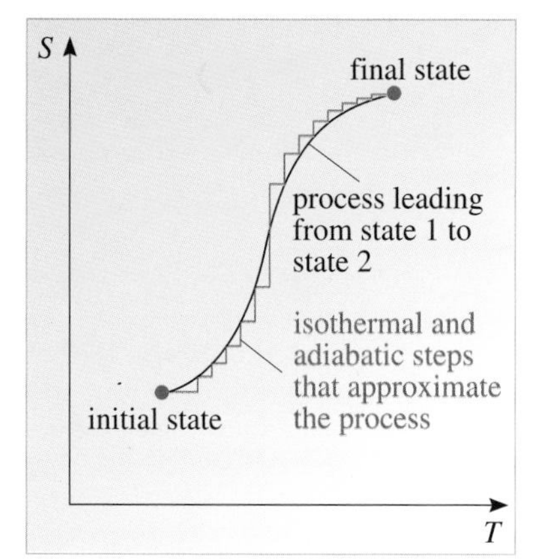

Figure 3.26 The entropy difference between any two states can be determined by considering a sequence of reversible isothermal and adiabatic steps that together approximate a reversible process leading from one state to the other.

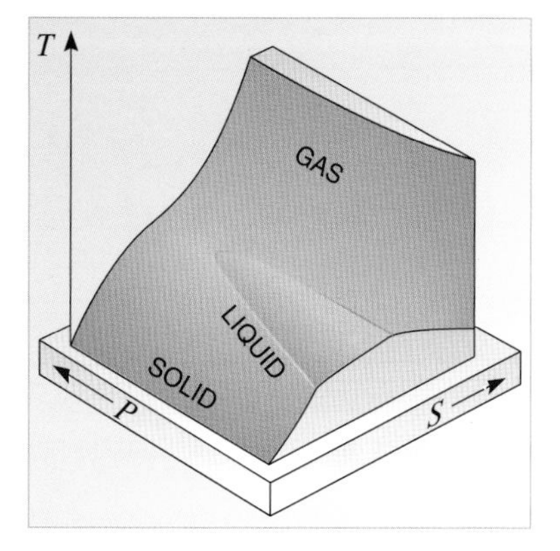

Figure 3.27 The *TSP* surface for a fixed quantity of a pure substance. The variables *P* and *T* determine the state of the substance. The surface shows that the entropy *S* is a function of state, as the second law of thermodynamics implies.

Of course, the likelihood that we will want to know the difference in entropy between two states at the same temperature is not particularly great. But fortunately we can use Equation 3.27 as the basis of a general definition of entropy difference. The essential idea behind this generalization is indicated in Figure 3.26. Given any two states of the system and a reversible process that links them, then it is always possible to find a process composed of reversible isothermal and adiabatic steps which approximates that reversible process. The reversible adiabatic steps in this process do not involve any change in the entropy of the system, so the total entropy change results from the reversible isothermal steps and is given by

$$\Delta S = \sum_i \frac{Q_{\text{rev},\,i}}{T_i} \tag{3.28}$$

where $Q_{\text{rev},i}$ is the heat reversibly transferred to the system in the ith step, which takes place at absolute temperature T_i, and the summation sign (Σ) tells us to add together the value of $Q_{\text{rev},i}/T_i$ for all of the steps involved in the process. Equation 3.28 is, essentially, the general definition of the entropy difference between two states of a system. (It can be further generalized by use of the concept of definite integration introduced in *Predicting motion*, but we shall not take that step here.) This definition is particularly attractive because it implies that entropy is additive: if two smaller systems are joined together to form a larger system, and no further changes take place as a result of this joining, the entropy of the whole is the sum of the entropies of its parts.

Now, with this general definition in mind, you can see how it would be possible experimentally to start from a given state of a system and perform a series of experiments that would determine the entropy change needed to reach any other state of the system. If you assigned an arbitrary value, S_0 say, to the entropy of the initial state and carried out enough experiments, you could eventually build up a picture of how entropy varied as you explored all the possible states of the system. Figure 3.27 shows what the result of such an investigation might look like for a system consisting of a fixed quantity of a pure substance confined in a cylinder of volume V at pressure P and temperature T.

As you can see, states belonging to the solid phase of the substance, broadly speaking, have a relatively low entropy, while states belonging to the liquid phase have intermediate entropy, and states belonging to the gas phase have a high entropy. Later, you will see that entropy can be interpreted as a measure of disorder, so a solid phase is less disordered than a liquid phase which in turn is less disordered than a gas phase. But what are the values of entropy, and in what units are they measured? The second of these questions is easily answered; you can work out the units of entropy for yourself by responding to the following question.

● Given that there are processes in which $\Delta S = Q_{\text{rev}}/T$, what is a suitable SI unit of entropy?

○ Since heat transfers are measured in joules (J), and absolute temperature is measured in kelvin (K), it follows that entropy (and entropy difference ΔS) may be measured in J K^{-1}. ■

Finding typical values of entropy is less straightforward since we have to arbitrarily assign a value for the entropy to some particular state. However we can get a good idea of the entropy difference between some states as the following example shows:

Example 3.3

(a) 1.00 kg of ice (pure H_2O) at 0 °C, contained in a bucket, is allowed to gradually melt while the pressure is maintained at standard atmospheric pressure. Under these conditions, the specific latent heat of melting of ice is $3.33 \times 10^5\,\mathrm{J\,kg^{-1}}$. By how much will the entropy of the H_2O have increased when it has all turned into liquid water, but its temperature is still 0 °C?

(b) Perform a similar calculation to find the entropy change when 1.00 kg of boiling water at 100 °C is converted to steam at the same temperature. Take the relevant specific latent heat of vaporization to be $2.26 \times 10^6\,\mathrm{J\,kg^{-1}}$.

Solution

(a) The heat transferred to ice in order to melt it at fixed temperature is simply the product of the mass of ice and its specific latent heat of melting. Hence

$$Q_{\mathrm{rev}} = 1.00\,\mathrm{kg} \times 3.33 \times 10^5\,\mathrm{J\,kg^{-1}} = 3.33 \times 10^5\,\mathrm{J}.$$

A temperature of 0 °C is equivalent to $T = 273.15\,\mathrm{K}$, so the required entropy change is

$$\Delta S = Q_{\mathrm{rev}}/T = (3.33 \times 10^5\,\mathrm{J}) / (273\,\mathrm{K}) = 1.22 \times 10^3\,\mathrm{J\,K^{-1}}.$$

(b) Similarly, for vaporization,

$$Q_{\mathrm{rev}} = 1.00\,\mathrm{kg} \times 2.26 \times 10^6\,\mathrm{J\,kg^{-1}} = 2.26 \times 10^6\,\mathrm{J}.$$

In this case, the relevant temperature is $T = 373.15\,\mathrm{K}$, so

$$\Delta S = Q_{\mathrm{rev}}/T = (2.26 \times 10^6\,\mathrm{J}) / (373\,\mathrm{K}) = 6.06 \times 10^3\,\mathrm{J\,K^{-1}}.$$

As a check we note that the units are correct; in particular, we have remembered to convert the temperature from the Celsius scale to the absolute scale before carrying out the division. (Forgetting to do this is a common source of error.)

Having obtained a general definition for the entropy of a system, we can return to the special case of an ideal gas, and take a closer look at the relationship between entropy and the adiabatic accessibility index, A. The relationship is easily stated.

Consider a fixed mass of ideal gas. Let S_1, A_1, P_1, V_1 and T_1 be the entropy, adiabatic accessibility index, pressure, volume and temperature in the initial state of the gas. Let S_2, A_2, P_2, V_2 and T_2 be the corresponding quantities in the final state of the gas. Then the change in entropy between the initial and final states is:

$$S_2 - S_1 = C_V \log_e \left(\frac{A_2}{A_1} \right). \tag{3.29}$$

We shall not prove this result, but can explore its implications for different values of A_1 and A_2. Three cases can be distinguished:

1. When the adiabatic accessibility index rises, $A_2 > A_1$, and $\log_e (A_2/A_1) > 0$. Equation 3.29 then shows that $S_2 > S_1$, so the entropy rises.
2. When the adiabatic accessibility index remains constant, $A_2 = A_1$, and $\log_e (A_2/A_1) = 0$. Equation 3.29 then shows that $S_2 = S_1$, so the entropy remains constant.
3. When the adiabatic accessibility index falls, $A_2 < A_1$ and $\log_e (A_2/A_1) < 0$. Equation 3.29 then shows that $S_2 < S_1$, so the entropy falls.

Because the entropy rises, stays constant, and falls in tandem with the adiabatic accessibility index, we are entitled to say that they both play a similar role in determining the accessibility of different states.

It is worth recasting Equation 3.29 in a slightly different form. Using the adiabatic condition, $A = PV^{\gamma}$, and the following general properties of the logarithm function:

$$\log_e(xy) = \log_e(x) + \log_e(y)$$

$$\log_e(y^z) = z\log_e(y)$$

we can re-express Equation 3.29 first as

$$S_2 - S_1 = C_V \log_e\left(\frac{P_2V_2^{\gamma}}{P_1V_1^{\gamma}}\right)$$

and then as

$$S_2 - S_1 = C_V\left(\log_e\left(\frac{P_2}{P_1}\right) + \gamma\log_e\left(\frac{V_2}{V_1}\right)\right).$$

Recalling that $\gamma = C_P/C_V$, we finally have

$$S_2 - S_1 = C_V \log_e\left(\frac{P_2}{P_1}\right) + C_P \log_e\left(\frac{V_2}{V_1}\right) \tag{3.30a}$$

where C_P and C_V are the heat capacities of the gas at constant pressure and constant volume. This formula describes the *entropy change* in an ideal gas. Notice that the change in entropy has more direct physical significance than the entropy itself. To define the entropy of a system, we must make an arbitrary choice of reference state with pressure P_0, volume V_0 and entropy S_0. Then,

$$S = S_0 + C_V \log_e\left(\frac{P}{P_0}\right) + C_P \log_e\left(\frac{V}{V_0}\right). \tag{3.30b}$$

(This is reminiscent of the definition of potential energy, which always involves an arbitrary choice of energy zero.)

The existence of an explicit formula helps to emphasize the 'concrete' nature of entropy. Remember, though, that Equation 3.30b applies only to ideal gases. A graphical representation of this formula for the entropy is shown in Figure 3.28.

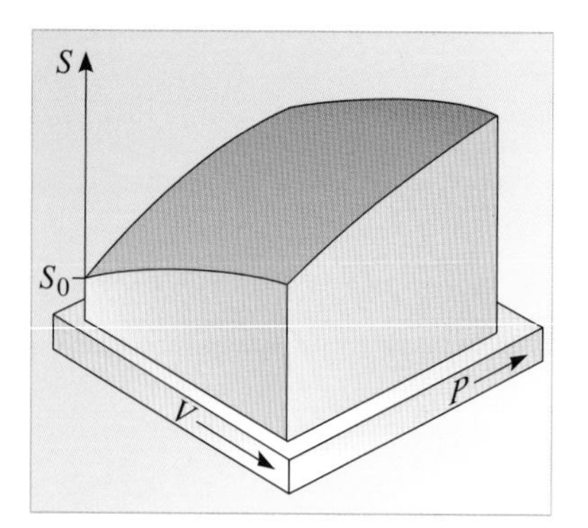

Figure 3.28 The *SVP* surface for a fixed quantity of ideal gas. Different choices for the value of S_0 would have the effect of raising or lowering the entire surface, but would leave its shape unchanged.

In other systems, the entropy will be a different, usually more complicated function. But it remains true, in any system, that the entropy is constant for all reversible adiabatic processes and increases for all irreversible adiabatic processes. Since an isolated system is adiabatic, we can be sure that the entropy of an isolated system will not decrease.

Recalling that all naturally occurring processes are to some extent irreversible, we can state the following principle which some authors use in place of the second law:

The principle of entropy increase

There exists a function of state, known as entropy, which increases during any naturally occurring change of an isolated thermodynamic system.

Note that this principle is restricted to an *isolated* thermodynamic system, i.e. one that does not exchange energy or matter with its environment. Systems that are not isolated may decrease their entropy, but not without other consequences, as you will see when we discuss refrigerators. In fact, exchanging heat with the environment entails altering the entropy of the system *and* altering the entropy of its environment. What happens in such cases is that the combined entropy of the system and its environment never decreases, even though the entropy of the system itself may decrease if it is not isolated. This leads to another way of stating the second law:

Boltzmann's statement of the second law of thermodynamics

The entropy of the Universe tends to a maximum.

The word 'Universe' is used here in its technical sense of 'system plus environment.'

We shall not provide detailed arguments to support the various assertions that have just been made about the consequences of the second law, they require too much mathematical detail. However, it is instructive to compare the implications of the second law with those of the first law as stated in Section 2.3. The first law had three direct implications: the existence of internal energy, its property of being a function of state, and the fact that the energy of the Universe is constant (i.e. the conservation of energy). In the case of the second law, we are again presented with a new function of state — the entropy of the system — but this time, instead of a conservation principle we are led to a principle of no decrease or a tendency towards a maximum. It is this principle that underlies the tendency of natural processes to follow an 'arrow of time.' If you see a film of breaking an egg or of almost any other natural phenomenon, it is generally easy to tell if the film is running backwards or forwards. This is because natural processes tend to increase entropy and therefore distinguish the future from the past. The second law is almost unique amongst physical laws in providing insight into the distinction between past and future that is such an important feature of the physical world.

Question 3.9 Boltzmann's statement of the second law is sometimes expressed mathematically by saying that in any process $\Delta S_{\text{Univ}} \geq 0$, where S_{Univ} represents the entropy of the Universe, and the symbol $\geq$ is read as 'is greater than or equal to'. Under what circumstances would you expect the equality to hold, rather than the inequality?

Question 3.10 Suppose the entropy of one mole of monatomic ideal gas at a certain pressure and volume is $500\,\text{J}\,\text{K}^{-1}$. (a) What is the entropy of the same gas at the end of a process that doubles both the volume and the pressure? (b) What significance do you attach to the fact that the initial entropy is $500\,\text{J}\,\text{K}^{-1}$?

Question 3.11 (a) Use Equation 3.30a together with the equation of state of an ideal gas and the relation $C_P - C_V = nR$ to show that, for an ideal gas,

$$S_2 - S_1 = C_V \log_e\left(\frac{T_2}{T_1}\right) + nR \log_e\left(\frac{V_2}{V_1}\right). \qquad (3.31)$$

(b) Use this result, and an entry in row 3 of Table 3.3, to show that Equations 3.30a and 3.27 are consistent in the special case of an isothermal process in an ideal gas. ■

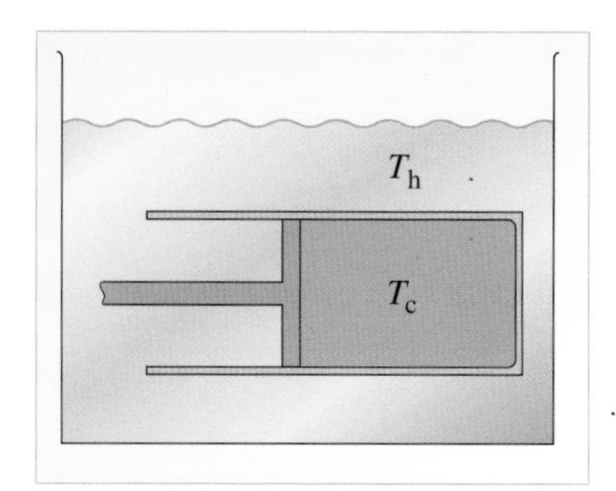

Figure 3.29 A system at temperature T_c in contact with its environment at temperature T_h.

3.4 Entropy and heat flow

When two bodies at different temperatures are placed in thermal contact, energy flows spontaneously from the hotter to the colder. As far as the first law of thermodynamics is concerned, there is no reason why energy should not pass from the colder to the hotter body, but it certainly does not. Why not?

The answer, of course, is that the second law of thermodynamics does not permit it. To understand why, it is easiest to start from Boltzmann's statement of the second law, that 'the entropy of the Universe tends to a maximum'.

Figure 3.29 shows a system, consisting of a cylinder of gas at temperature T_c, in thermal contact with its environment, a heat bath at temperature T_h. The temperature of the heat bath is slightly higher than the temperature of the cylinder so $T_h > T_c$. Because these two parts of the Universe are in thermal contact, heat can flow from one to the other. The second law allows us to establish the *direction* of heat flow.

First, suppose that as a result of the temperature difference a small quantity of heat Q is transferred reversibly *from* the heat bath *to* the system. Note that, at this stage, Q may be either positive or negative. If Q is positive, heat flows from the heat bath to the system; if Q is negative, heat flows from the system to the heat bath. So the direction of energy transfer has not yet been settled. Our task will be to determine the sign of Q.

The transfer of heat Q causes the entropy of the system to change. Since the heat transferred is small, it has a negligible effect on the temperatures T_h and T_c of the heat bath and the system. Thus, we can use Equation 3.27 to calculate the change in the entropy of the system:

$$\Delta S_c = \frac{Q}{T_c}.$$

The transfer of heat also causes the entropy of the heat bath to change. Since the heat transferred *from* the heat bath is Q, the heat transferred *to* the heat bath is $-Q$. It follows that the change in the entropy of the heat bath is

$$\Delta S_h = \frac{-Q}{T_h}.$$

Again, we stress that no assumptions have been made about the signs of these entropy changes: they will depend on the sign of Q. Whatever the sign of Q, we can say that the total change in the entropy of the Universe as a result of the heat exchange between the system and the heat bath is

$$\Delta S_{\text{Univ}} = \Delta S_c + \Delta S_h = \frac{Q}{T_c} - \frac{Q}{T_h}$$

i.e.
$$\Delta S_{\text{Univ}} = Q\left(\frac{1}{T_c} - \frac{1}{T_h}\right).$$

Now, since T_c is less than T_h, it follows that $1/T_c$ is greater than $1/T_h$, so the term in brackets must be positive. Since Boltzmann's statement of the second law requires that ΔS_{Univ} cannot be negative, it follows that Q cannot be negative: heat cannot flow from the cold system to the hot heat bath.

It is possible for Q to be zero, but that would correspond to perfect insulation between the system and the heat bath. We are assuming that the system and heat bath are in thermal contact — in this case, Q is positive, showing that heat flows from the hot heat bath to the cold system. As a result, the entropy of the system increases and the entropy of the heat bath decreases, but these entropy changes are of different sizes and their sum is positive, indicating an increase in the entropy of the Universe.

In summary, Boltzmann's statement of the second law, which requires that

$$\Delta S_{\text{Univ}} \geq 0 \tag{3.32}$$

(where the equality holds for reversible changes and the inequality for irreversible ones), combined with the relation expressing the additivity of entropy

$$\Delta S_{\text{Univ}} = \Delta S_{\text{c}} + \Delta S_{\text{h}} \tag{3.33}$$

and the general expression for entropy transferred in a reversible isothermal process

$$\Delta S = Q_{\text{rev}}/T$$

leads directly to the conclusion that heat flows from hot bodies to cold ones.

Question 3.12 Suppose that the temperature of the system discussed above was slightly higher than that of the surrounding heat bath. Use thermodynamics to show that positive heat will flow from the system to the environment. ■

3.5 But what is entropy anyway?

You have now seen that there is a function of state, the entropy S, that determines the possibility or impossibility of natural processes such as heat flow between bodies at different temperature (and much else besides). You have also seen how the difference in entropy between two states of a system can be determined (using $\Delta S = Q_{\text{rev}}/T$ or Equation 3.28), and that, in the particular case of an ideal gas,

$$\Delta S = C_V \log_{\text{e}}\left(\frac{P_2 V_2^{\gamma}}{P_1 V_1^{\gamma}}\right).$$

In spite of this detail, many students feel that entropy is a 'slippery' concept, far less tangible than the concept of internal energy. They would like a clear answer to the question: 'but what is entropy, anyway?'

One of the main reasons why students may feel that they have a better grasp of internal energy than entropy is that, thanks to kinetic theory, they have some insight into the *microscopic nature* of internal energy. They can picture the internal energy of a gas, for example, as the sum of the kinetic and mutual potential energies of the molecules that comprise the gas. In this section, we shall try to provide an equivalent insight, based on statistical mechanics, into the microscopic nature of entropy.

Suppose you have a fixed mass of gas, initially at pressure P_0 in a container of volume V_0. This container (A) is connected via a tap to an evacuated container (B) also of volume V_0, as shown in Figure 3.30. When the tap is opened, gas passes from A to B until the pressures in the two containers are equal. We will assume that the gas is in good thermal contact with a heat bath so that its temperature remains fixed. At the end of the process, the gas then occupies a volume $2V_0$ and has pressure $0.5P_0$. Because the temperature is fixed, the internal energy of the gas remains unchanged, but (according to Equation 3.31) its entropy will have been increased.

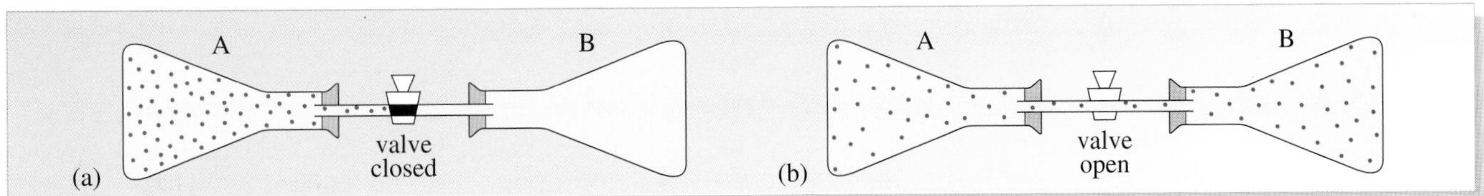

Figure 3.30 Increasing the entropy of a gas by allowing it to expand at constant temperature. (a) Valve closed; (b) valve open.

Now, if we could reverse the velocity of every molecule in the gas, the whole process would go into reverse. All the gas would then return to container A and we could close the tap. Since the entropy of the gas increased during the expansion, it would decrease during the contraction, and the second law of thermodynamics would be broken. Of course, you would not expect to see this happening, but why not? Since the reversed velocity state is a perfectly reasonable state for the system; why doesn't it occur naturally? Why do we never find all the gas returning to container A? A little thought soon gives the answer — it could happen, but it is incredibly unlikely.

You saw in Chapter 2 that a detailed microscopic description of the state of a gas can be given in terms of phase cells: each phase cell defines the position and velocity of a particle to within a chosen precision. The state of the gas is specified by stating *which* molecule is in *which* phase cell. Such a detailed description is known as a **configuration** (or microstate). It is clear that relatively few of the conceivable configurations correspond to arrangements in which all the molecules happen to be in container A (see Figure 3.31). We can even make an estimate of the likelihood of this happening. If we consider one molecule moving randomly through the whole volume, it will, on average, spend half of its time in container A; that is, the probability of finding the molecule in container A is 0.5. For two molecules, the probability that they are both in A is $(0.5)^2$; for three molecules it is $(0.5)^3$, and so on. If there are N molecules in the system, the probability that they are all in A is $(0.5)^N$. Now N is a very large number for any macroscopic quantity of gas; if there is one mole of gas in the system, then N will be equal to Avogadro's number, i.e. about 6×10^{23}. Consequently, the probability of finding all the molecules in A will be a really minuscule number; if we were to try to write it down as a decimal, we would need to start with about 2×10^{23} zeros!

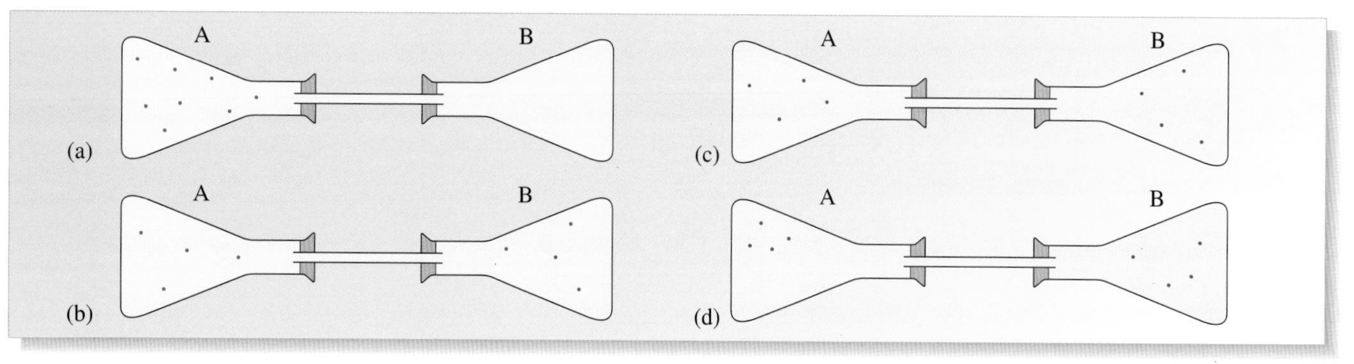

Figure 3.31 The number of configurations in which the molecules are spread throughout the combined volume is much greater than the number in which they are confined to container A.

What has this to do with entropy? Well, in the course of his pioneering studies in statistical mechanics, Ludwig Boltzmann (1844–1906) concluded that the entropy S associated with any given (macroscopic) equilibrium state of a system was a measure of the number of different configurations, W, that corresponded to that particular equilibrium state. Specifically,

$$S = k \log_e W \qquad (3.34)$$

where $k = 1.381 \times 10^{-23}\,\mathrm{J\,K^{-1}}$ is Boltzmann's constant. This important result, linking the macroscopic concept of 'entropy of an equilibrium state' with the microscopic concept of 'number of corresponding configurations', is **Boltzmann's equation for**

entropy. Its significance is such that a version of it appears on the famous monument to Boltzmann (Figure 3.32) dedicated by the city of Vienna, where he was born and spent much of his life.

Figure 3.32 The monument to Boltzmann in Vienna.

Boltzmann's equation makes it easy to understand why entropy always tends to increase to its maximum possible value; it is just the tendency of a system to spend most of its time in the macroscopic state that corresponds to the largest number of configurations. For example, when two bodies at different temperatures are in thermal contact, heat will flow from the hotter to the cooler because there are more microscopic states that correspond to arrangements in which the temperatures of the two bodies are closer together than to arrangements in which the temperatures are further apart. The condition of thermal equilibrium in which the two bodies have the same temperature corresponds to the largest number of configurations and therefore represents the state towards which the system will evolve. From the microscopic point of view, the system does not have any 'preference' for equal temperatures: it is simply that the unbiased exploration of all possible configurations means that the observed (macroscopic) state will be that which corresponds to the greatest number of configurations. You may be surprised that this results in such a well-defined macroscopic state, but that is because of the enormously large number of microscopic entities in any macroscopic sample.

This microscopic view of entropy also provides another insight into its nature. Consider the outcome of mixing two different gases, such as the oxygen and nitrogen in the air around us. At any given temperature and pressure these gases will always tend to mix to the greatest degree possible. This is because the thoroughly mixed state corresponds to a greater number of configurations than any comparable unmixed state. Unmixed macroscopic states correspond to rather orderly configurations that are relatively rare; mixed macroscopic states correspond to disordered configurations that are common. Thus there is always a tendency for states to become as mixed (i.e. disordered at the microscopic level) as possible. For this reason, entropy is often described as a measure of disorder.

4 Entropy, engines and refrigerators

The science of thermodynamics was developed during the nineteenth century, and was initially concerned with practical matters. A particular preoccupation, as we noted earlier, was the efficiency of engines such as the steam engine. Another concern is with refrigerators, which are a bit like engines in reverse in that they use work to transfer heat from cooler to hotter bodies. In this section, we will see what the second law tells us about the behaviour of engines and refrigerators, and this will lead us to two of the earliest statements of the second law. As a tailpiece, you will also meet the third law of thermodynamics.

4.1 Heat engines

Heat engines are devices that convert heat into work. They include steam engines and internal combustion engines, but not electric motors (though they too are subject to the laws of thermodynamics). We said in Section 1 that the efficiency of real engines is limited — none of them manage to convert 100% of the heat they absorb into useful work. We also said that this limitation was not just a matter of friction and other such effects, but was a more fundamental requirement that could not be evaded. Our aim in this section is to show why the laws of thermodynamics demand that this should be so and to quantify the limitations in a few simple cases.

In order to do this, we start by considering an especially simple kind of heat engine, almost an 'ideal' one you might say, called a **Carnot engine**. Once we have described the operation of the Carnot engine, we shall use the second law of thermodynamics to work out its maximum efficiency, and then argue that any other heat engine operating under the same conditions as the Carnot engine cannot possibly be more efficient. Thus by starting with a specific but ideal case, we shall arrive at a general result of great practical importance. This is essentially the path that Carnot himself followed in the 1820s, though unlike us he knew nothing of the first law of thermodynamics or of entropy.

A particular version of the Carnot engine is shown schematically in Figure 3.33. It consists of hot and cold thermal reservoirs (heat baths) at temperatures T_h and T_c, and a cylinder containing a fixed quantity of ideal gas confined by a frictionless piston. The system is arranged in such a way that the ideal gas may be in thermal contact with either of the thermal reservoirs, or it may be isolated from them. (This might be done by raising or lowering insulated barriers between the cylinder and the thermal reservoirs, though the details are unimportant.) When the cylinder is in thermal contact with one or the other of the thermal reservoirs, heat may be transferred to or from the gas, but there is no other way for heat to enter or leave the gas. As usual we shall assume that the thermal reservoirs are so large that any heat entering or leaving them has no perceptible effect on their respective temperatures. We shall also suppose that all the changes that take place happen quasi-statically and reversibly.

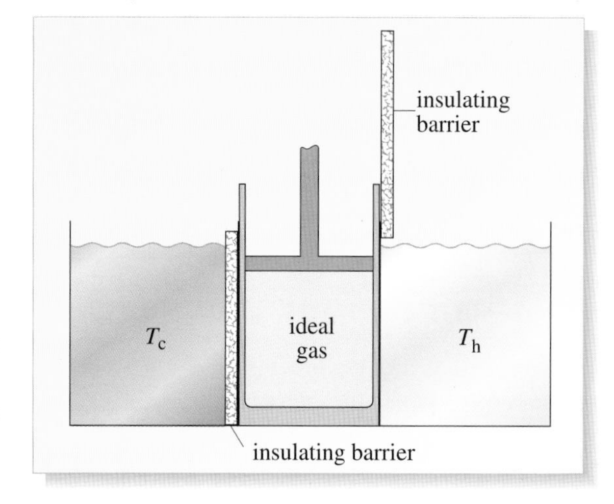

Figure 3.33 A Carnot engine. An ideal gas, confined in a cylinder, may be in thermal contact with either of two thermal reservoirs at temperatures T_h and T_c (where $T_h > T_c$), or it may be isolated from both of them.

The Carnot engine operates in a cycle, called a **Carnot cycle**, that consists of four distinct processes; these are described below and illustrated in Figure 3.34.

1 Start with the gas at the temperature of the cooler reservoir, T_c, in the state represented by point A on the *P–V* diagram shown at the centre of Figure 3.34. Establish thermal contact between the gas and the cooler reservoir, and then compress the gas quasi-statically by pushing the piston until the volume of the gas is reduced to that corresponding to point B on the *P–V* diagram. The thermal contact with the reservoir ensures that this will be a reversible *isothermal* compression. A positive amount of work will be done *on* the gas and an equal amount of heat Q_c will be transferred to the reservoir. (These quantities of heat and work must be equal because there is no overall change in the internal energy of an ideal gas during an isothermal process.)

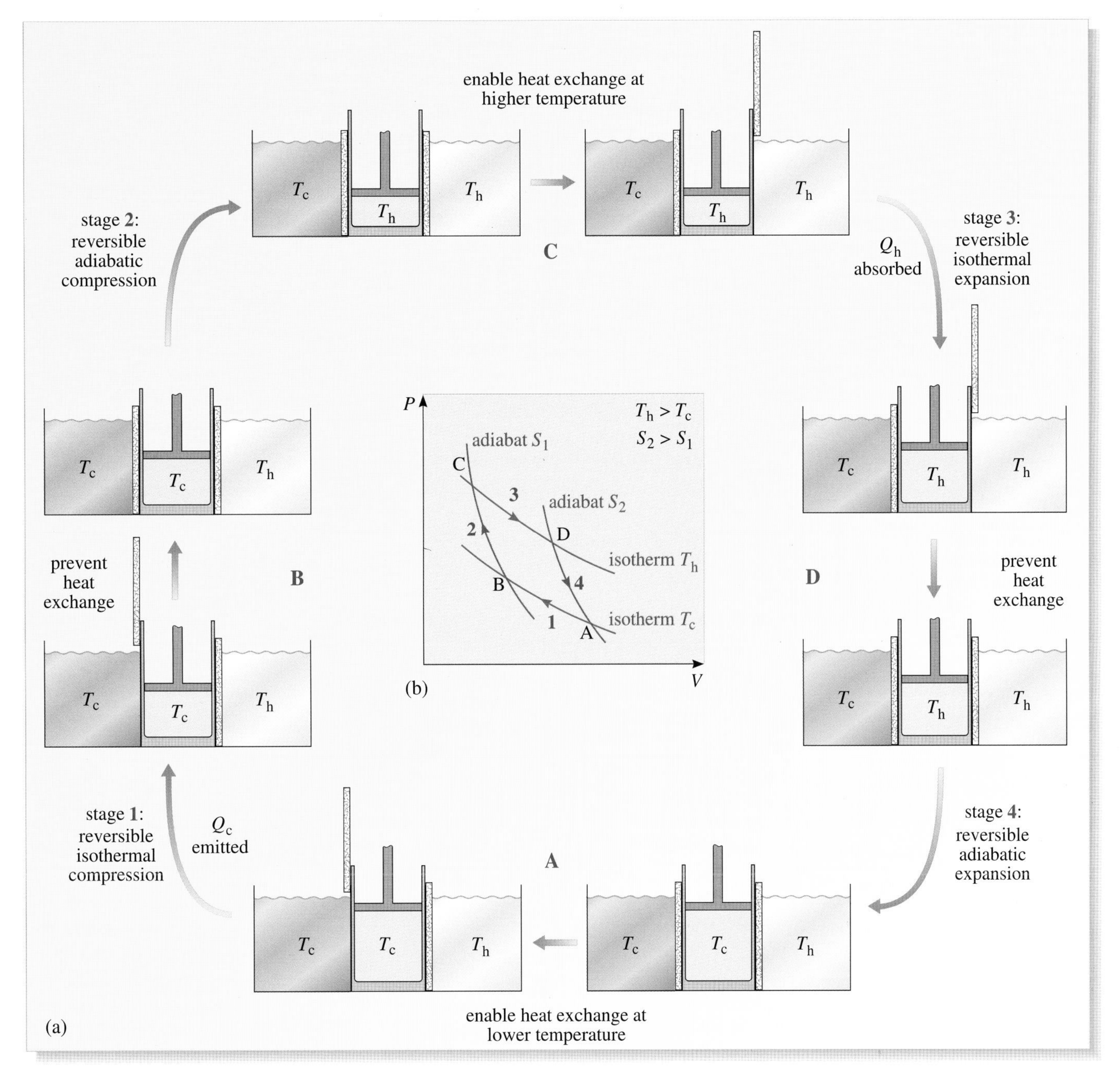

Figure 3.34 The four stages of the Carnot cycle, shown as (a) physical processes, and (b) pathways on the *P–V* diagram of the gas. Description of the cycle starts at point A.

2 With the gas in the state represented by point B on the *P–V* diagram, isolate the cylinder from the thermal reservoir and then continue to reduce its volume quasi-statically until it reaches the value corresponding to point C on the *P–V* diagram. The thermal isolation ensures that this will be a reversible *adiabatic* compression. A positive amount of work will be done *on* the gas but no heat will be transferred to or from the gas. As a result, the temperature of the gas will rise during the adiabatic compression. The point C has been chosen so that the temperature of the gas will be equal to that of the hotter reservoir, T_h, at the end of this stage.

3 With the compressed gas having reached temperature T_h, establish thermal contact with the hotter reservoir and then allow the gas to expand quasi-statically until its volume is that corresponding to point D on the P–V diagram. This will be a reversible isothermal expansion, and this time the gas will do positive work *on* its environment while an equal amount of heat, Q_h, will be transferred *to* the gas.

4 With the gas now in the state represented by point D on the P–V diagram, isolate it from the thermal reservoir and allow it to continue expanding quasi-statically until it returns to its initial state represented by point A. This will be a reversible adiabatic expansion in which the gas will do work on its environment, but no heat will be transferred to the gas. At the end of this process the gas will have returned to its original temperature T_c and be ready to repeat the cycle all over again.

Considering the cycle as a whole, the gas transfers heat Q_c to the cooler reservoir during the isothermal compression, and absorbs heat Q_h from the hotter reservoir during the isothermal expansion. It also performs a net amount of work W_{out} on its surroundings. However, the gas returns to its initial state, without any overall change in its internal energy. Applying the first law of thermodynamics to this situation, and noting that the net amount of heat transferred *to* the gas is $Q_h - Q_c$ and the net work done *on* the gas is $-W_{out}$, we have $\Delta U = -W_{out} + Q_h - Q_c = 0$. So,

$$W_{out} = Q_h - Q_c\,. \tag{3.35}$$

Using the second law of thermodynamics, we can then show that W_{out} is positive which implies that the Carnot engine is successful in converting heat into work. To achieve this, we note that since the gas returns to its original state at the end of a cycle, the overall change in entropy of the gas over the complete cycle must be zero ($\Delta S = 0$). Nonetheless, there are entropy changes within the cycle.

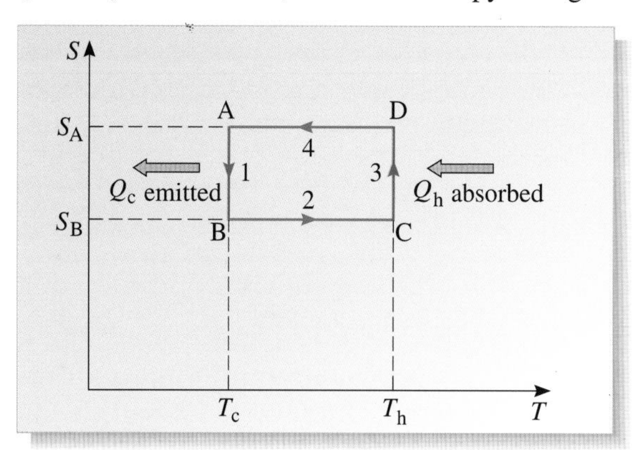

Figure 3.35 The entropy–temperature diagram of the Carnot cycle.

● What are the entropy changes of the gas in the four stages of the Carnot cycle?

○ In stage 1, heat Q_c is transferred *from* the gas at temperature T_c, so the change in entropy of the gas is $\Delta S_1 = -Q_c/T_c$.
In stage 2, no heat is transferred, so $\Delta S_2 = 0$.
In stage 3, heat Q_h is absorbed at temperature T_h, so $\Delta S_3 = Q_h/T_h$.
In stage 4, as in stage 2, no heat is transferred, so $\Delta S_4 = 0$. ■

It follows from the above answer that the total change in the entropy of the gas is

$$\Delta S = \Delta S_1 + \Delta S_2 + \Delta S_3 + \Delta S_4 = -Q_c/T_c + 0 + Q_h/T_h + 0$$

i.e.
$$\Delta S = Q_h/T_h - Q_c/T_c\,. \tag{3.36}$$

But we already know that $\Delta S = 0$, so $Q_h/T_h = Q_c/T_c$

i.e. $$Q_h/Q_c = T_h/T_c. \qquad (3.37)$$

Since T_h is greater than T_c, Equation 3.37 implies Q_h is greater than Q_c, and that $W_{out} = Q_h - Q_c$ is positive, as claimed above. The relationship between W_{out}, Q_h and Q_c is indicated schematically in Figure 3.36.

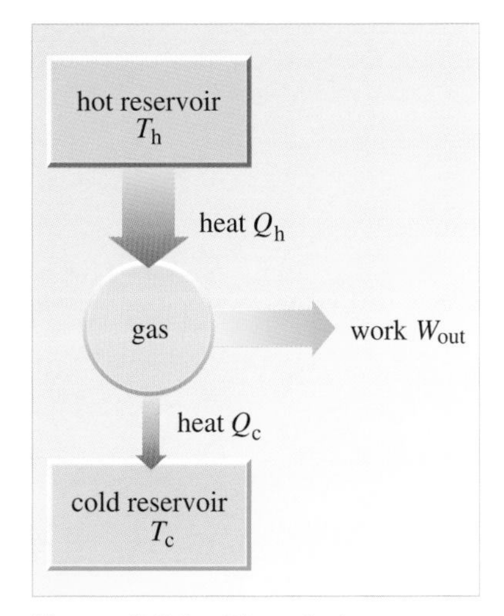

Figure 3.36 The relation between W_{out}, Q_h and Q_c for a Carnot cycle engine. Note that the width of the arrows has been used to indicate that $W_{out} = Q_h - Q_c$.

Broadly speaking, the reversible transfer of heat Q_h to the gas at temperature T_h brings with it an amount of entropy $\Delta S_{in} = Q_h/T_h$. The need to ensure that there is no change in the entropy of the gas over a complete cycle therefore requires that a sufficient amount of heat Q_c must be transferred from the gas at the lower temperature T_c to ensure that $\Delta S_{out} = Q_c/T_c$ is equal to ΔS_{in}. So the conversion of heat to work must be accompanied by the 'rejection' of some of the heat to ensure compliance with the second law. Only the difference $W_{out} = Q_h - Q_c$ is available to be converted to work.

Now, the **efficiency** η of a heat engine is defined as

$$\eta = W_{out}/Q_h \qquad (3.38)$$

and represents the fraction of the heat supplied that is converted into work. Complete conversion of heat to work corresponds to $\eta = 1$, and is usually described as 100% efficiency.

In the case of the Carnot engine, where $W_{out} = Q_h - Q_c$,

$$\eta = \frac{Q_h - Q_c}{Q_h} = 1 - \frac{Q_c}{Q_h}$$

and using Equation 3.37 (which expresses the implications of the second law in this case), we see that:

$$\eta = 1 - \frac{T_c}{T_h}. \qquad (3.39)$$

This result was quoted without proof in Section 1. You can now see where it comes from.

Thus, the efficiency of a Carnot engine is determined by the (absolute) temperatures of the thermal reservoirs that transfer heat to and from the gas. Whatever those temperatures may be, provided T_c is not 0 K, the ratio T_c/T_h will be between 0 and 1, so η will be less than 1 (i.e. the engine will be less than 100% efficient).

Now, without going into detailed arguments about different kinds of engine that operate between the same two temperatures, but which include additional or different processes in their operating cycle, I hope you will accept the correctness of the claim that no engine operating between the same two maximum and minimum temperatures (T_h and T_c) can be more efficient than a Carnot engine. Remember, a Carnot engine is a reversible engine, in which no energy is wasted by friction or any other dissipative effects, so you would expect it to have a relatively high efficiency.

Apart from a loophole (which we shall close later) about what happens when $T_c = 0$ K, we have now arrived at the conclusion that it is not possible to construct any heat engine that will absorb a positive amount of heat and completely convert it to work. This, in a somewhat generalized form, constitutes one of the earliest formulations of the second law of thermodynamics:

Kelvin's statement of the second law of thermodynamics

No cyclic process is possible which has as its sole result the complete conversion of a quantity of heat into work.

Notice the presence of the words 'sole result'. All heat engines convert heat to work — that's why they are called heat engines — but Kelvin's statement tells us that none of them can do it ideally, with 100% efficiency. The conversion of heat to work must always be accompanied by some other effect, typically the transfer of part of the heat to the environment.

We have arrived at Kelvin's statement as the conclusion of a lengthy argument based on entropy and the Carnot engine, but it is important to realize that we could turn the whole argument on its head. If we had started with Kelvin's statement as a credible generalization of everyday experience, and investigated its implications for the Carnot engine, we could have deduced the existence of a function of state (the entropy) that never decreased in an isolated system. We did not do this because it provides a rather tortuous route to the concept of entropy. Nonetheless, this is essentially what Kelvin did in the early days of thermodynamics, and it was implicit even earlier in the work of Carnot.

Kelvin himself regarded Equation 3.39 as quite fundamental, and used it to *define* a temperature scale. In other words, T_h was defined to be twice as large as T_c if an ideal heat engine working between these temperatures had an efficiency of 0.5. This is entirely consistent with the absolute scale used throughout this book, but has the merit of defining temperatures without referring to the properties of any particular substance (such as mercury or an ideal gas). This offers a fundamental macroscopic interpretation of temperature, to set alongside the fundamental microscopic interpretation given by the Boltzmann distribution law.

Question 3.13 A certain engine has an efficiency of 45%. Supposing it to be a Carnot engine with its cooler reservoir at a temperature of 295 K, what is the temperature of its hotter reservoir? ■

The efficiency of real engines

Real engines never have efficiencies anywhere near as high as those of the Carnot engine, for several reasons. First, they do not operate on a Carnot cycle, which, as we have just seen, is the most efficient possible, and secondly because of all the losses that inevitably occur due to friction, or due to the difficulty of making heat transfers or doing work in an ideal manner. However, the limiting thermodynamic efficiency is still an important point to be taken into account when designing an engine. From Equation 3.39, we see that the higher the temperature difference between the input and output, the higher the efficiency will be. Since the lowest temperature we can use is the temperature of the environment, the only one we can profitably change is the input temperature. So one of the most important things one can do is to raise the input temperature as high as possible.

An early (eighteenth century) steam engine used steam at atmospheric pressure, i.e. at 373 K, and emitted waste heat at about 300 K, with a limiting efficiency of $(373\,\mathrm{K} - 300\,\mathrm{K})/373\,\mathrm{K} = 0.20$. A modern steam turbine (Figure 3.37), with an input temperature of 850 K, has a limiting efficiency of

(850 K – 300 K)/850 K = 0.65. The actual efficiency achieved is about 0.38. However, the latest power station technology using gas as fuel has a turbine with an input temperature of 1450 K; the waste heat from this then goes to heat steam for steam turbines, and the exhaust gas leaves at a temperature of about 370 K. The temperature range is thus considerably greater, leading to a limiting thermodynamic efficiency of 0.74, and an actual efficiency of 0.56. The amount of useful work obtained from a given quantity of fuel is thus increased considerably.

Figure 3.37 A modern steam turbine.

4.2 Refrigerators

A **refrigerator** maintains objects at a lower temperature than their surroundings. In the case of a domestic refrigerator, energy that 'leaks' into the refrigerated compartment is extracted by means of a circulating fluid and transferred to the surrounding air behind the refrigerator. The details are quite complicated but the important point is that the circulating fluid undergoes a cyclic process. As you will see shortly, this process is essentially the reverse of that found in a heat engine.

Figure 3.38 indicates the main energy transfers that occur in a refrigerator. The circulating fluid operates at two different temperatures, T_c (below room temperature) and T_h (above room temperature). Heat Q_c is transferred *to* the fluid at the lower temperature, T_c; this cools the contents of the fridge (the cold reservoir). Heat Q_h is transferred *from* the fluid at the higher temperature, T_h; this warms the surroundings of the fridge (the hot reservoir). Just as for a heat engine, we shall show that Q_h is greater than Q_c, so more heat is transferred to the surroundings than is absorbed from the interior of the fridge. The difference is accounted for by work done, W_{in}, on the fluid by an external power source, usually provided by the domestic electricity supply.

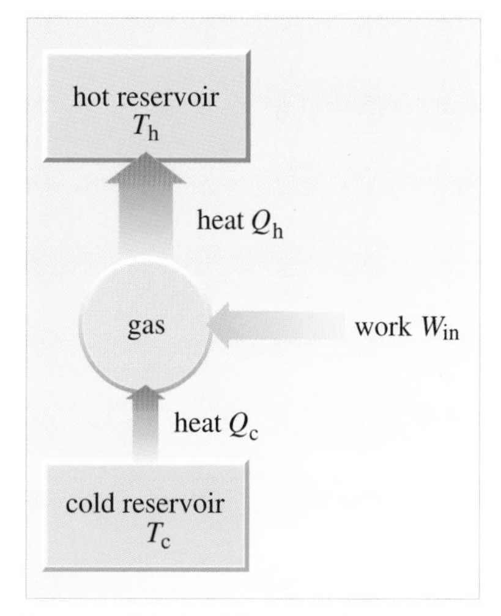

Figure 3.38 The relation between W_{in}, Q_h and Q_c for a refrigerator.

- Could you cool your kitchen by leaving the fridge door open?

Figure 3.39 A domestic refrigerator.

❍ No. The pipes at the back of the fridge (Figure 3.39) would transfer more energy to the air in the kitchen than those inside the refrigerator would absorb. As a result, leaving the fridge door open is a way of heating your kitchen, though not a particularly good one. ■

As in the case of an engine, we can ask how efficient a refrigerator can be. How much work must be supplied to transfer a given quantity of heat from the interior of a refrigerator to its environment? In particular, could we construct a refrigerator that simply transferred heat from the cool interior to the relatively hot exterior without requiring any external work to be performed? You will not be surprised to learn that the second law of thermodynamics implies that this is not possible. Let us see why.

We shall approach this practical problem by again considering the Carnot cycle. When dealing with the Carnot engine, we assumed that the Carnot cycle was performed in such a way that the expansion took place at the higher temperature, and compression at the lower temperature. This ensured that heat would be absorbed by the gas at the higher temperature, and emitted at the lower temperature. However, since the Carnot cycle is reversible, it can also be followed in the reverse direction, as indicated in Figure 3.40. Heat Q_c will then be *absorbed* by the gas at the lower temperature, and heat Q_h will be *emitted* at the higher temperature. Clearly, what this reversed Carnot cycle provides is a **Carnot refrigerator** rather than a Carnot engine.

Following exactly the same argument as given for a Carnot engine, we can show from an audit of entropy changes round a complete cycle that

$$Q_h/Q_c = T_h/T_c. \tag{3.37}$$

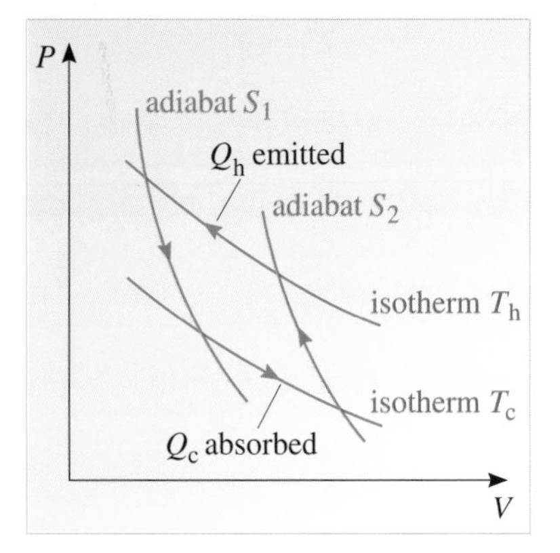

Figure 3.40 The four stages of a reversed Carnot cycle.

Since T_h is greater than T_c, it follows that Q_h must be greater than Q_c. So, the transfer of heat from a cooler to a hotter body by a Carnot refrigerator must be accompanied by the performance of work $W_{in} = Q_h - Q_c$, as indicated on Figure 3.38.

As in the case of engines, we can now argue that no other refrigerator operating between the temperatures T_h and T_c could possibly perform better than a reversible Carnot refrigerator. This brings us to the conclusion that it is not possible to construct any refrigerator that will absorb a positive amount of heat from a cooler body and transfer it to a hotter body without requiring some additional effect, such as a supply of work from an external source. The construction of a perfect refrigerator that transfers heat from a cooler to a hotter body without requiring any external work is therefore an impossible dream.

A generalization of this conclusion provides yet another form of the second law of thermodynamics:

Clausius's statement of the second law of thermodynamics

No cyclic process is possible which has as its sole result the transfer of heat from a cooler body to a hotter one.

Note the reference to 'sole result'. Clausius's statement does *not* say that it is impossible to transfer heat from a colder to a hotter body; refrigerators do that all the time. However, the statement does imply that such an 'unnatural' transfer of heat must be accompanied by some additional effect, such as the performance of work.

As in the case of the Kelvin statement, we could have started from Clausius's statement and, via a fairly lengthy argument, deduced the existence of entropy. Many authors prefer to treat thermodynamics in this way, taking the view that Clausius's statement is the most intuitively appealing of all the statements of the second law. It is notable that Clausius himself was the first to explicitly develop the concept of entropy and appreciate its significance. The term 'entropy' was coined by him in 1865, though his most important work on the subject was published 15 years earlier, in 1850.

Question 3.14 I've just had a great idea! When my electrically powered refrigerator is operating, I can easily feel the hot air rising from the grill at the back. If I captured this hot air, I'm sure I could use it alone to power a small generator and thus obtain the electricity needed to make the refrigerator work. I could then unplug the refrigerator and continue to use it without having to pay for the electricity it consumes. What do you think of my idea? Will it work, and if not, why not? ■

Real refrigerators

Real domestic refrigerators are quite complicated. They usually involve evaporation and condensation of the circulating fluid, as indicated in Figure 3.41.

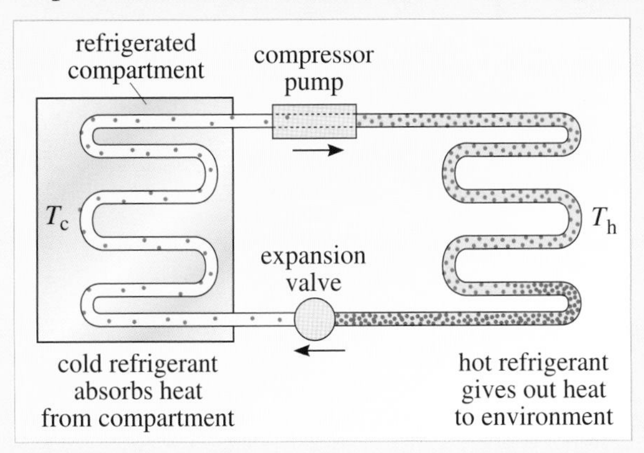

Figure 3.41 Schematic view of a refrigerator. A refrigerant, pumped around a closed system of tubes, passes through an expansion valve into a low pressure section where it evaporates and becomes cold. The cold fluid absorbs heat Q_c from the refrigerated compartment; it then passes through a compressor, into a high pressure section where it warms up. As it gives off heat Q_h to the environment, it condenses back into a liquid, ready to begin the cycle again. The operation of the pump requires that work is performed by some external agency.

The **performance** κ of a refrigerator is defined as

$$\kappa = Q_c/W_{in}$$

and represents the heat absorbed per unit work done by the pump. The greater the value of κ, the better the refrigerator, at least as far as its energy consumption is concerned. For a Carnot refrigerator, in which $W_{in} = Q_h - Q_c$ and $Q_h/Q_c = T_h/T_c$, it can easily be shown that

$$\kappa = \frac{T_c}{T_h - T_c}.$$

This implies that the limiting performance of a typical domestic refrigerator, operating between room temperature (20 °C, say) and 0 °C, will be 273 K/(293 K − 273 K) = 13.7. This implies that for every watt of electrical energy used, 13.7 W of heat will be transferred from the refrigerated compartment. In practice, the performance of a real refrigerator is considerably lower, typically less than half of this.

Figure 3.42 Walther Nernst (1864–1941), winner of the 1920 Nobel Prize for chemistry in recognition of his discovery of the third law of thermodynamics.

4.3 The third law of thermodynamics

The efficiency of a Carnot engine approaches its maximum value of 1 as T_c approaches absolute zero, and would reach that ideal value when $T_c = 0\,K$. Is it possible then, even if impractical, that perfect engines might be constructed with a thermal reservoir maintained at 0 K?

Sadly, it is not. There is a *third* law of thermodynamics, discovered by the German physical chemist Walther Nernst (Figure 3.42) in 1906, which says (in one of its formulations):

> **The third law of thermodynamics**
>
> It is impossible to reduce the temperature of any system to absolute zero by a finite number of operations.

Discussing this law in detail would take us far beyond the intended scope of this chapter, but you can see that it closes the loophole concerning ideal engines with $T_c = 0\,K$. With the addition of the third law, you will now appreciate the light-hearted summary of thermodynamics that appeared in *The American Scientist* in March 1964:

'1st law You can't win, you can only break even.

2nd law You can break even only at absolute zero.

3rd law You cannot reach absolute zero.'

5 Closing items

5.1 Chapter summary

1. A quasi-static process is one that happens sufficiently slowly that at every stage of the process it is a good approximation to regard the system as being in an equilibrium state.
2. Heat is the term used to refer to any quantity of energy transferred between a system and its environment as a result of a temperature difference. The transfer may be achieved by conduction, radiation or convection.
3. Work is the term used to refer to any quantity of energy transferred between a system and its environment that is not transferred as heat. The work done *on* a gas when its volume increases by ΔV while its pressure P remains constant is
$$W = -P\,\Delta V. \tag{3.5}$$
4. The first law of thermodynamics states that: 'When a system undergoes a change from one equilibrium state to another, the sum of the heat transferred *to* the system and the work done *on* the system depends only on the initial and final equilibrium states and not on the process by which the change is brought about'.
5. The first law implies the existence of a function of state called the internal energy U, and may be represented by the equation $\Delta U = Q + W$, where Q is the heat transferred to the system and W is the work done on the system. One consequence of this is that the internal energy of any isolated system must be constant. This provides the basis for a general law of energy conservation, which is amongst the most important of all scientific principles.

6 The molar heat capacity of a substance is the amount of energy required to raise the temperature of 1 mol of that substance by 1 K. It is measured in units of $\mathrm{J\,K^{-1}\,mol^{-1}}$. In the case of a gas, the heat capacity may be determined at constant volume or constant pressure, and for an ideal gas it can be shown that the ratio of these heat capacities $\gamma = C_P/C_V$ is given by

$$\gamma = 1 + \frac{2}{f} \tag{3.22}$$

where f is the effective number of degrees of freedom (3 for a monatomic gas, 5 for a diatomic gas under moderate conditions). For n moles of an ideal gas, it is also the case that

$$C_P = C_V + nR\,. \tag{3.18}$$

7 An isothermal process is one in which there is no change of temperature. In the case of a given quantity of ideal gas (which always obeys the equation $PV = nRT$), any particular isothermal process is described by the condition

$$PV = I \tag{3.25}$$

where $I = nRT$ is a parameter that has a constant value for any particular isothermal process, but which has different values for isothermal processes that take place at different temperatures.

8 An adiabatic process is one in which no heat is transferred. In the case of a given quantity of ideal gas, any reversible adiabatic process may be described by the condition

$$PV^{\gamma} = A \tag{3.26}$$

where A, the adiabatic accessibility index, is a parameter that has a constant value for any particular adiabatic process, but which has different values for different adiabatic processes.

9 A thermodynamic process is said to be reversible if, following the process, both the system and its environment can be returned to the states they were in prior to the process.

10 The second law of thermodynamics (Carathéodory's version) states that: 'In the neighbourhood of any equilibrium state of a thermodynamic system there are equilibrium states that are adiabatically inaccessible'.

11 The second law implies the existence of a function of state called the entropy S. This is defined in such a way that the reversible transfer of an amount of heat Q_{rev} to a system at a fixed absolute temperature T increases the entropy of the system by an amount $\Delta S = Q_{\text{rev}}/T$. In the case of an ideal gas with pressure P and volume V, this implies that

$$S = C_P \log_e\left(\frac{V}{V_0}\right) + C_V \log_e\left(\frac{P}{P_0}\right) + S_0. \tag{3.30b}$$

where S_0 is the entropy of an arbitrarily chosen reference state with pressure P_0 and volume V_0.

12 The entropy of an *isolated* system can never decrease, so the change in entropy of such a system, as a result of any process, is given by $\Delta S \geq 0$, where the equality holds when the process is reversible. The entropy of a system that is not isolated may decrease, but the entropy of its environment will then increase by an equal or greater amount. This is expressed in Boltzmann's statement of the second law: 'The entropy of the Universe tends to a maximum'.

13 Boltzmann's equation, $S = k \log_e W$, relates the entropy of any given equilibrium state of a thermodynamic system to the number of configurations, W, that correspond to the given equilibrium state. This justifies the view that entropy is a measure of (microscopic) disorder.

14 Kelvin's statement of the second law of thermodynamics is that: 'No cyclic process is possible which has as its sole result the complete conversion of a quantity of heat into work'.

15 An example is provided by the reversible Carnot engine which converts heat to work but which additionally emits heat as well. This can be regarded as an inevitable consequence of its need to absorb and emit equal amounts of entropy over the course of any complete cycle of operation.

16 Clausius's statement of the second law of thermodynamics is that: 'No cyclic process is possible which has as its sole result the transfer of a quantity of heat from a cooler body to a hotter one'.

17 An example is provided by the reversible Carnot refrigerator which transfers heat from a cooler to a hotter body but which requires an external supply of work to do so. This can again be regarded as an inevitable consequence of its need to absorb and emit equal amounts of entropy over the course of any complete cycle of operation.

18 One version of the third law of thermodynamics states that: 'It is impossible to reduce the temperature of any system to absolute zero by a finite number of operations'.

5.2 Achievements

Now that you have completed this chapter, you should be able to:

A1 Understand the meaning of all the newly defined (emboldened) terms introduced in this chapter.

A2 Distinguish between internal energy, work, and heat, and evaluate those quantities in various circumstances (particularly for simple processes involving an ideal gas).

A3 State the first law of thermodynamics, explain its significance, and use it in simple calculations and arguments.

A4 Define and calculate heat capacities, distinguish between the values obtained for a gas under different conditions, and recognize various relations between the heat capacities of ideal gases.

A5 Recall and use the equations that describe the behaviour of an ideal gas under adiabatic and isothermal conditions, and relate those equations to relevant graphs and surfaces (such as the *PVT* surface).

A6 State the second law of thermodynamics (four versions), explain its significance, and use it in simple calculations and arguments (e.g. those concerning the direction of heat flow).

A7 Explain what is meant by the entropy of a system, describe how the difference in entropy between two states of a system may be evaluated, and comment on the extent to which such differences influence the accessibility of one state from another.

A8 Calculate the entropy changes that occur when an ideal gas undergoes a change of state (you are not expected to remember the formula for this).

A9 Provide a microscopic interpretation of entropy that accounts for its tendency to increase.

A10 Show how the behaviour of entropy limits the efficiency of a heat engine and the performance of a refrigerator.

A11 Explain in outline the working of a Carnot engine and a Carnot refrigerator.

A12 State the third law of thermodynamics and comment on its significance.

A13 Outline the historical development of thermodynamics and describe the contributions of some of its founders and pioneers.

5.3 End-of-chapter questions

Question 3.15 A professional male athlete eats a diet with an energy content of 3800 Cal (i.e. kcal). If he is not to put on weight, what must his average power consumption be?

Question 3.16 If the athlete lets things slip and decides he needs to lose weight by weightlifting, how many times would he need to lift a 60 kg weight through 1 m to lose 1 kg of fat, assuming that the energy content of this is 15 000 kJ? (For the purposes of this question, you may assume that the conversion of fat to mechanical energy is 100% efficient and that the energy used in each lift is the increase in gravitational potential energy of the weight.)

Question 3.17 A sample of gas expands from 1.00 m^3 at a pressure of 10.00 Pa to 6.00 m^3 at a pressure of 2.00 Pa. How much work is done by the gas if the expansion proceeds by each of the two paths ABC and ADC shown in Figure 3.43?

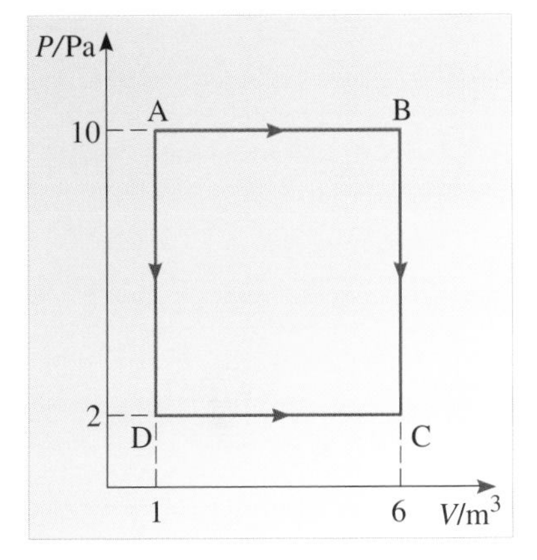

Figure 3.43 Diagram for Question 3.17.

Question 3.18 1.00 m^3 of a monatomic ideal gas initially at 300 K is compressed adiabatically and reversibly from a pressure of 1.00×10^5 Pa to eight times that pressure. What are the final volume and temperature? By how much does the entropy of the Universe change as a result of the process? What is the change in the entropy of the gas alone?

Question 3.19 State the second law of thermodynamics in four different ways and comment briefly on the relative merits of the various formulations.

Question 3.20 Show that the area enclosed by the S–T diagram of a Carnot cycle (Figure 3.35) is equal to the work done by the gas in that cycle.

Question 3.21 Modern gas-fired power stations operate a two-stage process: Stage 1 uses a gas turbine with a high input temperature T_H and an output temperature T_M. Step 2 employs a steam turbine which uses the waste heat from the gas turbine at temperature T_M and discards heat at a lower temperature T_L. Derive an expression for the maximum overall efficiency that could be obtained using such a system by assuming that each stage can be represented by a Carnot cycle. ■

Chapter 4 The physics of fluids

1 Introduction

This chapter is about fluids. In physics, a fluid is any apparently continuous medium that is able to flow. Fluids have no set shape, but adopt the form of any container into which they are placed. For example, before it sets, a jelly is a fluid; afterwards it is not. Both liquids and gases can be treated as continuous media that are able to flow, so both are classified as fluids.

Fluids are all around us (Figure 4.1). The Earth is 75% covered by water and 100% surrounded by air. Currents in the oceans and winds in the atmosphere largely control our climate. Occasionally, violent motions spring up, wreaking havoc as hurricanes or tornadoes. We breathe air, while blood circulates in our veins and arteries. Gas and water are pumped through pipes into our homes. In spite of the downward pull of gravity, a massive jumbo jet soars into the air by deflecting air around its wings. Ships, made of steel far denser than water, float on the sea and cars are streamlined to minimize the effects of air resistance. Even the Sun and the stars are made of fluids, in a state of continual motion.

Figure 4.1 (a) Planet Earth is 75% covered by oceans up to 10 km deep, and enveloped by an atmosphere a few tens of km thick. (b) The atmosphere is in constant motion, revealed by the motion of clouds; (c) waves break as they approach the shoreline; (d) cheers!

The motion of fluids is full of surprises, much more so than the motion of particles. For example, take a glass of water at room temperature and put a spoonful of sugar in it. Stir vigorously and then leave it to flow by itself. You will see the sugar forming a little pile in the centre at the bottom of the glass (Figure 4.2). How surprising! Wouldn't you expect the sugar to be flung to the outside of the glass, like the clothes in a tumble dryer? This is just the sort of unexpected behaviour that arises as soon as one starts to look closely at fluids. We know that it should be explicable using Newton's laws, but the explanation is by no means obvious.

Figure 4.2 When a glass of water containing sugar is stirred vigorously, the sugar collects together at the centre of the bottom of the glass.

The task of applying Newton's laws to fluids turns out to be quite hard. Relatively simple questions remained unanswered 200 years after Newton's *Principia*, and it was not until the twentieth century that a modern view of fluid motion emerged. A comprehensive study of fluids requires several advanced mathematical techniques. Perhaps for this reason, undergraduate mathematicians often learn far more about fluids than undergraduate physicists. Nevertheless, the study of fluids is an important part of any survey of the physical world. Our aim here is to explore a few basic ideas — using mathematics no more complicated than elsewhere in the course — and to give a non-mathematical description of other key topics such as turbulence and flight.

The study of fluids within the framework of Newtonian mechanics is known as **fluid mechanics**. This subject is generally divided into two areas: **fluid statics** is concerned with fluids at rest, while **fluid dynamics** is concerned with fluids that are moving. The structure of this chapter reflects this division. Section 2 considers fluids that are at rest. You will see how pressure increases when a diver descends into a lake, and decreases when a mountaineer climbs Everest. You will also see why some objects float and others do not. Sections 3 and 4 consider fluids that are in motion. The phenomena here are much more diverse and depend on the type of fluid motion under consideration. Section 3 deals with a special type of motion, known as *ideal flow*, which is simple enough for detailed analysis. Non-ideal flow is the subject of Section 4. Detailed analysis is much harder in this case, but you will see how scale models can be used to deduce vital information about the system being studied. Finally, Section 5 deals with the special topic of flight, discussed with the support of a video band.

2 Fluids at rest

2.1 Pressure in a fluid

You have met the idea of pressure several times in this book. In Chapter 1, you saw that a gas exerts a pressure on the walls of its container, and that this pressure depends on other macroscopic quantities such as the temperature and volume of the gas. In Chapter 3, you saw that the work needed to compress a system depends on the pressure of the system. In both cases, pressure was treated as an equilibrium property, with a single value describing the whole system. We did not think of pressure as varying throughout a system, or consider how pressure might be measured inside a system, far away from the container walls. It is now necessary to do this because, in order to think about the motion of a fluid, we need to discuss the forces that influence this motion, and pressure is one of the prime sources of force in a fluid.

To measure the pressure inside a fluid, imagine introducing a sensor in the form of a tiny evacuated cylinder, with a piston at one end supported by a spring (Figure 4.3). The compression of the spring indicates the magnitude of the force exerted along the axis of the piston by the fluid. Then, if the cross-sectional area of the piston is A,

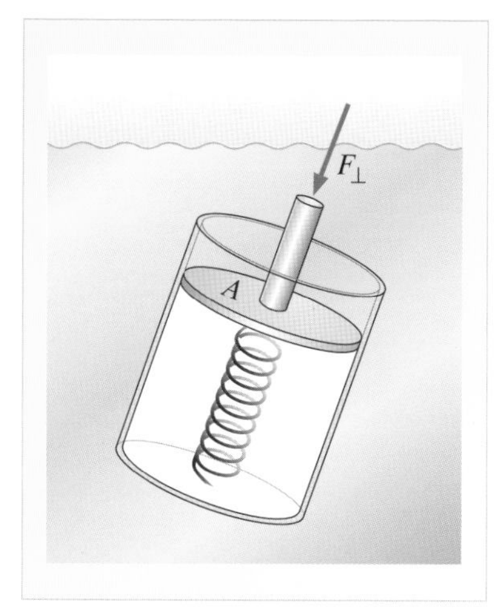

Figure 4.3 A sensor for measuring pressure inside a fluid. The fluid exerts a force of magnitude $F_\perp$ along the axis of the piston, causing it to compress. The ratio of $F_\perp$ to the area A of the piston defines the pressure.

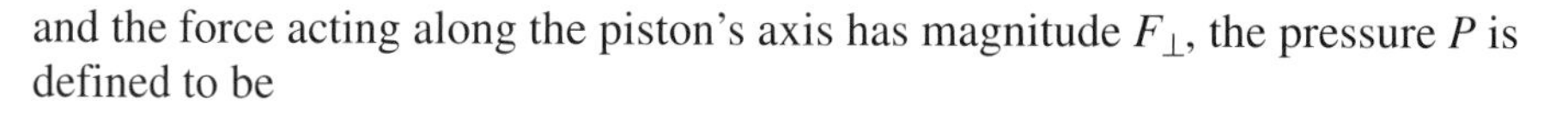

and the force acting along the piston's axis has magnitude $F_\perp$, the pressure P is defined to be

$$P = \frac{F_\perp}{A}. \tag{4.1}$$

It is assumed that the act of introducing the sensor has no significant effect on the pressure in the fluid. If the fluid is static this is a fair assumption, but if the fluid is flowing, like a river in spate, we should be more careful. It is essential for the sensor to drift with the fluid, so that it experiences exactly the same pressure as the fluid itself. We can imagine introducing a large number of sensors and letting them drift with the flow, allowing the pressure to be measured at different places and times throughout the fluid. The expense and practical difficulty of putting this plan into action need not concern us — the important point is simply that pressure at any point in the fluid can be given a precise definition.

Our definition of pressure is based on an important observation. In a fluid, the reading on a pressure sensor does not depend on the orientation of the sensor. If the sensor is turned around so that it points in a different direction, the same pressure will be recorded. This fact allows us to use Equation 4.1 as the definition of the pressure at a given point, with no mention of the orientation of the sensor. In mathematical terms we can say that pressure is a scalar quantity, with no associated direction, in contrast to force which is a vector and acts in a definite direction. Both the magnitude of $F_\perp$ and the cross-sectional area are positive quantities, so pressure is itself a positive quantity. Zero pressure corresponds to a perfect vacuum.

In a solid, the stresses acting are more complicated than the pressure alone, and may depend on direction.

In this section, we will consider the pressure inside a fluid that is at rest. Two rather different examples will be discussed: the pressure experienced by a diver and the pressure experienced by a mountaineer.

2.2 The pressure experienced by a diver

One of the main problems confronting a diver is the rapid increase in pressure that occurs with increasing depth. This is noticeable even for shallow dives, such as those to the bottom of a swimming pool. A serious problem arises because the human body is not completely solid: internal spaces like the lungs are filled with air, so deep diving introduces the danger of being crushed by the pressure of the surrounding water, like an empty aluminium can being crushed underfoot. Breathing special mixtures of high-pressure gases, scuba divers have dived to 300 metres; but unfortunately, air becomes toxic at very high pressures, so special diving suits or submarines become essential beyond this depth (Figure 4.4).

In order to learn about the increase of pressure with depth, we will consider a fluid that is at rest, with no currents flowing. We will also assume that the fluid is incompressible, so that it has a fixed density, independent of the pressure. Both these assumptions are reasonable because currents have almost no effect on the pressure experienced by a diver, and water is practically incompressible.

Suppose a cube of water is at rest somewhere within a lake. If you like, you can imagine the cube to be surrounded by a thin plastic membrane, so that it is clearly separated from the rest of the water, although this is not really necessary. What are

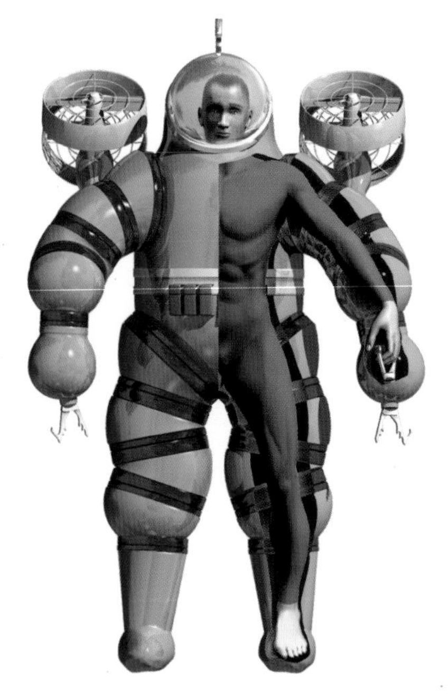

Figure 4.4 Divers can swim in surface waters without special protection, but heavy diving suits are needed to work at greater depths.

the forces acting on the cube? There is the weight of the cube acting downwards and there are inward forces acting on each face of the cube due to the pressure of the surrounding water.

First, consider the forces acting along the horizontal x-axis. These are due solely to the external pressure on the two shaded faces in Figure 4.5a. Suppose the area of one face of the cube is A, the pressure on the left-hand face is P_L, and the pressure on the right-hand face is P_R. Then the x-component of the total external force on the cube is

$$F_x = P_L A - P_R A\,.$$

Notice how the signs appear here: the force on the left-hand face is positive because it acts along the x-axis, while that on the right-hand face is negative because it acts in the opposite direction. Finally, remember that the cube is supposed to be at rest so F_x must be zero, leading to the conclusion that $P_L = P_R$. More generally, we can say that any two points at the same depth in a static fluid must be at the same pressure — if they were not, currents would flow, driven by the pressure difference.

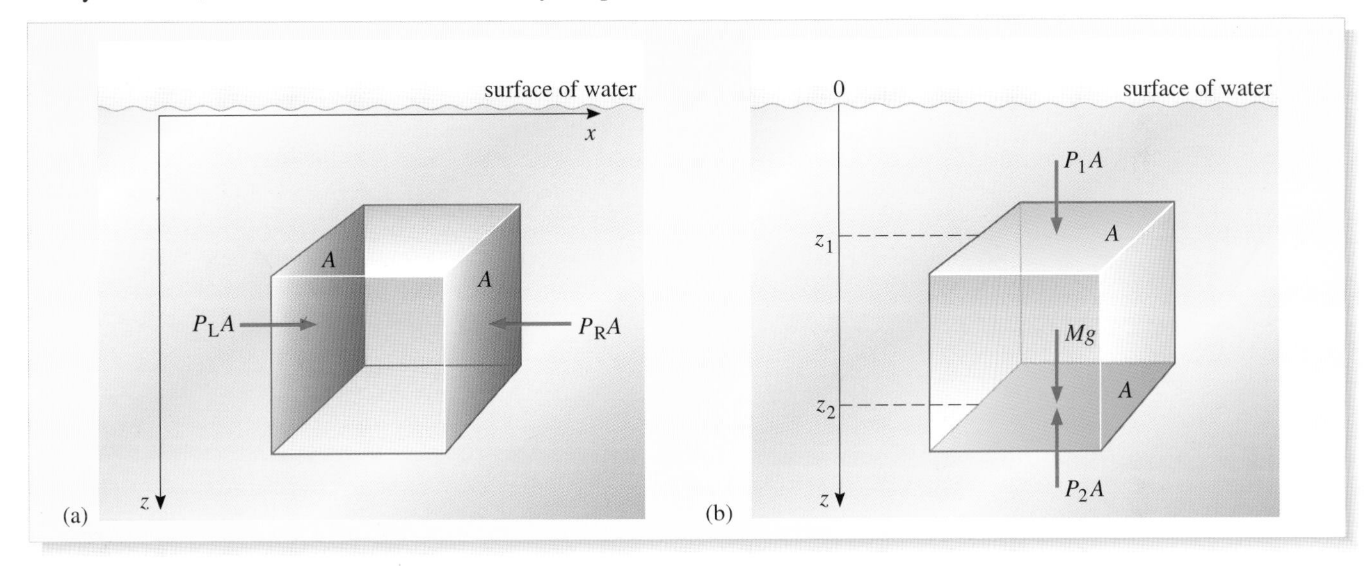

Figure 4.5 Forces on an imaginary cube of water in a lake. (a) Horizontal forces on the cube, acting in the x-direction, are determined by the pressure on the two faces shaded blue, each of area A. (b) Vertical forces on the cube, acting in the z-direction, are determined by the weight of the cube and by the pressure on the two faces shaded red, each of area A.

In order to see how pressure depends on depth, we must consider the vertical forces acting on the cube (Figure 4.5b). It is convenient to choose a z-axis that points vertically downwards, with its origin at the surface of the water. Suppose the cube has mass M, the pressure on the upper face is P_1 and the pressure on the lower face is P_2. Then the z-component of the total external force on the cube is

$$F_z = Mg + P_1 A - P_2 A\,.$$

Again, we can argue that F_z must be zero in order for the cube to remain at rest, so it follows that

$$P_2 = P_1 + \frac{Mg}{A}\,.$$

The pressure on the lower face is greater than the pressure on the upper face, allowing the weight of the cube to be supported. It is worth expressing this result in a slightly different way by noting that the mass of the cube is given by

$$M = \rho V$$

where ρ is the density of the water and V is the volume of the cube. Since $V = A(z_2 - z_1)$, we have

$$M = \rho A(z_2 - z_1).$$

This allows us to write

$$P_2 = P_1 + \rho g(z_2 - z_1)\,. \qquad (4.2)$$

If we now place the cube with its upper face at the surface of the water, z_1 will be zero and P_1 will be equal to atmospheric pressure, P_A. On the lower face of the cube, at depth z_2, the pressure is then

$$P_2 = P_A + \rho g z_2\,.$$

Now there was nothing special about z_2. It could have been any point below the surface, so the subscript 2 can be dropped, giving a simple formula for the pressure P at depth z below the surface:

$$P = P_A + \rho g z\,. \qquad (4.3)$$

The quantity $\rho g z$, which is the contribution of the liquid to the total pressure, is known as the **gauge pressure**, since it is the pressure that would be registered on a diver's pressure gauge, normally calibrated to read zero at the surface. It is interesting to calculate the rate of increase of pressure with depth. This can be done by rearranging Equation 4.2 to obtain

$$\frac{P_2 - P_1}{z_2 - z_1} = \rho g\,.$$

The left-hand side can be written as $\Delta P/\Delta z$ so, taking the limit as Δz becomes very small,

$$\frac{\mathrm{d}P}{\mathrm{d}z} = \rho g\,. \qquad (4.4a)$$

- ● Equation 4.4a can also be obtained directly from Equation 4.3. How?
- ❍ By differentiating both sides of Equation 4.3 with respect to z, remembering that P_A, ρ and g are constants. ■

The right-hand side of this equation, ρg, is the force magnitude per unit volume due to gravity. The left-hand side of the equation is the (vertical) **pressure gradient**, which can be interpreted as the force magnitude per unit volume due to pressure. Notice that the pressure gradient, $\mathrm{d}P/\mathrm{d}z$, has the *constant* value ρg. Inserting appropriate values for ρ, the density of pure water, and g, the magnitude of the acceleration due to gravity close to the Earth's surface, we obtain

$$\frac{\mathrm{d}P}{\mathrm{d}z} = 10^3\,\mathrm{kg\,m^{-3}} \times 9.8\,\mathrm{m\,s^{-2}} \approx 10^4\,\mathrm{kg\,m^{-2}\,s^{-2}} = 10^4\,\mathrm{Pa\,m^{-1}}.$$

Comparing this rate of increase of pressure with standard atmospheric pressure (10^5 Pa), we see that each metre of depth corresponds to about 10% of atmospheric pressure. So, if you dive to a depth of 10 metres, you will experience roughly twice the pressure as at the surface.

Before leaving the subject of pressure under water, it is worth trying to represent our findings in a more visual way. The most obvious method is to plot a graph of pressure against depth, as in Figure 4.6. This is a straight line of slope $10^4\,\mathrm{Pa\,m^{-1}}$. An alternative method is shown in Figure 4.7. This shows a vertical slice through the water, with lines connecting all points at the same pressure. For example, points A, B and C are at the same pressure because they are on the same line. Lines drawn in this way are called **isobars**. Since all points at the same depth have the same pressure, the isobars are horizontal, and the uniform increase of pressure with depth is revealed by the constant spacing of the isobars.

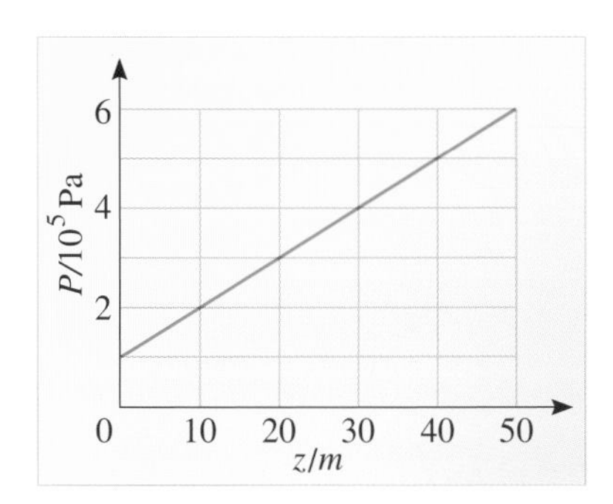

Figure 4.6 Graph showing the increase of pressure with depth below the surface of water. The graph is a straight line, showing that pressure increases uniformly with depth.

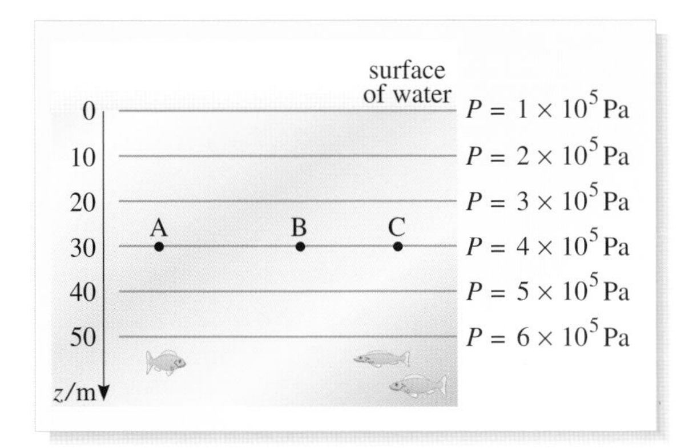

Figure 4.7 The isobars (i.e. contour lines for pressure) beneath the surface of water. The isobars are horizontal and evenly spaced, showing that all points at the same depth have the same pressure and that pressure increases uniformly with depth.

Question 4.1 The crew of a damaged submarine need to escape from their vessel, 50 m below the surface of the sea. What upward force must be applied to a circular escape hatch of diameter 0.75 m to push it open? (The pressure of air inside the submarine hull is 1 atmosphere and the density of the seawater is $1025\,\mathrm{kg\,m^{-3}}$.) ■

2.3 The pressure experienced by a mountaineer

The problem experienced by a mountaineer is the opposite of that faced by a diver. As the mountaineer climbs, the pressure falls. Above five kilometres, oxygen is in such short supply that mountain sickness is normal. Above eight kilometres, the body begins to die, so that any attempt to scale the highest peaks without artificial supplies of oxygen becomes a desperate race against time. Unfortunately, lack of oxygen means that this race must be carried out with an exhausted body and a fuzzy mind.

Figure 4.8 Near the summit of Everest, a mountaineer is breathing air at very low pressure and should not delay the descent for too long.

To see how pressure decreases with height we shall again take the atmosphere to be a static fluid. However air, unlike water, is easily compressed so it is not safe to assume that it has a constant density. As shown in Figure 4.9, the lower layers of the

Figure 4.9 Schematic diagram showing how air density decreases with height.

atmosphere are much denser than the tenuous upper layers. This creates a new situation. Consider two cubes of atmosphere of the same size, one near ground level and the other near the top of Everest. The cube near ground level contains denser gas, so it has a greater mass, and a greater pressure drop between its bottom and top faces is needed to maintain equilibrium. This means that the pressure of the atmosphere must drop rapidly close to the ground, and less rapidly at high altitude. The pressure drop will not be uniform.

To investigate this behaviour, we shall assume that the atmosphere is an ideal gas *at constant temperature*. This is an approximation, but not an unreasonable one, bearing in mind that the physical properties of a gas are largely determined by its *absolute* temperature and that, up to a height of 100 km, the absolute temperature of the atmosphere varies by less than 25%. We will also assume that the atmosphere is *thin* compared to the radius of the Earth, so that the force of gravity can be taken to be independent of height within the atmosphere. Because of these assumptions, the model is called a **thin isothermal atmosphere**.

Now let's examine the forces on a small cube of a thin isothermal atmosphere (Figure 4.10). Just as before, the horizontal forces on the cube balance, so the pressure is the same at all points with the same height. To examine the vertical forces on the cube, we choose a z-axis that points vertically *upwards*, and has its origin at ground level. As shown in Figure 4.10, the bottom face of the cube is at height z_1, where the pressure is P_1, and the top face of the cube is at height z_2, where the pressure is P_2. Adding up all the vertical forces acting on the cube then gives

$$F_z = -Mg - P_2A + P_1A\,.$$

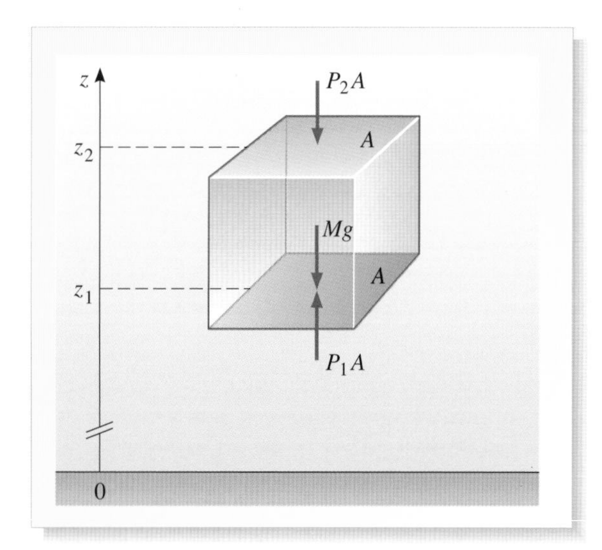

Figure 4.10 Vertical forces on a small cube of a thin isothermal atmosphere.

We can again argue that F_z must be zero in order for the cube to remain at rest, so

$$P_2 = P_1 - \frac{Mg}{A}.$$

Provided the cube is small enough, we can neglect any variation of density within the cube, and take its density to be constant. The mass of the cube is then

$$M = \rho A(z_2 - z_1)$$

so $$P_2 = P_1 - \rho g(z_2 - z_1)$$

and $$\frac{P_2 - P_1}{z_2 - z_1} = -\rho g\,.$$

Taking the limiting case of a very small cube, the rate of change of pressure with height is found to be

$$\frac{\mathrm{d}P}{\mathrm{d}z} = -\rho g\,. \qquad (4.4\text{b})$$

All of this looks very similar to the discussion given for water, except for the appearance of a minus sign on the right-hand side. The minus sign is easily explained. It arises because we have chosen the z-axis to point upwards and pressure *decreases* with height in the atmosphere (whereas it *increases* with depth in a lake).

In spite of the similarity, there is one crucial difference — the density of the atmosphere is not constant, but varies with height in some unknown way. At first sight, this looks like a serious setback because Equation 4.4b provides a relationship between the pressure and the density, *neither of which is known*.

But wait! We are dealing with an ideal gas, so it should be possible to express the density in terms of the pressure. For an ideal gas, with N molecules in volume V at temperature T,

$$PV = NkT$$

so the number of molecules per unit volume is

$$\frac{N}{V} = \frac{P}{kT}\,.$$

If the mass of one molecule is m, the mass of N molecules is mN so the density of the gas is

$$\rho = \frac{\text{mass}}{\text{volume}} = \frac{mN}{V} = \frac{mP}{kT}\,. \qquad (\text{Eqn } 1.13)$$

We can therefore rewrite Equation 4.4b as

$$\frac{\mathrm{d}P}{\mathrm{d}z} = -\left(\frac{mg}{kT}\right)P\,.$$

The quantities m, g, k and T are all constants (k is Boltzmann's constant, m is a constant provided that the atmosphere has a fixed composition, g is constant for a thin atmosphere and T is constant for an isothermal atmosphere). We can therefore write

$$\frac{\mathrm{d}P}{\mathrm{d}z} = -\frac{1}{\lambda}P \qquad (4.5)$$

where $\lambda = kT/mg$ is a constant. (The reason for expressing the equation in terms of λ is that this quantity turns out to be a length with an immediate physical significance, as you will see below.)

Equation 4.5 shows that the rate of decrease of pressure at any height is proportional to the pressure at that height. To solve this equation, we need to find a function $P(z)$ that obeys the following rule — differentiating the function with respect to z must simply give the original function, multiplied by the constant, $(-1/\lambda)$.

In looking for a function that fits the bill, it is worth noting that an equation very similar to Equation 4.5 has appeared earlier in the course. In *Predicting motion* (Eqn 2.51), you saw that the equation:

$$\frac{dv}{dt} = -\frac{1}{\tau}v$$

has the solution

$$v(t) = v_0 e^{-t/\tau}$$

where v_0 is a constant, independent of t. Equation 4.5 has a very similar form to this so, if we make the replacements $v \rightarrow P$, $t \rightarrow z$ and $\tau \rightarrow \lambda$, we can conclude that its solution is

$$P(z) = P_0 e^{-z/\lambda} \quad (4.6)$$

where P_0 is a constant, independent of z. At ground level, $z = 0$, so we can use the fact that $e^0 = 1$ to deduce that $P_0 = P(0)$. The pressure as a function of height is therefore:

Note that $\exp(-z/\lambda)$ is an alternative way of writing $e^{-z/\lambda}$.

$$P(z) = P(0) \exp(-z/\lambda) \quad (4.7)$$

where z is the height above ground level and $P(0)$ is the pressure at ground level ($z = 0$). This equation is generally called the **barometric formula** because it gives the pressure that would be recorded on a barometer (a meter for pressure) at any given height. It tells us that the pressure decreases exponentially with height.

The quantity

$$\lambda = \frac{kT}{mg} \quad (4.8)$$

is called the **scale height** of the atmosphere. It is a distance whose significance can be appreciated in terms of energy. At the scale height, λ, the gravitational potential energy of a molecule, $mg\lambda$, is equal to the energy kT. On general grounds, we would not expect many molecules to have much more energy than kT so the atmosphere is expected to thin out above the scale height. In retrospect, the exponential decrease of pressure with height is not too surprising. Combining Equations 4.7 and 4.8 gives

$$P(z) = P(0) \exp(-E_{pot}/kT),$$

where E_{pot} is the gravitational potential energy, mgz. This is reminiscent of the Boltzmann distribution law, which tells us that the probability of finding a molecule in a phase cell of energy E is proportional to the Boltzmann factor, $\exp(-E/kT)$. In fact, it is possible to derive the barometric formula directly from the Boltzmann distribution law. We have chosen to follow a more macroscopic derivation here, sticking closer to the spirit of fluid mechanics, but it is reassuring to note that two very different branches of physics — fluid mechanics and statistical mechanics — lead to the same conclusion.

The Earth's atmosphere is largely made up of nitrogen and oxygen molecules (of average mass 5.0×10^{-26} kg) and has an average temperature of about 270 K. Its scale height is therefore estimated to be

$$\lambda = \frac{1.38 \times 10^{-23}\,\text{J K}^{-1} \times 270\,\text{K}}{5.0 \times 10^{-26}\,\text{kg} \times 9.8\,\text{m s}^{-2}} = 7.6\,\text{km}$$

which is slightly lower than the height of Everest. This highlights the dangers facing a Himalayan mountaineer. Breathing air at a pressure below 40% of the ground-level value, we can say that the mountaineer has almost climbed out of the atmosphere!

The exponential function in Equation 4.7 leads to a simple pattern for the decrease of pressure with height. The pressure falls by a given factor for a given increase in height, no matter what the starting point. For example, the pressure falls by a factor of $1/e = 0.368$ between sea-level and the scale height; it falls by further factor of $1/e = 0.368$ between the scale height and twice the scale height, and so on.

In order to visualize the barometric formula more directly, we can plot a graph of pressure against height, as in Figure 4.11. Alternatively, we can plot isobars, as in Figure 4.12. Notice that the isobars are not equally spaced: they are tightly packed close to the ground, where the pressure is falling rapidly and less tightly packed at high altitude where the pressure is decreasing more slowly.

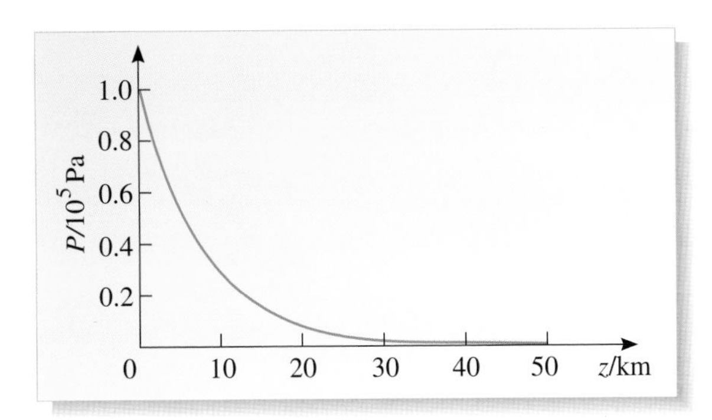

Figure 4.11 Graph of pressure against height for a thin isothermal atmosphere.

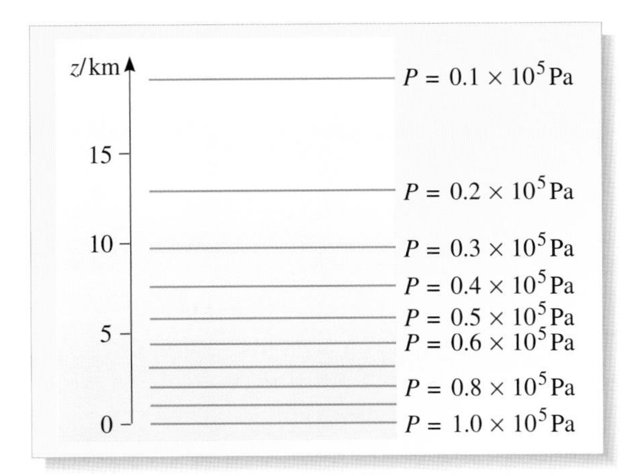

Figure 4.12 Isobars for a thin isothermal atmosphere.

Box 4.1 Morals drawn from the barometric formula

It is worth standing back for a moment to consider the route that led to Equation 4.7. The situation was not straightforward because the pressure and density of the atmosphere are non-uniform, which means that equilibrium is reached under different conditions at each level.

We considered a small cube, which could be anywhere in the atmosphere and related the pressure drop between its bottom and top faces to the mass of air in the cube. Because the cube was very small, we neglected any variation of density within the cube, and took its mass to be simply density times volume. This allowed us to link the rate of change of pressure at any given height to the density at that height (Equation 4.4a). Then, using the model of a thin isothermal atmosphere, we expressed density in terms of pressure, and so obtained a differential equation for the *rate of change of pressure* as a function of height (Equation 4.5). The solution of this equation gave the barometric formula (Equation 4.7).

The remarkable point about this derivation is that a complicated situation was made much easier by considering a small cube (or 'element') in which the density was taken to be constant. This simplification, valid on a small scale, but invalid on a large scale, allowed us to relate a small change in height to a small change in pressure. The beauty of the approach emerges when we imagine the cube becoming infinitesimally small. In this limit the approximation becomes exact and, at the same time, we obtain an equation involving derivatives — a differential equation.

This way of simplifying complicated situations by considering infinitesimal elements was introduced into physics in the eighteenth century, and is particularly associated with the name of a remarkable genius, Leonhard Euler (Figure 4.13). Born in Switzerland, he graduated at the age of 16, and became a professor of physics in St. Petersburg at 23. He steadily accumulated a body of work which secured his reputation as one of the most prolific mathematicians and scientists in history. For the last 15 years of his life, Euler was almost totally blind but was able to calculate so fluently in his head that his output scarcely decreased. The awe of his friends may be discerned from the curious inscription on his tombstone: 'on the 18th day of September 1783 Euler ceased to calculate'.

Figure 4.13 Leonhard Euler (1707–1783).

The impact of differential equations was enormous, as previously vague problems could be specified in precise mathematical terms. Even if solutions were not immediately forthcoming, it was at least clear what mathematical help was required. If there was a drawback, it was that the gap between everyday speech and the language of physics was irretrievably widened, cutting non-specialists off from the details of the subject.

Example 4.1

The atmosphere of a certain planet can be modelled as a thin isothermal atmosphere at temperature 150 K composed of molecules of mass 8.3×10^{-26} kg. The pressure at the surface of the planet is 4.5×10^{7} Pa and the pressure 50 km above the surface is 1.5×10^{5} Pa. What is the magnitude of the acceleration due to gravity near the planet's surface?

Solution

Preparation *Useful equations:*

$$P(z) = P(0)\exp(-z/\lambda) \qquad \text{(Eqn 4.7)}$$

$$\lambda = \frac{kT}{mg} \qquad \text{(Eqn 4.8)}$$

Known values:

$P(z) = 1.5 \times 10^{5}$ Pa when $z = 50$ km

$P(0) = 4.5 \times 10^{7}$ Pa

$m = 8.3 \times 10^{-26}$ kg

$T = 150$ K

$k = 1.38 \times 10^{-23}\ \text{J K}^{-1}$.

Brief plan: We can use the barometric formula with $z = 50$ km to find λ, and then use the formula for the scale height to find g.

Working The barometric formula can be written as

$$\frac{P(z)}{P(0)} = \exp(-z/\lambda).$$

In order to find the scale height λ, we take logs on both sides (remembering that the $\log_e$ function is the inverse of the exponential function). This gives

$$\log_e\left(\frac{P(z)}{P(0)}\right) = -\frac{z}{\lambda}$$

so $$\lambda = -\frac{z}{\log_e\left(\dfrac{P(z)}{P(0)}\right)} = -\frac{50\,\text{km}}{\log_e\left(\dfrac{1.5\times10^{5}}{4.5\times10^{7}}\right)} = 8.77\,\text{km}.$$

Rearranging the equation for the scale height then gives the magnitude of acceleration due to gravity for this planet:

$$g = \frac{kT}{m\lambda} = \frac{1.38\times10^{-23}\,\text{J K}^{-1}\times150\,\text{K}}{8.3\times10^{-26}\,\text{kg}\times8.77\times10^{3}\,\text{m}} = 2.8\,\text{m s}^{-2}.$$

Checking (i) The units of g explicitly check because $1\,\text{J} = 1\,\text{kg m}^2\,\text{s}^{-2}$ so $(\text{J K}^{-1}\,\text{K})/(\text{kg m}) = \text{m s}^{-2}$, which are the units of acceleration. (ii) Our equations show that, if $P(z)$ were smaller, λ would be smaller and g would be larger. This makes good sense because a stronger gravitational pull will keep the atmosphere more closely bound to the planet's surface. ■

Question 4.2 What is the pressure 76 km above the Earth's surface (i.e. at an altitude that is ten times the scale height)? How many molecules would you expect to find in one cubic millimetre at this height? (You may use the thin isothermal atmosphere model and take the temperature to be 270 K throughout.)

Question 4.3 A spacecraft approaches an unknown planet in a distant star system. It finds that the pressure 100 km above the surface is 2.1×10^5 Pa, while the pressure 75 km above the surface is three times higher (6.3×10^5 Pa). Use the model of a thin isothermal atmosphere to estimate the pressure at the surface of the planet. Should the spacecraft attempt to land if it is designed to withstand a maximum external pressure of 1.5×10^7 Pa? ■

2.4 Buoyancy and Archimedes' principle

Figure 4.14 Floating in the Dead Sea.

You have probably experienced your weight becoming less burdensome when you are immersed in water. In the bath, or when swimming in fresh water, your *mass* remains unaltered — it is just as difficult to accelerate as on dry land — but your apparent weight is decreased considerably. This effect is even more pronounced in seawater; in fact, your buoyancy in seawater depends on where you are in the world. You will feel slightly more buoyant in the Caribbean than in British seas, and in the Dead Sea you can float without any effort at all. Floating becomes easier the more dense the water is; seawater is more dense than freshwater, and the water of the Dead Sea is the densest of all (Figure 4.14).

It is not just humans who can float, of course; anything made of wood will float, and it is even possible for hollow objects made of iron, such as ships, to float, although a solid lump of iron will sink to the bottom of the water at once. Most of these facts have been known to humanity since the dawn of history, but they were put on a systematic and quantitative basis by the ancient Greek scientist and mathematician Archimedes, who answered the questions:

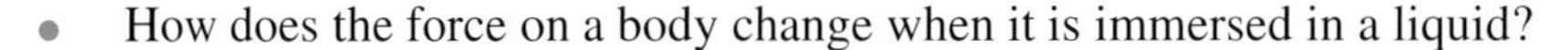

- How does the force on a body change when it is immersed in a liquid?

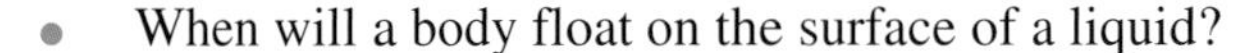

- When will a body float on the surface of a liquid?

These questions are answered by **Archimedes' principle**:

Archimedes' principle

If a body is immersed, partly or wholly, in a fluid, it appears to lose weight compared to when it is surrounded by a vacuum. The apparent loss of weight has a magnitude equal to that of the weight of the fluid displaced by the body.

To see why Archimedes' principle applies, we can consider an element of fluid, surrounded by more fluid. In a state of equilibrium there is no net force acting on the element so its weight must be exactly balanced by an upward force due to the surrounding fluid. This upward force is called the **upthrust**, and you have seen that it arises because the pressure on lower surfaces is greater than the pressure on upper surfaces. If the fluid has density ρ_0 and the element has volume V_0, equilibrium is achieved provided the upthrust force has magnitude

$$F = \rho_0 V_0 g\,.$$

Archimedes argued that, if the element of fluid is replaced by some other object of the same shape, *the upthrust will remain exactly the same*: it remains equal to $\rho_0 V_0 g$. Note that it is equal to the weight of the fluid displaced by the object irrespective of the composition or mass of the object. We can see what this means in various circumstances.

Suppose, first, that a block of concrete of density ρ and volume V is totally submerged in water of density ρ_0 (Figure 4.15). The submerged block displaces its own volume of water so the volume of fluid displaced is $V_0 = V$. With the z-axis pointing downwards, the total downward force is the weight of the block minus the upthrust due to the surrounding fluid so

$$F_z = \rho V g - \rho_0 V g = (\rho - \rho_0) V g . \tag{4.9}$$

Since the density of concrete, ρ, is greater than that of water, ρ_0, the value of F_z is positive and the total force is downwards. The concrete block sinks, but with apparently diminished weight.

What happens if a block of wood is submerged in water? Exactly the same arguments apply, again leading to Equation 4.9. Now, though, the density of wood, ρ, is less than that of water, ρ_0, so the value of F_z is negative, corresponding to an upward force. The submerged block of wood heads upwards for the surface. According to Archimedes' principle, the wood rises until the weight of the fluid displaced is equal to the weight of the wood. Because the density of water is greater than the density of wood, this is achieved by having the wood only partly immersed in the water — the wood floats (Figure 4.16).

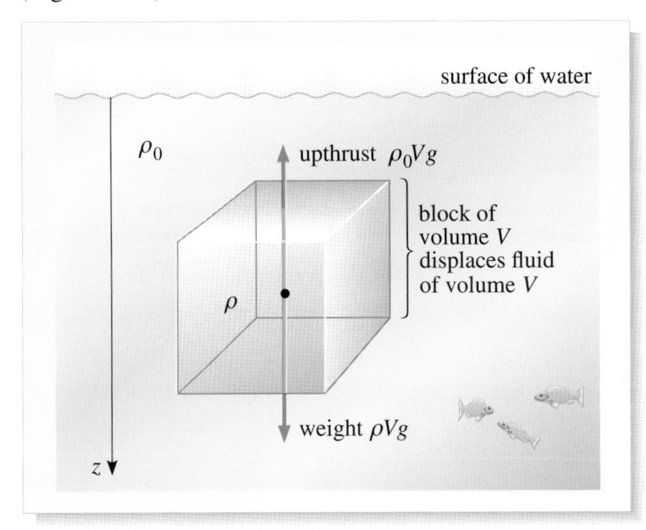

Figure 4.15 Forces on a concrete block of density ρ and volume V, totally submerged in water of density ρ_0.

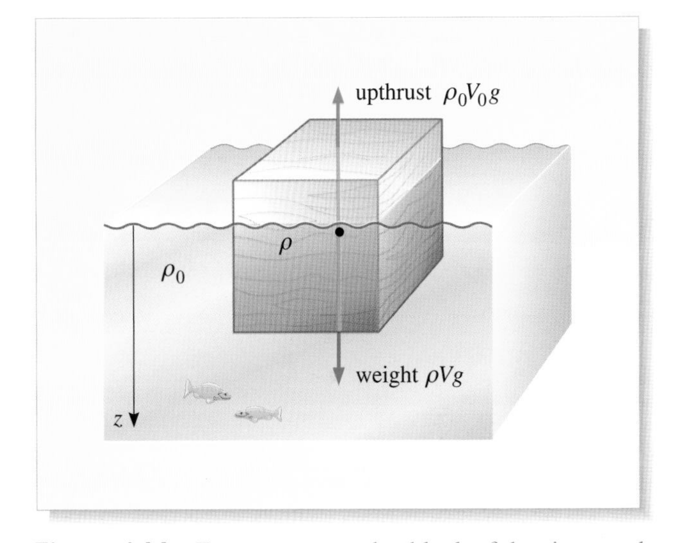

Figure 4.16 Forces on a wooden block of density ρ and volume V, partly submerged up to volume V_0 in water of density ρ_0.

Equation 4.9 cannot be used to discuss a floating block of wood because its volume V is not equal to the volume of water V_0 displaced. Instead, the total force is

$$F_z = \rho V g - \rho_0 V_0 g.$$

In a state of mechanical equilibrium, the total force is zero so the fraction of the block of wood immersed is

$$\frac{V_0}{V} = \frac{\rho}{\rho_0}. \tag{4.10}$$

If, for example, the block has 80% of the density of water, it is 80% immersed.

Since it is only the total volume of displaced fluid which is significant, it is not necessary for the immersed object to be uniform; indeed most interesting objects, whether floating or submerged, are distinctly non-uniform, whether they are ships,

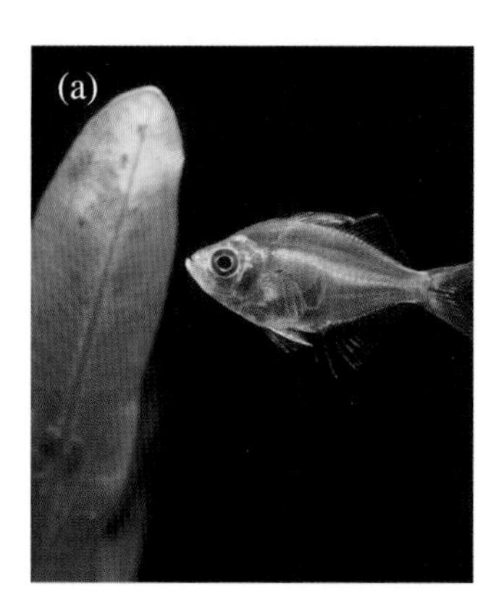

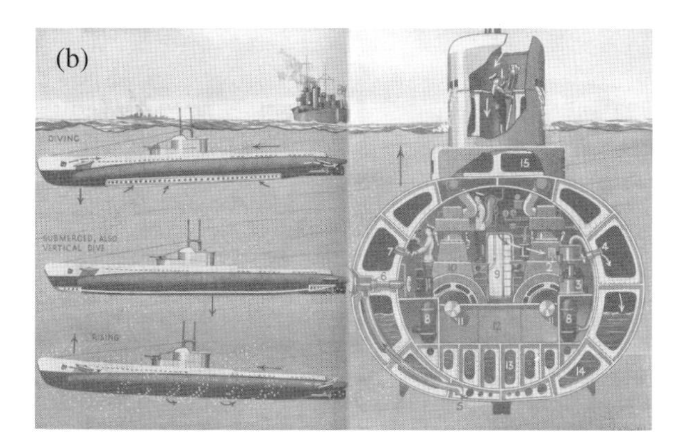

Figure 4.17 (a) Fish have swim-bladders and (b) submarines have ballast tanks in order to control their buoyancies.

bottles, fish, submarines or humans. All ships are hollow, and so will displace a much greater volume of water than would a solid object of the same mass, and even for a wooden ship, the majority of its buoyancy comes from the hollow interior rather than the low density of the wood used in its construction.

Other interesting examples of floating bodies are those for which it is possible to control the buoyancy. Fish are able to control their own average density by expanding and contracting the size of their swim-bladders, which contain air. These enable them to move between different depths at sea, where variations in salt content lead to small variations in density with depth. The same principle is used in submarines, which contain tanks which can be filled with air or water, thus allowing the boat either to float on the surface or sink out of sight (Figure 4.17).

Archimedes' principle applies to objects surrounded by air in exactly the same way as for objects immersed in water. A balloon or airship stays aloft because it contains a large volume of a gas that is less dense than air. The gas is either hydrogen, helium, or hot air, all of which have a lower density than that of the surrounding atmosphere.

- A spherical balloon has a volume of $4000\,\mathrm{m}^3$. What payload can the balloon lift when it is filled with hot air at 100 °C, while the surrounding air is at 0 °C? The air inside and outside the balloon is at a pressure of 1 atmosphere. At 0 °C the density of air is $1.29\,\mathrm{kg\,m^{-3}}$. (Ignore the weight of the skin of the balloon.)

- ❍ Suppose the balloon carries a payload of mass m, and that the densities of the hot air in the balloon and the surrounding air are ρ_b and ρ_0. Choosing the x-axis to point upwards, the total force acting on the balloon and its payload is

$$F_x = (\rho_0 - \rho_b)Vg - mg\,.$$

The maximum payload that can be lifted is found by setting the total force equal to zero, giving

$$m = (\rho_0 - \rho_b)V\,.$$

Using the ideal gas laws, the density of a gas at a fixed pressure is inversely proportional to the absolute temperature (see Equation 1.13 derived earlier). So

$$\rho_b = \frac{273\,\mathrm{K}}{373\,\mathrm{K}}\rho_0$$

and $$m = (1 - \frac{273\,\mathrm{K}}{373\,\mathrm{K}})\,\rho_0 V.$$

Substituting for the density and volume, the maximum payload has a mass of 1.38×10^3 kg. ■

Question 4.4 'Only the tip of the iceberg …' is a common phrase. If the density of an iceberg is 860 kg m^{-3}, and the density of the surrounding seawater is 1025 kg m^{-3}, what proportion of the iceberg is beneath the water?

Question 4.5 A cubical block of wood floats partly submerged in water. It is not in equilibrium, but bobs up and down, whilst always remaining partly submerged. Let x be the vertical displacement of the block, measured downwards from the equilibrium position. When x is positive, the block is over-submerged and experiences a net upwards force. When x is negative, the block is under-submerged and experiences a net downwards force. Show that, in both cases, these forces are proportional to x. What type of motion does the block perform? ■

Figure 4.18 An iceberg in the Southern Ocean. Only a small fraction of the iceberg is visible above the surface.

3 Ideal fluids in motion

The study of fluids in motion, or **fluid dynamics**, is more complicated than that of fluids at rest. Although its history goes back to the beginning of the eighteenth century, some problems remain unsolved, even today. To see why fluid motion is difficult to predict, imagine looking down on a river from a bridge. To describe the flow of water, you would need to specify the velocity *at each point of the flow*: the velocity of water directly below the centre of the bridge is different from that near the river banks, and the velocity at the surface is different from that near the river bed. The flow may also depend on time, being larger in times of flood and smaller in times of drought. A complete theory of fluid dynamics would have to explain the entire fluid velocity as a function of position and time at every point in the river.

How could this be done? The most obvious approach is to use Newton's laws. You saw earlier that gradients in pressure generate forces, and you know that forces produce accelerations, which are rates of change of velocity. Thus, at each point in the fluid, Newton's second law will give a relationship between pressure gradients (such as $\mathrm{d}P/\mathrm{d}x$, $\mathrm{d}P/\mathrm{d}y$ and $\mathrm{d}P/\mathrm{d}z$) and acceleration components (such as $\mathrm{d}v_x/\mathrm{d}t$, $\mathrm{d}v_y/\mathrm{d}t$ and $\mathrm{d}v_z/\mathrm{d}t$). The crucial point is that derivatives with respect to position (e.g. $\mathrm{d}P/\mathrm{d}x$) are intimately linked to derivatives with respect to time (e.g. $\mathrm{d}v_x/\mathrm{d}t$). Equations involving mixtures of different types of derivative (known as *partial differential equations*) are notoriously difficult to solve, although modern computers have made the task somewhat easier. We shall simply take this as a cue to avoid the direct use of Newton's laws. Fortunately, there are many alternative techniques to use. In some cases, the laws of conservation of mass and conservation of energy provide valuable information. In others, it is possible to use 'scaling' arguments so that the behaviour of a fluid near a large system, such as a jumbo jet, can be inferred from observations made on a small-scale model in a wind tunnel.

3.1 Ideal flow

Figure 4.19 shows a familiar example of fluid flow — water flowing from a tap. You can probably reproduce something like this in your kitchen sink (taking care to run the tap gently so as to avoid turbulence and the mixing of air and water.) Notice that the jet of water narrows as it descends from the tap. Why should this happen?

Figure 4.19 Water flowing from a tap.

In order to explain this phenomenon and many others, we can adopt a simple model of fluid flow known as **ideal flow**.

Ideal flow

A fluid is said to exhibit ideal flow if:

1. The fluid is *incompressible*. That is, the density of the fluid is constant, with equal volumes having equal masses.
2. The fluid has *no viscosity*. Viscosity in a fluid is the analogue of friction between two solids. It resists the flow of one layer of fluid past a neighbouring layer, and dissipates the kinetic energy of flow as heat. Treacle is a very viscous fluid; water and air are much less viscous.
3. The flow is *irrotational*. It has no eddies or vortices (i.e. whirlpools).
4. The flow of fluid is *steady*. The speed of flow at any fixed point does not vary with time.

In common with all models, ideal flow is an approximation, but it is one that works well under a wide range of circumstances. Even the flow of air past an obstacle can often be regarded as ideal (including the assumption of constant density) provided the flow is not too violent. For brevity, we will use the term **ideal fluid** to describe a fluid exhibiting ideal flow. Note however, that a given substance may be an ideal fluid in some circumstances and non-ideal in others.

It is helpful to introduce a way of visualizing fluid flow, based on the idea of a streamline. A **streamline** is the path taken by a particle drifting in the fluid. For example, Figure 4.20a shows streamlines drawn for the case of water falling from a tap. When drawing streamlines, it is normal to include arrows to indicate the direction of flow — in this case, of course, downwards. Figure 4.20b shows streamlines in rapidly flowing air, made visible by shining light on drifting smoke. Streamlines have a number of properties, which follow directly from their definition.

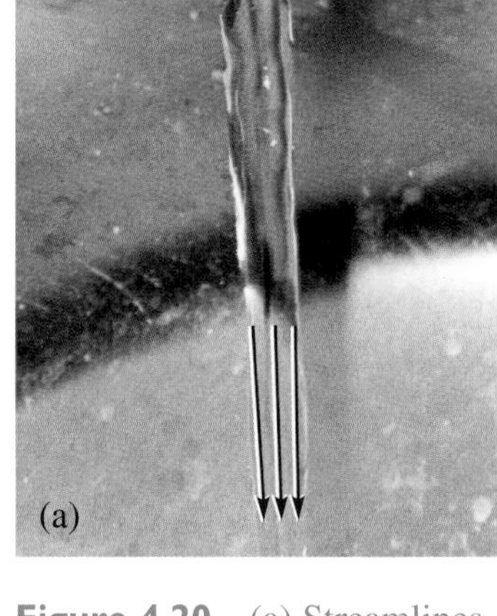

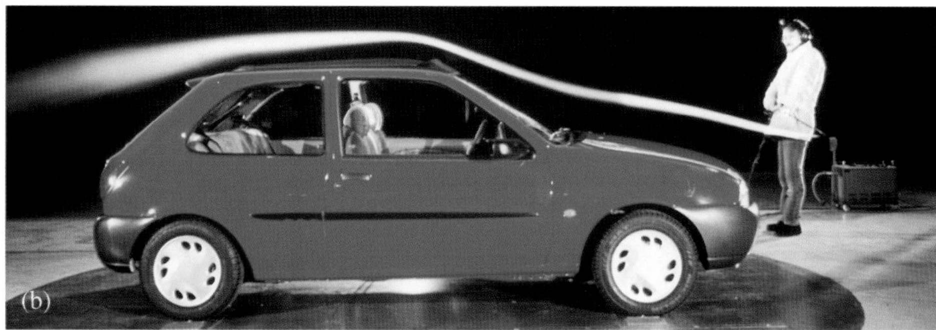

Figure 4.20 (a) Streamlines drawn for water flowing from a tap. (b) Streamlines shown by smoke in a wind tunnel.

Properties of streamlines

1. Streamlines are continuous during steady flow; they never appear or disappear. A particle in the fluid must have somewhere to go, and must have come from somewhere.
2. Streamlines cannot cross; at a crossing point, a particle would have no unique path to follow.
3. At each point, the velocity of the fluid is in the direction of a tangent to the streamline.
4. In ideal flow, the streamlines cannot form closed loops because no eddies or vortices are allowed.

3.2 Conservation of mass and the equation of continuity

We will now apply the law of conservation of mass to ideal fluid flow. If we consider a closed system of fluid, such as the water in a bath, we know that, if no fluid enters or leaves the system, the total mass of fluid will be constant. This way of using the law of conservation of mass is said to be *global* because it applies to the system as a whole. However, more interesting information emerges if we use the law of conservation of mass in a *local* way, to analyse a small region within the fluid. To do this, we consider again an imaginary cube. Fluid cannot appear from nowhere or disappear into oblivion, so any change in the mass of fluid inside the cube must be due to fluid flowing into the cube or fluid flowing out of it. In general we can write

change in mass of fluid = mass flowing into cube – mass flowing out of cube.

However, an ideal fluid is incompressible. Because the density of the fluid is constant, the mass of fluid inside the cube cannot change. Thus, ideal fluid flow is characterized by the fact that

mass flowing into cube = mass flowing out of cube.

And, because the density is constant, this also means that

volume flowing into cube = volume flowing out of cube.

To take a definite case, consider the flow of water along a pipe of varying cross-sectional area, as shown in Figure 4.21.

The streamlines are practically perpendicular to the cross-sectional areas shown as A_1 and A_2. In this situation the volume of water entering the shaded region in time Δt is $v_1 A\,\Delta t$, while the volume of water leaving the shaded region is $v_2 A_2\,\Delta t$. Thus,

$$v_1 A_1\,\Delta t = v_2 A_2\,\Delta t$$

so $$\frac{v_2}{v_1} = \frac{A_1}{A_2}. \tag{4.11}$$

This equation is sometimes called the **equation of continuity**. Since A_1 is greater than A_2, it follows that v_2 is greater than v_1: the flow is most rapid where the pipe is narrowest. In fact, the speed of flow is inversely proportional to the cross-sectional area of the pipe. Expressed in terms of streamlines, we can say that the flow is fastest where the streamlines are closest.

Now, let's return to the problem of water flowing from a tap. Since the water accelerates as it falls, the speed v_2 further down the jet will be greater than the speed v_1 at the tap. Equation 4.11 then shows that that the cross-sectional area A_2 must be smaller than A_1. The narrowing of the jet is therefore explained by the equation of continuity, which follows from the conservation of mass.

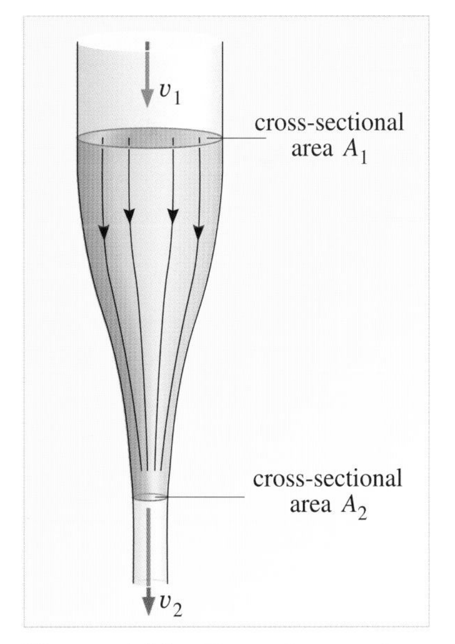

Figure 4.21 Flow of water along a pipe of varying cross-sectional area.

Question 4.6 A water pipe of internal diameter 19 mm is divided at a Y-junction into two pipes each of diameter 12 mm. What is the ratio of the flow speed in the 19 mm pipe and the flow speed in one of the 12 mm pipes? ■

3.3 Conservation of energy and Bernoulli's equation

An ideal fluid has no viscosity so no energy can be dissipated as heat. This allows us to apply the law of conservation of energy in a relatively straightforward way.

We again consider ideal flow through a pipe of varying cross-sectional area. The pipe bends upwards so that fluid enters and leaves at different heights. Let's take as our system a fixed quantity of fluid, shaded purple in Figure 4.22. This fluid is initially between points A and B in Figure 4.22a. At a slightly later time, this fluid has moved along the pipe to occupy the volume between C and D in Figure 4.22b. Comparing these two parts of the figure, and remembering that the flow is steady, the net effect of the flow can be characterized as follows: it is *as if* the volume between A and C had been removed from the fluid in Figure 4.22a, and the volume between B and D added. Because the fluid is incompressible, these volumes are equal. Let them both be ΔV.

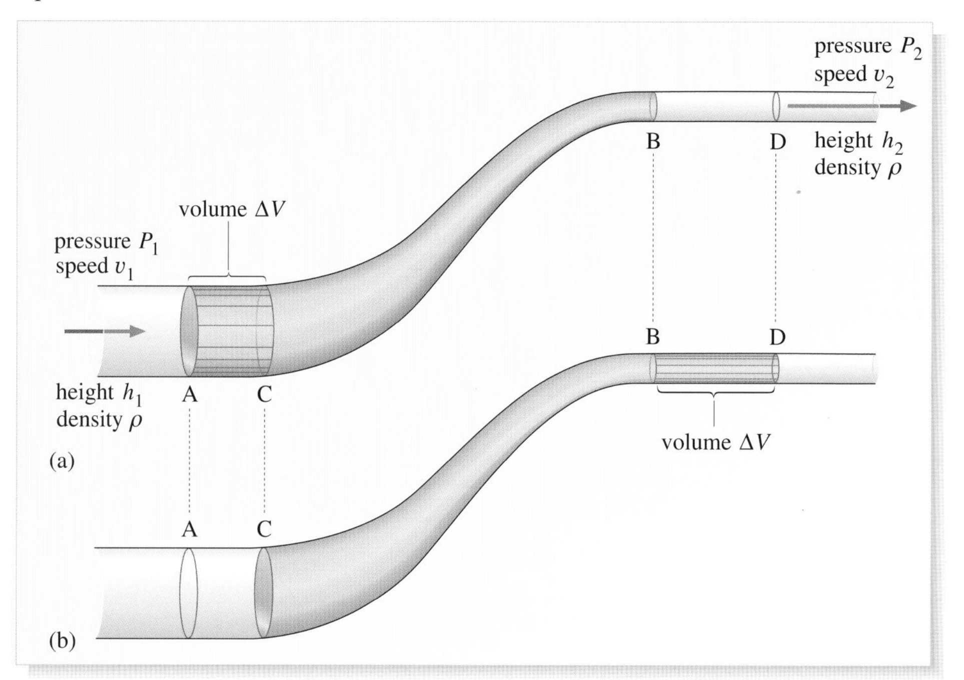

Figure 4.22 Applying the law of conservation of energy to the flow of an ideal fluid. Remember that (b) shows the same bit of pipe as (a), but at a slightly later time

Using the notation indicated in Figure 4.22, the portion of fluid between A and C has mass $\rho\,\Delta V$, speed v_1 and height h_1; it therefore has kinetic energy $\frac{1}{2}(\rho\,\Delta V)v_1^2$ and gravitational potential energy $(\rho\,\Delta V)gh_1$. Similarly, the portion of fluid between B and D has mass $\rho\,\Delta V$, speed v_2 and height h_2, and therefore has kinetic energy $\frac{1}{2}(\rho\,\Delta V)v_2^2$ and gravitational potential energy $(\rho\,\Delta V)gh_2$. We conclude that the net change in energy of the purple-shaded fluid is

$$\Delta U = \tfrac{1}{2}(\rho\,\Delta V)v_2^2 + (\rho\,\Delta V)gh_2 - \tfrac{1}{2}(\rho\,\Delta V)v_1^2 - (\rho\,\Delta V)gh_1.$$

Where does this change of energy come from? The answer is *from the surrounding fluid*. You can think of the fluid surrounding the purple region as behaving like a piston doing work on the system. At the bottom left-hand corner of the pipe in Figure 4.22, the pressure is P_1 and the change in volume of the system is $-\Delta V$; at the top right-hand corner, the pressure is P_2 and the change in volume of the system is ΔV. The total work done on the purple-shaded fluid is therefore

$$W = [-P_1(-\Delta V)] + [-P_2\Delta V] = (P_1 - P_2)\,\Delta V.$$

Using the law of conservation of energy, we can equate the work done, W, to the change in energy, ΔU. Dividing through by the volume ΔV and rearranging slightly, we obtain

$$P_1 + \tfrac{1}{2}\rho v_1^2 + \rho g h_1 = P_2 + \tfrac{1}{2}\rho v_2^2 + \rho g h_2. \tag{4.12}$$

Both sides of this equation must be equal to the same constant. Since we could have chosen our input and output points to have been anywhere in the flow, we can finally state that

$$P + \tfrac{1}{2}\rho v^2 + \rho g h = \text{constant} \tag{4.13}$$

throughout the entire length of the tube. The value of the constant depends on the nature of the flow, but provided we know the flow at one point in the tube we can deduce the flow elsewhere.

Equation 4.13 is known as **Bernoulli's equation**. It applies to the flow of an ideal fluid along a tube or, more generally, to flow along any streamline. Very often, we don't need to consider changes in height, so it is sometimes quoted in the form

$$P + \tfrac{1}{2}\rho v^2 = \text{constant}. \tag{4.14}$$

In other words, where the velocity is high, the pressure is low, and *vice versa*. This fact is known as **Bernoulli's principle**.

Figure 4.23 Daniel Bernoulli (1700–1782) was a member of a famous Swiss family containing at least 11 noted mathematicians in four generations. He researched in many areas of science, including the kinetic theory of gases and fluid dynamics.

● Is there any relationship between the pressure in a fluid and the spacing of the streamlines?

❍ We already know that the closer the spacing of the streamlines the higher the flow speed. Bernoulli's equation shows that the higher the flow speed, the lower the pressure. Thus, closely spaced streamlines indicate a low pressure. ■

Example 4.2

Figure 4.24 shows a tank of water which has developed a tiny hole near the bottom at depth $d = 0.6$ m below the water surface. At what speed will the water escape?

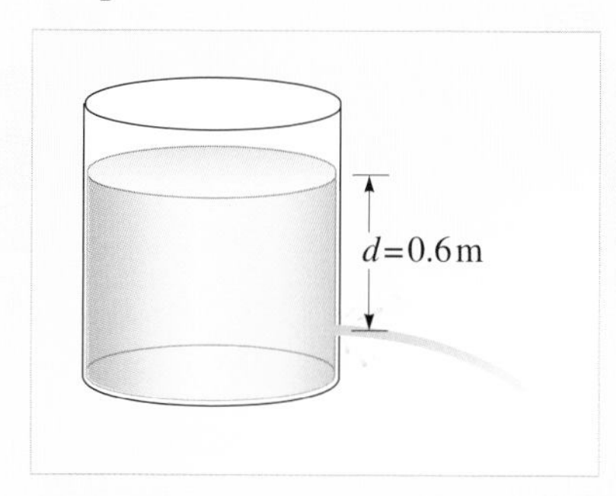

Figure 4.24 A leaking tank of water. A cylindrical tank has been shown but the shape does not affect the answer to Example 4.2.

Solution

Preparation We can use Bernoulli's equation to compare conditions at the water surface in the tank and in the jet of water *just as it emerges from the tank*. Both are open to the atmosphere and so they are both are at atmospheric pressure, P_0.

Useful equations:

$$P + \tfrac{1}{2}\rho v^2 + \rho g h = \text{constant}. \qquad \text{(Eqn 4.13)}$$

Known values:

at the surface: $P = P_0$, $v = 0$, $h = d$, where $d = 0.6\,\text{m}$

at the outlet: $P = P_0$, $h = 0$.

Working Inserting the known values into Bernoulli's equation gives

$$P_0 + \rho g d = P_0 + \tfrac{1}{2}\rho v^2$$

so $$\tfrac{1}{2}\rho v^2 = \rho g d$$

and $$v = \sqrt{2gd} = \sqrt{2 \times 9.8\,\text{m}\,\text{s}^{-2} \times 0.6\,\text{m}} = 3.4\,\text{m}\,\text{s}^{-1}\,.$$

Checking The units check explicitly. According to the last equation, increasing g and increasing d both lead to an increase in outlet flow speed, v. This seems reasonable.

Question 4.7 Water is moving through a pipe of cross-sectional area $6.0\,\text{cm}^2$ at a rate of $2.0\,\text{m}\,\text{s}^{-1}$. The pipe descends 50 cm and broadens out to a cross-sectional area of $12\,\text{cm}^2$. What is the final speed of the water, and what is the pressure drop between start and finish? (Density of water = $1000\,\text{kg}\,\text{m}^{-3}$.) ■

3.4 Bernoulli's principle in action

Direct confirmation in a pipe

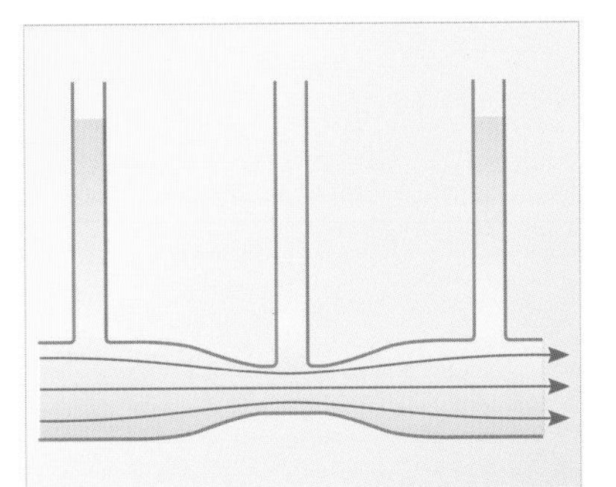

Figure 4.25 Direct demonstration of Bernoulli's principle.

The equation of continuity tells us that an ideal fluid moves most rapidly in the narrowest part of a pipe. This is a direct consequence of the conservation of mass. Bernoulli's principle gives us fresh insight into this phenomenon. It tells us that the pressure is low where the flow is fast. This makes good sense, as the low pressure can be thought of as providing an influence that causes the fluid to accelerate.

Nevertheless, some people find this counter-intuitive, falsely expecting the pressure to be greatest in the narrowest part of a pipe. It is therefore worth carrying out an experiment. Figure 4.25 shows the results. The height of the water in each vertical tube is proportional to the pressure at the point where it joins the horizontal pipe. Clearly, the pressure is lowest in the narrow part of the pipe, where the flow is fastest, just as Bernoulli's principle predicts.

Here are some examples of Bernoulli's principle in action:

A strip of paper

Figure 4.26 A paper strip rises when an air current passes over its upper surface.

Take a strip of paper and hold it by one end, so that the end is horizontal just below your lips (Figure 4.26). Now blow along the strip. You should find that the paper rises up towards the horizontal.

● How would you explain this effect?

❍ When an air current is passing along one side of the strip, the pressure is lower than on the other side, where the air speed is zero. This pressure difference pushes the paper upwards. ■

Chimney design

The chimneys of coal fires are generally built with a constriction (Figure 4.27).

- Why is this a good idea?

- To a first approximation, smoke rising up the chimney behaves like an incompressible fluid in that it moves faster as it passes through the constriction. According to Bernoulli's principle, this leads to a pressure drop in the vicinity of the constriction which reduces the likelihood of smoke being blown back into the room. ■

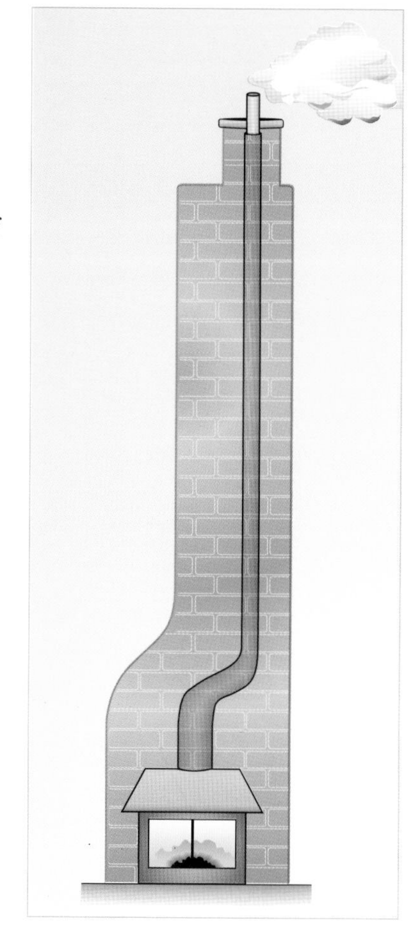

Figure 4.27 A well-designed chimney with a constriction (where the cross-section narrows).

Storm damage

When a wind blows past a surface such as the roof of a building or a window, which has stationary air on the other side, a pressure difference develops which may be enough to take the roof off, or to blow the window out, rather than blow it in, as one might naively expect. These effects are relatively common in countries where hurricanes occur. But the effect may also happen closer to home on a smaller scale.

Question 4.8 A dustbin lid has a mass of 1.0 kg and a diameter of 50 cm. What is the minimum wind speed which might blow the lid off? You may treat the lid as merely resting on the dustbin, neglect any frictional forces that might hold it on and take the density of air to be $1.2\,\mathrm{kg\,m^{-3}}$. ■

Flight

Perhaps the most famous application of Bernoulli's principle is in the context of flight. We will explore this subject later, but one crucial idea can be mentioned here. Suppose you could arrange things so that the air flowing over the top surface of an aircraft wing were moving faster than the air flowing over the lower surface. What would this imply? Because of its higher speed, the air flowing above the wing would be at a *lower* pressure than the air flowing below the wing. This would provide an upward force acting on the wing. If you are reading this book on an aircraft, you have every reason to be grateful for this effect, and might like to consider the magnitude of force being generated — sufficient to support many tons of aircraft, fuel, cargo and passengers.

Question 4.9 An aircraft has a mass of 60 tonnes (i.e. $6.0 \times 10^4\,\mathrm{kg}$) and is in steady, level flight. Each wing is 20 m long and 2 m wide. The speed of air just below the wing is $300\,\mathrm{m\,s^{-1}}$. Calculate the speed of the air just above the wing. (Again, take the density of air to be $1.2\,\mathrm{kg\,m^{-3}}$.) ■

4 Non-ideal fluids in motion

4.1 Viscosity

So far we have dealt with ideal flow, involving no compression of the fluid and no dissipation of energy. The ideal flow model is certainly useful in understanding many aspects of flow, but there are some situations where it is misleading.

Many of the most important deviations from ideal flow occur as a result of viscosity. When one solid surface slides over another solid surface, the frictional force acts to reduce the relative speed, dissipating kinetic energy as heat. An analogous effect occurs in fluids, though the details are not quite the same. It is useful to consider the situation shown in Figure 4.28, where two parallel plates are a small distance Δx apart.

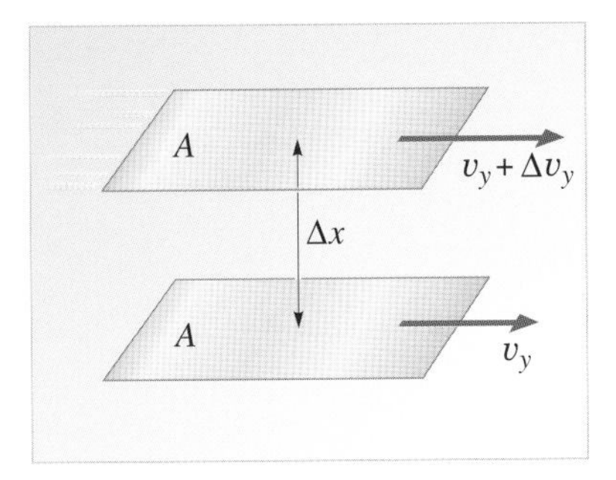

Figure 4.28 Two plates in a fluid experience viscous forces as a result of their relative sideways motion.

The lower plate is moving sideways with velocity v_y and the upper plate is moving sideways at a slightly greater velocity $v_y + \Delta v_y$. If the space between the plates is filled with fluid, the fluid next to the lower plate will move along with the plate at velocity v_y, while that next to the upper plate will move along with the plate at velocity $v_y + \Delta v_y$. There will be a force acting on the upper plate tending to slow it down, and a force, equal in magnitude and opposite in direction, acting on the lower plate, trying to speed it up. The magnitude of the force, F, would be expected to increase with both the area of the plates, A, and the relative velocity Δv_y of the plates However, the force would be expected to *decrease* as the separation of the plates, Δx, increases. It is therefore plausible to suppose that the magnitude of the force on either plate obeys the relationship

$$F \propto A\left|\frac{\Delta v_y}{\Delta x}\right|.$$

η is the lower case Greek letter eta. An unrelated use of this symbol is for the efficiency of a heat engine (Chapter 3).

Equation 4.15 was first suggested by Newton and fluids that obey it are called *Newtonian fluids*.

This relationship turns out to be valid in a wide range of circumstances, provided Δv_y and Δx are not too large. It is generally written in the form of an equation, by introducing a constant of proportionality, η:

$$F = \eta A\left|\frac{\Delta v_y}{\Delta x}\right|. \tag{4.15}$$

The force F is called the **viscous force** and the constant of proportionality η is called the **coefficient of dynamic viscosity** (or just the **viscosity**, for short). The thicker and more viscous the fluid, the larger the viscosity and viscous force will be.

● What are the SI units of the viscosity η?

❍ From Equation 4.15,

$$\eta = \frac{F/A}{\Delta v_y/\Delta x}$$

so its units are $(\text{N m}^{-2})/(\text{m s}^{-1}/\text{m}) = \text{N s m}^{-2} = \text{kg m}^{-1}\,\text{s}^{-1}$. ■

The values of viscosity vary over a very wide range, as shown in Table 4.1.

Table 4.1 Viscosities for some common fluids.

	$\eta/\text{kg m}^{-1}\,\text{s}^{-1}$	Conditions
acetone	3.10×10^{-4}	25 °C
water	8.90×10^{-4}	25 °C
olive oil	6.70×10^{-2}	25 °C
glycerol	1.41	20 °C
air	1.82×10^{-5}	20 °C
carbon dioxide	1.47×10^{-5}	20 °C
water vapour	9.7×10^{-6}	20 °C

The above discussion has referred to the forces on solid plates. We could equally well replace the plates by layers of fluid moving past one another. The viscous force then tells us how much one layer pushes or pulls a neighbouring layer along the line of flow. Viscous forces always act in a direction that diminishes the velocity difference between neighbouring layers. Figure 4.29 shows a layer of fluid (marked

B) with layers A and C on either side. Layer A is moving slower than layer B, so it exerts a viscous force on B against the direction of flow, tending to slow it down. By contrast, layer C is moving faster than layer B, so it exerts a viscous force on B along the direction of flow, tending to speed it up.

At first sight, you might think that these two viscous forces would cancel out, but this is not true. Figure 4.30 is a graph showing how the velocity of a fluid varies in a pipe. Note that the velocity is zero at the walls of the pipe, and rises as we move towards the centre. Note too that the velocity gradient, given by the slope of this graph, is greater at the interface between B and A than at the interface between B and C. This means that the viscous force on layer B due to layer A is greater than the viscous force on layer B due to layer C, so the net effect of viscosity is to slow layer B down.

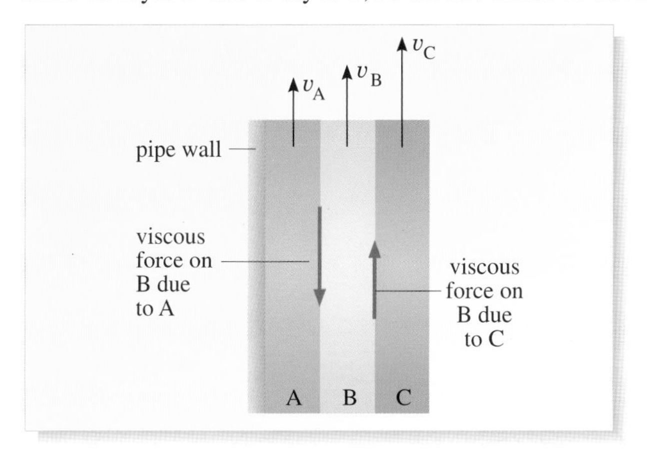

Figure 4.29 Viscous forces act between adjacent layers of a fluid.

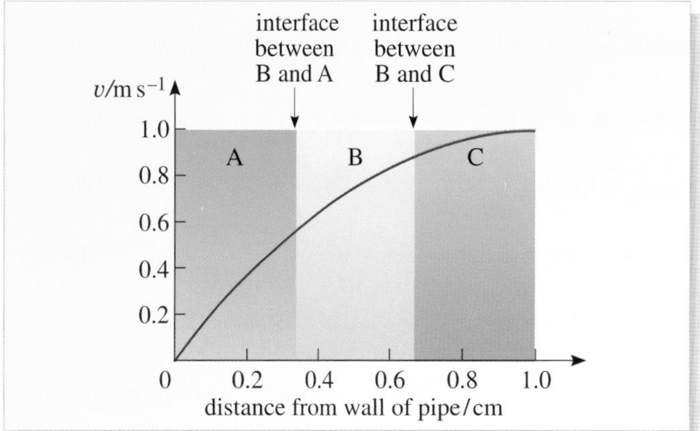

Figure 4.30 The velocity profile of a fluid flowing in a pipe.

In practice, there are other forces acting, all of which must be taken into account when analysing the flow. A situation that arises quite often is that of steady flow through a cylindrical pipe of radius R and length L. To maintain steady flow, the net retarding force due to viscosity must exactly balance the accelerating force due to pressure difference ΔP along the pipe. In this case, the speed of flow in the pipe turns out to be a function of the distance r from the axis of the cylindrical pipe:

$$v = \frac{\Delta P}{4\eta L}(R^2 - r^2). \tag{4.16}$$

According to Equation 4.16, the speed of flow is zero at the walls of the pipe (where $r = R$) and has its maximum value, $R^2 \Delta P / 4\eta L$, at the centre of the pipe (where $r = 0$). Clearly, the flow is nothing like uniform — the inner layers slip past the outer layers, and contribute most to the flow. Designers of pipelines know that to maximize throughput for a given pressure drop, it is better to build one large pipe rather than many smaller ones as this keeps more of the fluid away from the walls.

The type of flow described in Equation 4.16 is known as **Poiseuille flow** after the French physician and physiologist Jean-Louis-Marie Poiseuille (1799–1869) whose interest in fluids arose from studies of the circulation of blood. Although blood does not behave quite like an ideal fluid, Poiseuille's findings still have important consequences for health. As arteries clog, their radius decreases. You might expect the flow rate to be proportional to the cross-sectional area of an artery, which is proportional to the square of its radius. But the above discussion shows that the maximum speed of flow is also proportional to the radius squared. Taking both cross-sectional area and speed of flow into account, the throughput is proportional to the *fourth* power of the radius. This is a very strong dependence, making it all the more important to heed medical advice about taking exercise and maintaining a healthy diet.

Question 4.10 A water pipe has radius 1.5 cm and length 12 m (Figure 4.31). The portion of water that is in a thin cylindrical shell 1 cm from the centre of the pipe flows at a steady speed of $15\,\text{cm}\,\text{s}^{-1}$. What is the pressure drop over the length of the pipe? (Take the viscosity of water to be $1.0 \times 10^{-3}\,\text{kg}\,\text{m}^{-1}\,\text{s}^{-1}$.) ■

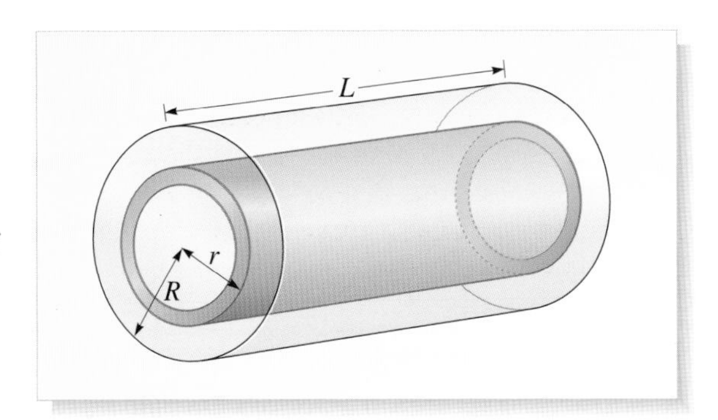

Figure 4.31 The cross-section of a cylindrical pipe of length L and radius R, with a thin cylindrical shell of water a distance r from the centre of the pipe.

4.2 Dynamical similarity and the Reynolds number

The complete equations of fluid flow are difficult to solve, especially when viscous forces are taken into account. Not surprisingly, scientists and engineers have searched for alternative ways of predicting phenomena and testing designs, based as much on observation as on mathematics. For example, small models of planes or cars are routinely placed in wind tunnels, where they are exposed to strong blasts of air. By observing the behaviour of these models, it is possible to gather information that is valid for full-scale planes or cars.

So, how can we use scale models to give us information about full-scale situations? To investigate this question, imagine observing a fluid flowing around a solid object — say the flow of water round a cylinder. For simplicity, suppose that the flow is incompressible, so that the fluid has a constant density; suppose too that gravity has a negligible effect on the flow. We can introduce a typical length, L_0, that characterizes the solid object. In the case of the cylinder, this could be its radius. Then all position coordinates can be scaled relative to this length. For example, the position coordinate, x, can be divided by L_0 to yield a scaled position coordinate, $x' = x/L_0$, which is *dimensionless* — that is, a pure number, without any units. Similarly, we can introduce a typical speed of flow, v_0. Note that this is not the actual speed of flow $v(\mathbf{r})$ (which will generally vary from point to point) but a *typical* value. We can also introduce a typical time interval, L_0/v_0, and a typical quantity with the units of pressure, ρv_0^2. These typical values can then be used to define scaled velocity, time and pressure variables, each of which is *dimensionless*.

Now, imagine taking the basic equation of fluid dynamics (i.e. Newton's second law, including forces due to pressure gradients and viscosity) and expressing it *entirely* in terms of our **scaled variables**. If this were done, we would get an equation of the form

Re is one symbol; it is not $R \times e$.

$$\text{(some combination of scaled variables)} = Re \qquad (4.17)$$

where Re is some parameter that depends on the properties of the fluid (ρ and η) and on the typical lengths and speeds, introduced above (L_0 and v_0). We shall not spell out the precise form of Equation 4.17 because we already have enough information to establish the principle of scaling.

So, let's look at Equation 4.17 more closely. The crucial point to notice is that the left-hand side depends only on dimensionless variables, so is itself dimensionless. It follows that Re must be dimensionless as well. This means that Re cannot be just *any* combination of ρ, η, L_0 and v_0. It must be a combination that is dimensionless. A suitable combination can be found by noting that:

- the units of ρ are $\mathrm{kg\,m^{-3}}$;
- the units of η are $\mathrm{kg\,m^{-1}\,s^{-1}}$;
- the units of L_0 are m;
- the units of v_0 are $\mathrm{m\,s^{-1}}$.

To obtain a combination of ρ and η that does not involve kg, we need to form the ratio of these quantities. The units of η/ρ are $\mathrm{m^2\,s^{-1}}$, and these are also the units of the product L_0v_0 . It then follows that the quantity $(L_0v_0)/(\eta/\rho) = (\rho L_0v_0)/\eta$ has the units $\mathrm{m^2\,s^{-1}/m^2\,s^{-1}} = 1$, and so is dimensionless. On these grounds, based purely on an analysis of units, it is reasonable to propose that

$$Re = \frac{\rho L_0 v_0}{\eta}. \tag{4.18}$$

In fact, any power of $\rho L_0v_0/\eta$, or any pure number times $\rho L_0v_0/\eta$, is dimensionless so the precise choice of Re is partly a matter of convention, closely linked to how we choose to write Equation 4.17. The universal choice is to take Equation 4.18 as the *definition* of Re, since this leads to the simplest form of Equation 4.17. There is some freedom in choosing the quantities L_0 and v_0. For the flow of fluid around a sphere, it would be natural to take L_0 to be either the radius or the diameter of the sphere, for example. This freedom does not matter, provided we make a consistent choice in any particular case.

Nowadays, the quantity Re is called the **Reynolds number**, after the Manchester engineer Osborne Reynolds (Figure 4.32) who popularized its use. You will see in the next section how Reynolds used Re to great effect in his studies of turbulence, but it is only fair to record that Re was introduced by George Stokes (1819–1903) in 1851, some 30 years before Reynolds saw its significance.

Figure 4.32 Osborne Reynolds (1842–1912) was the founder of the study of turbulence and popularized the use of the Reynolds number.

Our decision to use scaled variables may have seemed cosmetic, unlikely to yield any new knowledge, but it is actually a very powerful idea. The reason is as follows. Consider two fluid flows: one near a large-scale object and another near a geometrically similar, but much smaller, model. The two flows obey Equation 4.17 in exactly the same way *provided they both have the same Reynolds number, Re*. So, whatever phenomena occur in the small-scale flow (expressed in terms of its scaled variables) will also take place in the large-scale flow (but expressed in terms of its scaled variables). This allows us to use observations on a model aeroplane in a wind tunnel to predict the behaviour of a full-scale aeroplane. Two flows for geometrically similar situations with the same value of Reynolds number are said to be **dynamically similar**. The general principle can be stated as follows:

The principle of dynamic similarity

Two dynamically similar flows display the same patterns when expressed in terms of their own scaled variables.

- To investigate the flow of water in a river, a scientist builds a scale model with linear dimensions 50 times smaller than the river. Measurements are then taken on water flowing through the scale model. What precaution should the scientist take to ensure that the model yields information relevant for the real river?

○ Since the scientist is using water in the scale model, the density and viscosity are the same as in the river. The only way to ensure that the Reynolds number in the model flow is the same as in the river is to speed up the flow rate so that water flows through the model 50 times faster than in the real river. Note that dynamical similarity has nothing to do with visual appearances — a film taken of the model, with water hurtling through it, will not look at all like the stately flowing river, although the model and the river are dynamically similar. ■

The principle of dynamic similarity is usually applied by constructing small-scale models and increasing the flow rate. It has also been used the other way round. For example, studying the flight of insects poses problems because of their small size and rapid motion. Larger-than-life models have therefore been built, with the flow rate decreased. Either way, the product $L_0 v_0$ is held constant in order to maintain the same Reynolds number.

We may also choose to use a different fluid in the simulation to that present in the real situation. For example, at room temperature, the ratio η/ρ is about seven times higher for air than for water. So, systems involving air flow can be made dynamically similar to those involving water flow provided the air is made to flow seven times more rapidly than the water, or the length-scales in air are seven times longer than those in water.

Scaling and experimental design

The tactic of using scaled variables is a major tool in fluid mechanics which has saved billions of pounds in research effort. Suppose, for example, that you plan to investigate the magnitude of the drag force on a sphere — that is, the force exerted by the fluid on a sphere that is moving through it. This force opposes the motion of the sphere. We will use the symbol F_D to denote its magnitude. You could measure the effects of changing the radius, L_0, and speed, v_0, of the sphere and the density, ρ, and viscosity, η, of the fluid. This could be done by changing one variable at a time, keeping all the others fixed, so an enormous number of measurements would be needed.

A much smarter way of proceeding is to notice that the quantity $\rho L_0^2 v_0^2$, has the units of force ($\mathrm{kg\,m\,s^{-2}}$), so the scaled variable, $F_D/(\rho L_0{}^2 v_0{}^2)$, has no units — it is *dimensionless*. The great advantage of introducing a dimensionless quantity is that it can only depend on a *dimensionless* combination of ρ, η, L_0 and v_0, and there is only one such combination — the Reynolds number. In other words, $F_D/(\rho L_0{}^2 v_0{}^2)$ must be some function of the Reynolds number, $Re = \rho L_0 v_0/\eta$. We write

$$\frac{F_D}{\rho L_0^2 v_0^2} = f(Re).$$

The role of any experimental investigation is then to determine the form of the function *f*(*Re*). To do this, the Reynolds number would have to be varied over a suitable range. This could be achieved by, for example, taking 30 different readings for different flow rates around a given sphere in a given fluid. Think how much work has been saved compared with varying four variables independently, and taking $30 \times 30 \times 30 \times 30 = 810\,000$ readings!

Question 4.11 In the particular case described above of a sphere moving through a fluid, it is found that

$$f(Re) = \frac{6\pi}{Re}$$

provided $Re < 1$, but this relationship breaks down when $Re > 1$. Show that a small sphere, moving slowly through water, will experience a drag force that is proportional to its radius and speed. How slowly would a sphere 0.1 mm in radius have to move through water for this statement to remain valid? Use the data for water given in Table 4.1. (Assume that for water $\eta = 1.0 \times 10^{-3}\,\mathrm{kg\,m^{-1}\,s^{-1}}$.) ■

4.3 Turbulence and its onset

Steady flow is something that is understood reasonably well but, in studying it, we soon find that it only occurs in a limited range of conditions. There are many occasions when a flow starts off as steady, and then appears spontaneously to become unsteady. A good example is provided by the smoke from a cigarette (Figure 4.33). Initially, the smoke rises vertically and the flow appears to be steady, but then at a certain height the flow suddenly breaks up and becomes highly irregular. What we see here is the onset of **turbulence**, an unsteady flow whose details are unpredictable. There is no definitive theory of turbulence; it is in fact one of the areas of physics where there are still major unsolved problems. However, there are questions we can ask in order to reach a general understanding of what is happening.

Figure 4.33 Smoke rising from a cigarette, illustrating the spontaneous development of turbulent flow.

The onset of turbulence was first studied experimentally by Reynolds in 1883, using the apparatus shown in Figure 4.34. Water flows through a horizontal tube at a steady rate, and a dye is injected into the centre of the flow using a nozzle. If the flow rate is small, the dye appears as a single streamline along the length of the flow, and can be made to oscillate gently if the water in it is made to oscillate, but does not lose any of its definition. However, as the flow rate is increased, vortices begin to form at the outlet end of the apparatus, and the dye becomes dispersed throughout the whole width of the tube. As the flow rate is increased further, the vortices began to form nearer to the inlet end of the tube, until eventually the whole flow becomes turbulent.

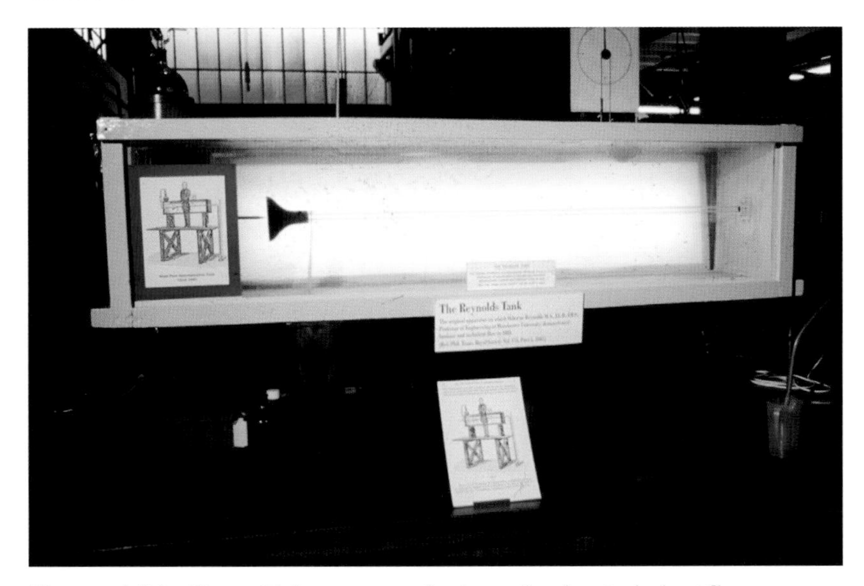

Figure 4.34 Reynolds's apparatus for investigating turbulent flow.

For a fluid of a given density and viscosity, flowing through a tube of fixed cross-section, turbulence is observed to set in at a critical speed of flow. From the principle of dynamical similarity, you might guess that this is part of a more general statement: for a fluid flowing through a tube, turbulence sets in at a critical value of the Reynolds number. Reynolds measured this critical value to be $Re \approx 2300$.

The precise details of the onset of turbulence depend on the geometry. Figure 4.35a–e shows the changes in flow pattern that occur around a cylinder as the Reynolds number is increased from very low values to very high values. The main features are as follows:

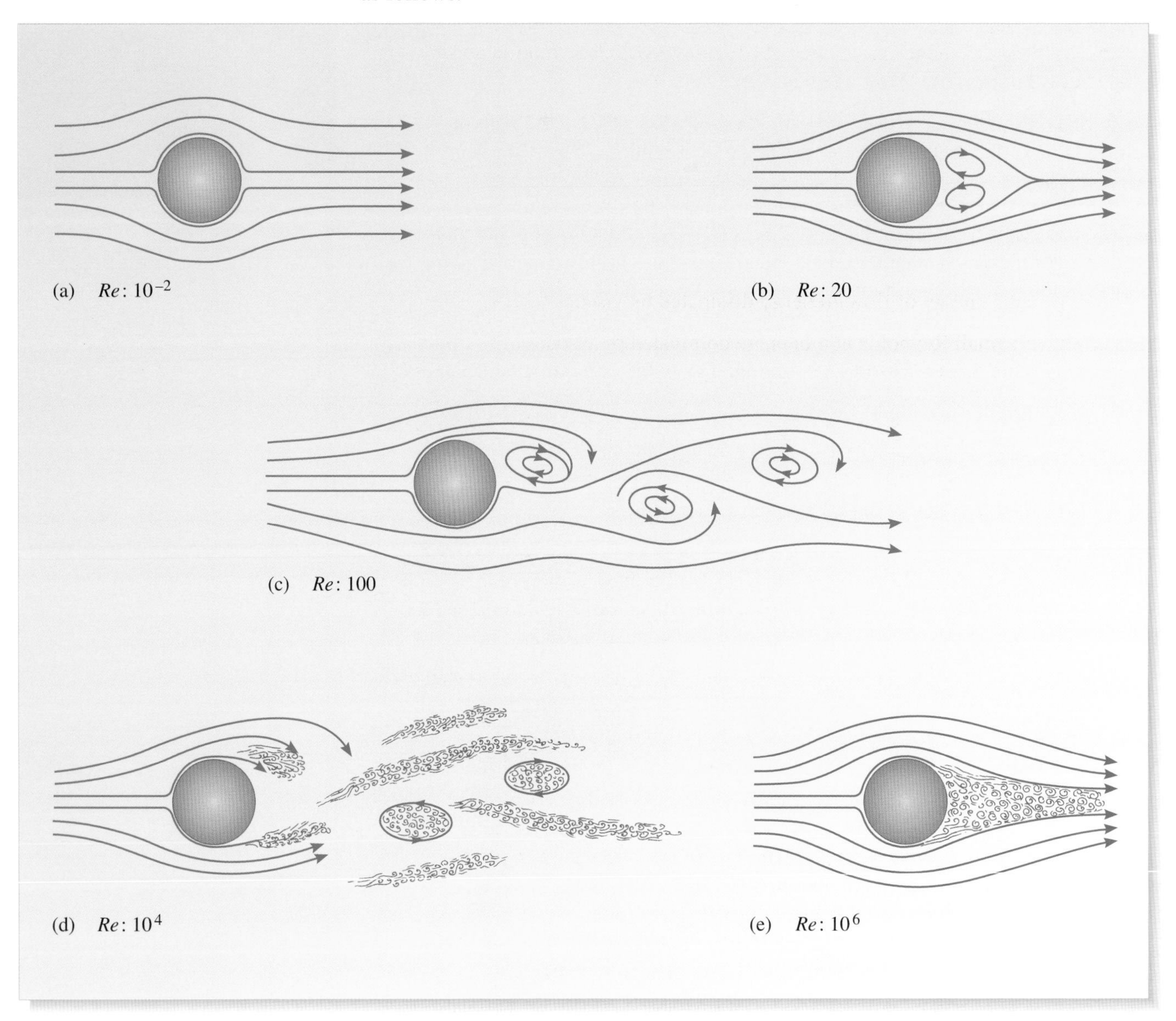

Figure 4.35 Flow of a fluid round a cylinder for various values of the Reynolds number.

(a) For Re below 1, the flow is very smooth.

(b) For Re between 1 and 30, the flow is still steady, but just behind the cylinder two vortices have developed; these stay and do not move away from the cylinder.

(c) For Re greater than about 40, the character of the motion changes suddenly. One of the vortices breaks away from the cylinder, and travels downstream; another is formed in its place, but in the meantime the other vortex breaks away, and the process is repeated. A double line of vortices is shed continually from the rear of the cylinder. The flow is no longer steady, but is varying in a regular periodic fashion.

(d) As Re is increased into the thousands, the periodicity starts to break down. This is the onset of turbulence as patches of irregular and chaotic motion appear which eventually destroy the regular vortex pattern.

(e) Finally, for Re greater than 10^5, the turbulent region has spread further forward, giving a highly turbulent wake just behind the cylinder. Wakes like this contain regions of very low pressure which tend to slow the cylinder down. They are just what designers of streamlined vehicles aim to avoid.

One way of interpreting these results is to note that the Reynolds number can be interpreted rather loosely as the ratio:

(kinetic energy of flow)/(energy dissipated by viscous forces).

Thus, flows with small Reynolds numbers are dominated by viscous effects and are very placid (Figure 4.35a). Flows with medium Reynolds numbers are influenced by viscosity, but in a subtle way, often involving the creation and shedding of vortices (Figure 4.35b,c). Finally, at high Reynolds numbers, the viscous forces are unable to damp down random motions and turbulence develops (Figure 4.35d,e).

4.4 The boundary layer

Try blowing the dust off the surface of a table. You will find it a frustrating experience. No matter how hard you blow, some dust will remain, although that dust could be easily removed with a cloth. Even the rotor blades of a helicopter gather dust, in spite of the strong currents of air that pass over their surfaces.

These observations illustrate an important fact: the fluid immediately in contact with a solid surface remains stuck to the surface. No matter what currents flow some distance away from the surface, there is no relative motion between the surface and the fluid immediately next to it. This fact is sometimes called the **no-slip condition**.

In 1904, Ludwig Prandtl (1875–1953) realized that it is possible to identify a thin layer of fluid next to any solid surface in which the fluid speed makes a rapid transition between the value expected for the main flow and the value of zero (relative to the surface) required by the no-slip condition. This layer is called the **boundary layer**. The abrupt change in velocity that occurs as we approach the surface is brought about by viscous forces which remain important in the boundary layer, no matter how large the Reynolds number characterizing the main flow.

For Reynolds numbers that are large compared to 1, the thickness of the boundary layer is of the order

$$\delta \approx \frac{L_0}{\sqrt{Re}} \qquad (4.19)$$

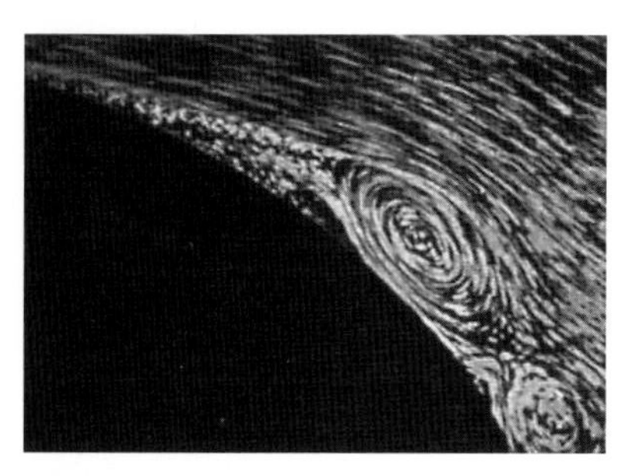

Figure 4.36 A boundary layer visible in fluid flowing around a cylinder. The boundary layer is seen next to the cylinder in the top left of the photo. A vortex is being shed from this layer.

where L_0 is a typical length in the system we are studying. If the Reynolds number is large, the thickness of the boundary layer, δ, can become very small, so it is perhaps not surprising that the significance of the boundary layer was missed until early in the twentieth century. Figure 4.36 shows a situation in which the boundary layer is visible.

The existence of a boundary layer helps resolve a paradox that arises from our study of turbulence. You have seen that increasing the Reynolds number leads to an increase in turbulence. An ideal fluid with no viscosity would have an infinite Reynolds number and so would be expected to display a lot of turbulence. But it is hard to see where this turbulence comes from. When the equations of fluid dynamics are solved for an ideal fluid, no turbulence is found. Even vortices cannot originate within an ideal fluid.

The resolution of this paradox is simply to note that ideal fluids do not exist. An *almost* ideal fluid would have a thin boundary layer. Within the boundary layer, there is a rapid change in fluid velocity and viscous effects play an important role. This means that vortices can develop within the boundary layer. Outside the boundary layer, the effects of viscosity are negligible and the flow is practically ideal, but vortices can drift in from the boundary layer.

In some circumstances, the boundary layer can become detached from the surface. This phenomenon, known as **boundary layer separation**, increases the rate of shedding of vortices, and may lead to turbulence in the region behind a moving object. This is usually regarded as something to avoid at all costs, as it leads to a low-pressure region behind a moving object, producing a marked increase in drag. Things can get even more complicated than this. It is possible for the boundary layer itself to become turbulent. Surprisingly, perhaps, this turns out to be beneficial because turbulent boundary layers remain stuck to surfaces at much higher flow rates than non-turbulent ones, so staving off the increase in drag.

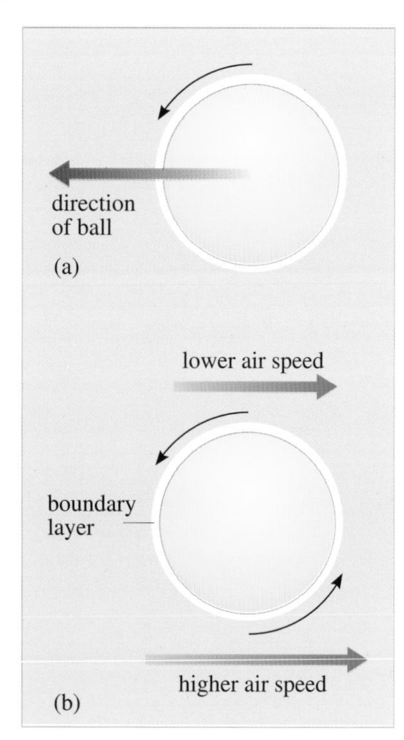

Figure 4.37 (a) A top-spun ball moving from right to left. (b) In the reference frame of the centre of mass of the ball, the air flow is from left to right; spin increases the flow on the lower surface and decreases the flow on the upper surface.

Spinning balls and the Magnus effect

Good ball players are able to apply spin to a ball in order to control its movement through the air. A tennis player, for example, can apply top-spin to make the ball dip sharply in its flight and so fall into the court rather than out. Table tennis players, footballers and golfers all make use of similar effects. The mechanisms by which these contestants achieve their aims are in general quite complex, but Bernoulli's principle gives us some understanding of what happens as the ball flies through the air.

Imagine a (top-spun) ball spinning about an axis at right angles to its direction of motion (Figure 4.37a). We can change our frame of reference so that the centre of mass of the ball is stationary, with air rushing past its surface (Figure 4.37b). The boundary layer, carried round with the spinning ball, will tend to carry other air with it, leading to a lower air speed on the upper side of the ball than the lower side. Bernoulli's principle then shows that there will be a higher pressure on the upper side than on the lower side, so the ball is deflected downwards. This has been described by saying 'the ball follows its nose'. The effect is known as the **Magnus effect** after the 19th century German physicist Gustav Magnus (1802–1870), although it seems to have been first recognized by the Englishman Benjamin Robins (1701–1751) 100 years earlier.

The case of a golf ball is particularly interesting since it is driven from the tee with considerable back-spin, so it experiences a very large Magnus force, and a markedly non-parabolic flight. Moreover, the surface of a golf ball is dimpled (Figure 4.38). You might intuitively expect a smooth ball to travel further than a dimpled one, but this is certainly not the case. The dimples cause the flow of air in the boundary layer to be turbulent, and so help to prevent its separation from the surface of the ball. This, in turn, reduces drag and increases the range of the shot.

Figure 4.38 A golf ball.

Answer to a puzzle

At this point we can return to the experiment described in the Introduction. Why does sugar, stirred into a glass of water, form a pile in the centre at the bottom of the glass?

To answer this question, first look at motion of the water well above the base of the glass. An element of fluid in the stirred water is in uniform circular motion, so accelerates towards the central axis of the glass. The force producing this acceleration must come from a variation in the pressure in the radial direction, with the pressure increasing along a radius drawn from the central axis to the edge of the glass. This ensures an inward radial force.

From Bernoulli's principle, we know that high pressures are accompanied by low flow speeds, so the speed of flow is predicted to be lower near the outside of the glass than near the central axis. This is not what you might expect if you thought of the water as rotating like a rigid body. However, all that matters for our present purposes is that the pressure increases towards the edge of the glass.

Now consider what happens as we approach the base of the glass. Just above the base there is a thin boundary layer in which the flow is much reduced. The pressure distribution does not change very much as we enter the boundary layer. For example, the large pressure towards the edge of the glass cannot suddenly decrease in the boundary layer, since that would produce a strong downward current which cannot exist given the unyielding nature of the glass base. In the boundary layer we therefore have practically the same inward force as elsewhere in the fluid, but a much reduced circular motion. The force must still produce the appropriate amount of acceleration and can only achieve this if the fluid in the boundary layer is driven towards the centre. The sugar collects in a pile and the current then rises towards the surface and circulates as shown in Figure 4.39a. A similar effect can be found at the bend in a river, where a spit of sand or other deposit may often be seen at the inside of the bend (Figure 4.39b).

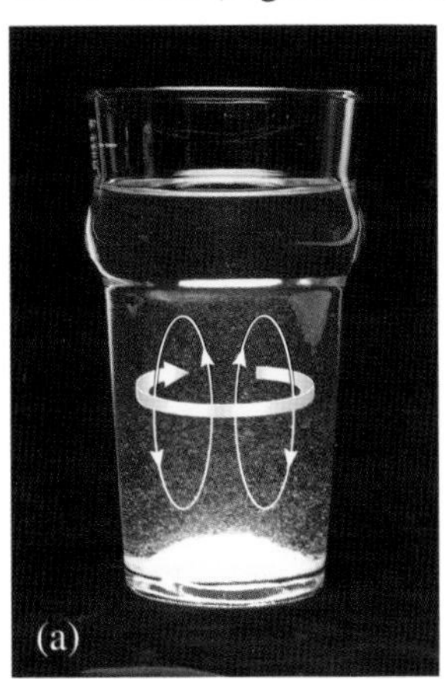

Figure 4.39 (a) Currents circulating in stirred water. (b) Sand deposited on the inside bend of a river.

5 Flight

Open University students should leave the text at this point and view Video 4 *Maiden flight*, which looks at some of the principles of flight, as discovered by a novice helicopter pilot. You should return to the text when you have finished viewing the video.

The simplest way of flying is to use balloons or airships, which exploit the buoyancy force implied by Archimedes' principle. It is worth noting that each cubic metre of air has a mass of 1.2 kg. By displacing several thousand cubic metres of air, upthrust forces of tens of thousands of newtons can be produced, allowing a balloon and its payload to rise and float.

A rocket, accelerating vertically upwards, gives another way of defying gravity. Exhaust gases are expelled from the rocket's tail with a certain downward momentum. They acquire this momentum thanks to chemical reactions in the rocket. By Newton's third law, the rocket and its remaining fuel must therefore experience an upward force. If the gases are expelled rapidly enough, the upward force will exceed the weight of the rocket and it will accelerate upwards. A few planes, such as the Harrier jump jet, use a similar principle to take off in confined areas.

The most popular method of flight, whether human or animal, is based on the concept of a wing. When a wing moves horizontally, air flows around it. The secret of a wing is to be shaped or oriented in such a way that it deflects some of this air downwards. Thus, the wing changes the momentum of the air, and exerts a force on it. By Newton's third law, the downward force on the air is accompanied by an upward force on the wing, known as **lift**. Note that, in order to carry out its function, a wing must move through the air. The engines on a plane are needed to maintain a high forward speed, so that air is deflected downwards at a sufficient rate for the lift on the wings to balance or exceed the plane's weight.

Although these arguments based on Newton's third law are complete, there is an alternative way of explaining the origin of lift, which is directly related to the theory of ideal fluid flow. A wing is shaped in such a way that the speed of flow over its top surface is greater than over its bottom surface (Figure 4.40a and Question 4.9). According to Bernoulli's principle, the pressure just above the wing must therefore be lower than the pressure just below the wing: this pressure difference is responsible for the lift.

A wing also experiences **drag**, a force that opposes forward motion through the air. An efficiently designed wing will minimize drag while maintaining a high lift. However, if the wing is tilted too steeply, a turbulent wake may be produced, leading to a region of low pressure just behind the wing, and a strong drag force (Figure 4.40b). Even worse, the lift force will be drastically reduced, and the wing will start to fall.

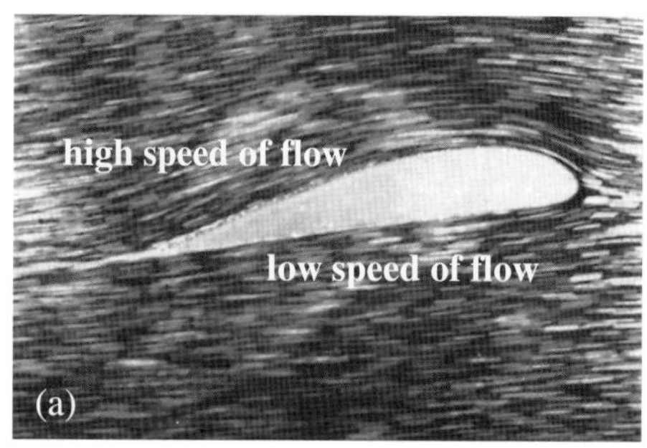

Figure 4.40 (a) Steady flow of air around a wing. (b) Turbulent flow of air around a wing.

6 Closing items

6.1 Chapter summary

1 The pressure in a fluid is defined as the magnitude of the normal component of the force exerted on a small element of area, divided by that area (Equation 4.1). The pressure is independent of the direction in which it is measured.

2 The pressure below the surface of a liquid is the pressure at the surface plus a contribution ρgz from the overlying fluid, which is proportional to the depth z.

3 According to the model of a thin, isothermal atmosphere, the pressure at height z above ground level is given by the barometric formula

$$P(z) = P(0)\exp(-z/\lambda) \qquad \text{(Eqn 4.7)}$$

where the scale height $\lambda = \dfrac{kT}{mg}$ (about 7.6 km for our atmosphere).

4 If a body is immersed wholly or partially in a fluid, buoyancy provides an upthrust force whose magnitude is equal to that of the weight of fluid displaced by the body (Archimedes' principle).

5 Ideal flow occurs at constant density and zero viscosity. The flow is steady and contains no vortices or eddies.

6 Applying the law of conservation of mass to ideal flow, we find that the speed of flow is inversely proportional to the cross-sectional area of the flow.

7 Applying the law of conservation of energy to ideal flow gives Bernoulli's equation:

$$P + \tfrac{1}{2}\rho v^2 + \rho gh = \text{constant}. \qquad \text{(Eqn 4.13)}$$

When differences in gravity can be neglected, this shows that the pressure is low where the speed of flow is high and *vice versa*.

8 Closely spaced streamlines indicate a high speed of flow and a low pressure.

9 The magnitude of the viscous force acting in a fluid is

$$F = \eta A\left|\frac{\Delta v_y}{\Delta x}\right| \qquad \text{(Eqn 4.15)}$$

where η is the coefficient of dynamic viscosity, A is the area of an interface between two neighbouring layers of fluid and $\Delta v_y/\Delta x$ is the velocity gradient at the interface.

10 Incompressible viscous flows are characterized by the Reynolds number Re, a dimensionless parameter given by

$$Re = \frac{\rho L_0 v_0}{\eta} \qquad \text{(Eqn 4.18)}$$

where L_0 is a typical length, v_0 is the flow speed, and ρ and η are the density and viscosity of the fluid.

11 Two flows for geometrically similar situations follow similar patterns when expressed in terms of their own scaled variables, provided both have the same Reynolds number. This is the principle of dynamic similarity.

12 The onset of turbulence occurs as Re rises into the thousands. Fully turbulent flow emerges when Re exceeds 10^5. Turbulent wakes lead to large drag forces and so designers try to avoid them.

13 There is no relative motion between a solid surface and the fluid in immediate contact with it. The effects of viscosity are often confined to a thin region near the solid surface, known as the boundary layer. The thickness of the boundary layer decreases as the Reynolds number increases.

14 The forces of drag and lift on a wing can be interpreted using Bernoulli's equation, in terms of the difference in speed of the air above and below it, or alternatively in terms of the change in momentum of the air deflected by the wing.

6.2 Achievements

Now that you have completed this chapter, you should be able to:

A1 Explain the meaning of all the newly defined (emboldened) terms introduced in this chapter.

A2 Calculate the forces on a cube of fluid, given the pressure acting on each of its faces.

A3 Use the barometric formula and the expression for the scale height in problems involving atmospheric pressure at various altitudes.

A4 Apply Archimedes' principle to calculate the upthrust due to buoyancy on submerged and floating objects.

A5 Apply the principle of conservation of mass to calculating the flow rates at different points in a fluid flow.

A6 Apply Bernoulli's principle to the explanation of various physical phenomena involving fluid flow.

A7 Apply Bernoulli's equation to calculate the speed of flow in terms of the pressure and height at different points in a fluid flow.

A8 Calculate the viscous force acting in simple fluid flows.

A9 Use the principle of dynamic similarity by calculating the Reynolds number in simple situations.

A10 Explain the significance of the boundary layer in a flow.

A11 Describe the origin of the forces of lift and drag on a wing.

6.3 End-of-chapter questions

These end-of-chapter questions give you further practice in using the definition of pressure, the barometric formula, Archimedes' principle, the equation of continuity, Bernoulli's equation and the principle of dynamic similarity.

Question 4.12 (a) A reservoir dam has a 500 m long vertical wall, which holds back an artificial lake of water. The depth of the water is 40 m. Sketch a graph of the pressure P exerted by the water on the wall against depth z below the water surface. (b) [*More difficult*] Use your graph to estimate the total magnitude of the force exerted by the water on the dam wall. (For simplicity, you may take g to be $10\,\mathrm{m\,s^{-2}}$.)

Question 4.13 Estimate the height above sea-level where the pressure of the Earth's atmosphere is 1% of the sea-level value. (The scale of height of the Earth's atmosphere is 7.6 km.)

Question 4.14 Consider a crude model of a wooden barge with a rectangular cross-section 20 m long and 3 m wide. Its unladen mass is 10^4 kg. What depth of freshwater will it draw when empty (i.e. how far is the bottom of the barge below the water surface), and what mass can it carry, if the depth of water drawn is not to exceed 1 m?

Question 4.15 A tree trunk conducts water to the upper branches of a tree at a rate of 10^{-2} kg s^{-1}. If the trunk behaves as a number of tubes of total cross-sectional area 10 cm^2, what is the speed of flow of the water?

Question 4.16 Serious damage can be caused to a pump if the liquid within it evaporates, forming small bubbles, or cavities, of vapour. This process, called cavitation, occurs when the pressure in the liquid becomes very close to zero. Suppose that a pump contains a tube of varying cross-section. The widest part of the tube has a cross-sectional area of 1.0 cm^2 and the pressure is 3 atmospheres. The narrowest part has a cross-sectional area of 0.2 cm^2. Use the equation of continuity and Bernoulli's equation to show that cavitation is inevitable if the speed of flow of water through the widest part of the tube approaches 5 m s^{-1}.

Question 4.17 A small bug can be modelled as a sphere of radius 1.0×10^{-6} m, moving through water at 1.0×10^{-4} m s^{-1}. Use Table 4.1 to calculate the Reynolds number for this motion. Explain why your answer means that the bug would perceive its surroundings as being very viscous. A sphere of radius 1.0 cm moves through a fluid with the same density as water, at a speed of 1.0 mm s^{-1}. What viscosity must this fluid have if the fluid flow around this sphere is to be dynamically similar to that of water around the bug? ■

Chapter 5 Consolidation and skills development

1 Introduction

This final chapter has two purposes. The first is to review some of the concepts that have been introduced in Chapters 1 to 4, to try to put into perspective the key ideas that are used to explain and predict the behaviour of matter using classical physics. The second purpose is to provide you with an opportunity to develop some of the skills that have been introduced in this book, and to continue the development of the problem-solving skills introduced in *Predicting motion*.

Section 2 provides an overview of the aims and approaches that have been taken in trying to understand the behaviour of matter. As well as reminding you about the important concepts that have been introduced, it should also prompt you to reflect on the methods that are used in trying to understand what can be very complex systems.

The remaining sections of this chapter are devoted to consolidation. Section 3 is devoted to skills development. One skill which is vital in physics is the interpretation and the formulation of written statements about the physical world. You have been introduced to many carefully worded statements which define concepts, principles and laws, and it is useful to review how to go about the business of writing such statements.

Section 4 consists of a series of questions which give you the opportunity to consolidate your understanding of the material covered in this book. Section 5 extends this consolidation by means of a set of questions which are provided on the interactive questions package for *Classical physics of matter*, while Section 6 invites you to try some of the questions contained in the *Physica* package.

2 Overview of Chapters 1 to 4

The aim of this book has been to provide an introduction to the physics of matter, using the concepts and methods of classical physics. After introducing some basic concepts and terminology, three different branches of physics were surveyed: statistical mechanics, thermodynamics and fluid mechanics.

Throughout, we emphasized the dual importance of understanding matter on the macroscopic scale of everyday life and the microscopic scale of atoms and molecules. Chapter 1 used both types of description for the phases of matter, looking at equations of state and *PVT* surfaces on a large scale, and the ordering and motion of molecules on a small scale. Chapter 2 used the methods of statistical mechanics to explore gases in more detail. By looking at the average behaviour of large numbers of molecules, subject to the rules of probability, we were able to derive the ideal gas equation of state, and so bridge the gap between the macroscopic and microscopic viewpoints. Chapter 3 discussed the laws of thermodynamics, which make no mention of atoms or molecules. Thermodynamics goes beyond ordinary mechanics by introducing thermal concepts such as temperature, heat and entropy, which have no precise definition in Newtonian mechanics. By concentrating on these quantities, and their relationships, thermodynamics is able to shed light on the phenomenon of irreversibility. This understanding is ultimately underpinned by statistical mechanics, which allows the concept of entropy to be interpreted in microscopic terms. Chapter 4 continued to take a macroscopic viewpoint, using fluid mechanics to explore the properties of fluids that are either at rest, or moving in various ways. Finally, it is worth remembering that this book has concentrated on aspects of the behaviour of matter that can be explained without resorting to quantum theory. The fascinating prospect of observing quantum mechanical effects directly through the properties of matter will be discussed in the final book of this course, *Quantum physics of matter*.

2.1 The atomic hypothesis

The single most important idea in developing an understanding of the material world is that matter is made from atoms. We can model macroscopic systems based on the properties of atoms and the interactions between atoms. In some senses the atomic world is remarkably simple, since only a limited number of types of atom exist — about one hundred or so, with each type corresponding to a different chemical element. Furthermore, physicists can develop models of matter without having a detailed knowledge of the internal workings of the atoms themselves.

As a starting point, the atomic hypothesis provides us with a simple picture of the three common phases of matter. We can envisage solid materials as consisting of atoms, vibrating about fixed positions. In liquids, the molecules are closely packed and are subject to intermolecular forces, but they retain the ability to escape from their immediate environment and so move throughout the liquid. Finally, in a gas, the molecules are free to move at random, and only interact by colliding with one another and with the walls of their container. The phase of matter obtained in a given situation is the result of a competition between the cohesive influence of molecular attraction and the disruptive influence of molecular kinetic energy. At low temperatures, cohesion wins and solids are stable; at high temperature, disruption wins and gases are formed.

The atomic hypothesis can also be used to make quantitative predictions about the macroscopic behaviour of matter. For instance, it is possible to derive an expression for the pressure of a gas using a simple model in which molecules are treated as point masses obeying Newton's laws. This simple gas model gives a quantitative description of the behaviour of a gas and also provides insight into the molecular origins of the pressure of a gas.

While we would hope that the behaviour of any type of matter could eventually be explained in terms of the behaviour of its constituent atoms, this is not always feasible. Historically, the starting point for theories of matter has been to attempt to relate macroscopic properties of matter to each other. This, in itself, is useful since these relations often have scientific and technological applications. However, the challenge to physicists is to explain these macroscopic relations from a knowledge of atomic properties. This challenge has been met with varying degrees of success, as we have seen from several examples in this book. The empirical gas laws, such as Boyle's law and Charles's law can be understood as special cases of the ideal gas law, which in turn may be derived from a statistical theory of molecules in gases. Similarly, thermodynamics, which successfully describes macroscopic energy changes in material systems, can now be understood on the basis of behaviour on an atomic level. However, in the case of thermodynamics, it is important to stress that the macroscopic relations are of immense importance in their own right. The subject of fluid flow, especially those aspects involving turbulence, is an area where the known macroscopic phenomena have not yet been fully linked to a microscopic explanation.

A final point to note about the atomic hypothesis is that once we accept that we are aiming to explain behaviour as being due to constituent atoms or molecules, we soon find that it is very convenient to consider quantities that refer not to a fixed mass of material but to a sample that consists of a fixed number of identical atoms (or identical molecules). This leads us to introduce the mole, which is an amount of a pure substance that contains Avogadro's number (6.022×10^{23}) of atoms (or molecules). Quantities defined as relating to one mole of material are referred to as *molar* quantities.

2.2 The macroscopic view

In attempting to understand the properties of matter and of material systems, an important reference point is given by observing macroscopic properties. There are several points to bear in mind about adopting this macroscopic approach.

First, we need to recognize the distinction between quantities which represent the state of a system and quantities which represent intrinsic properties of a material. For example, the temperature of a sample of solid metal is a variable which describes the state of the system, whereas the specific heat of that metal is a quantity which describes the properties of the metal. Similarly, the velocity of fluid at a given point in a pipe depends on the conditions in the flow (the state of the system), whereas the viscosity of the fluid is a quantity which describes the properties of the fluid.

The second point is that the study of the macroscopic properties should lead to relationships or laws between macroscopic variables. Examples include: the ideal gas equation, the laws of thermodynamics and Bernoulli's equation. The development of macroscopic relationships can also lead us to introduce new macroscopic variables whose meaning might not be immediately obvious. A good example is provided by the second law of thermodynamics, which leads us to introduce a new variable known as entropy, S. The behaviour of ideal gases similarly leads to the introduction of an absolute temperature, T.

In practice, we find that macroscopic relationships and laws depend on certain conditions in the system. For the case of ideal gases, we considered a situation where the macroscopic variables describing the gas were uniform throughout the sample and unchanging with time — a condition of equilibrium. This idea of equilibrium was also necessary in considering thermodynamic systems, and we found that these equilibrium states had a special property — they could be fully described by a small number of macroscopic variables. A variable whose value is determined by the equilibrium state (and not by the past history of a system) is called a state variable, or a function of state. Pressure, temperature, internal energy and entropy are functions of state; heat and work are not.

A different type of condition was imposed when we studied fluids. We found it useful to consider a system termed an ideal fluid, in which we could assume there would be steady incompressible flow, with no dissipation of energy and no eddies. We will return to these conditions in more detail below, but the key point to take from this is that in order to find useful relationships between macroscopic variables, it is often necessary to consider idealized systems.

A final point to make about macroscopic variables is a contrast between thermodynamic and fluid systems. In thermodynamic systems, we assume that the variables that define the state of the system are constant through some region of interest, and the aim of thermodynamics is to relate different equilibrium states to one another. In studying fluid systems, we are usually interested in macroscopic variables, such as pressure, that vary throughout the system, and the aim is to determine how such quantities vary with position. Fluid flow is not an equilibrium situation.

2.3 A statistical approach to physics

In accepting that the atomic view of matter should be the basis for physical models of material systems, we also have to recognize the problem inherent to such an approach, namely the difficulty of dealing with systems that contain a vast number of particles. We saw that, despite the power of Newtonian mechanics to predict the motion of individual particles, it is a practical impossibility to apply Newton's laws to each and every atom in a macroscopic system.

In order to tackle this problem, a radically different approach is used in which we describe a system in terms of probabilities. For example, consider a gas, in which the molecules behave as independent particles. The first step is to classify the microscopic states of the gas in a slightly crude way, in order to avoid the awkward fact that position and velocity are continuous variables. Rather than trying to specify the exact position and velocity of every molecule, we define positions and velocities to within a chosen precision, dividing phase space into a collection of tiny cells, known as phase cells. The microscopic state of the gas (known as a configuration) is specified by stating *which* molecules are in *which* phase cells. Then, rather than trying to follow the detailed history of individual molecules in the gas, we simply try to estimate the probability of one configuration compared to another. However, simply reformulating the problem in this way does not help much unless there is some way of assigning the probabilities. The insight that Boltzmann provided was to recognize that this could be done very simply, provided that the gas is in macroscopic equilibrium.

The idea of equilibrium is vital to the statistical approach to physics. It allows us to assume that, on a microscopic level, the internal energy of the system has been shared out randomly between the molecules of the gas, subject to one constraint – that the total energy of the system is a constant. This means that all configurations with the required total energy are equally likely, while other configurations (with other energies) are impossible. This leads to Boltzmann's distribution law, which describes the probability p of finding a given molecule in a given phase cell of energy E.

Boltzmann's distribution law provides the key to understanding systems of large numbers of particles in equilibrium, and is the basis for determining the distribution of speeds or energies of molecules in ideal gases. In order to determine the distribution functions, we need to know the probability of finding a molecule in a given phase cell and this is the information provided by Boltzmann's law. However, we also need to know the number of relevant phase cells in a given range of speed or energy. You saw in Chapter 2 that this approach leads to the Maxwell distribution for molecular speeds and the Maxwell–Boltzmann distribution for molecular energies. In this way, the perhaps rather abstract concept of a configuration can be translated into a measurable property of a gas, confirmed, for example, by the experimental determination of the molecular speed distribution using the rotating drum method.

A final point to note about statistical mechanics is that it provides a microscopic interpretation of entropy. In this book, we have concentrated on the thermodynamic role of entropy but, when we want an answer to the question 'what is entropy anyway?' we need to look to statistical physics. The answer, again provided by Boltzmann, is that entropy of a state is a measure of the number of allowed configurations corresponding to that state.

2.4 The laws of thermodynamics

Thermodynamics arose as a study of systems in which the primary interest was the relationship between the transfer of heat and mechanical work. It incorporates several important laws which are applicable in a wide variety of situations. Not only is it useful to know about these general laws before having to concern ourselves with a microscopic interpretation of a system, it is often the case that thermodynamic relations are also of great practical use in their own right.

The first law of thermodynamics tells us about the energy balance of any thermodynamic process. The precisely worded statement of the first law given in Chapter 3 allows us to define a quantity known as the internal energy, U, and to recognize that it is a function of state. This is in contrast to two other quantities of energy that are of interest in thermodynamic systems: the heat transferred and the work done. However, all three forms of energy are related by the first law, and we can often use knowledge of two of these quantities in order to determine the third.

While the first law tells us about the energy that is available in thermodynamic processes, it does not rule out some processes which are, in practice, never observed. For example, the first law does not exclude the possibility of a process whose only outcome is to cause heat to flow from a cooler to a hotter body. Yet, such a process is never seen. The second law of thermodynamics addresses the problem of which processes can actually occur.

Our starting point for discussing the second law was to investigate which states can be reached from an initial state by adiabatic processes (which involve no heat transfer into or out of the system). We found that there are always some states that are adiabatically inaccessible from the initial state, and used this fact to motivate the introduction of a function of state called entropy. In order to obtain a quantitative definition of a small entropy change, we relax the restriction to adiabatic processes and use the amount of heat reversibly transferred, divided by the absolute temperature, as a measure of the change in entropy. With this definition of entropy secured, we can state that:

- in any isolated system, the entropy cannot decrease;
- in practice, any naturally occurring process in an isolated system leads to an increase in entropy;
- the entropy of the Universe tends to a maximum.

The last statement is a form of the second law of thermodynamics due to Boltzmann, where the term 'Universe' is used in its technical sense of a system plus its environment. Different, but equivalent, forms of the second law arise in the context of cyclic thermodynamic processes.

'No cyclic process is possible which has as its sole result the complete conversion of heat into work' (Kelvin).

'No cyclic process is possible which has as its sole result the transfer of heat from a colder to a hotter body' (Clausius).

We considered an idealized cyclic process called the Carnot cycle. We found that an engine operating under a Carnot cycle is maximally efficient, but that its efficiency is only $1 - T_c/T_h$, where T_h is the absolute temperature at which heat enters the engine and T_c is the absolute temperature at which heat leaves the engine. Such an engine would only be capable of converting 100% of the heat input into useful work if T_c were equal to absolute zero. It turns out that it is impossible to reach absolute zero, so we can never have a heat engine which operates with perfect efficiency. The fact that we cannot decrease the temperature of a system to absolute zero is actually a separate law — the third law of thermodynamics.

2.5 The meaning of temperature

The concept of temperature has appeared throughout the first three chapters of this book. As our discussions have developed, our understanding of temperature has deepened well beyond our initial ideas, so this is an appropriate point to look back and review the various layers of meaning that have emerged.

To begin with, we introduced temperature as a variable that determines the direction of heat flow: heat flows spontaneously from a body at a higher temperature to a body at a lower temperature, but not in the opposite direction. You will now recognize that this is an informal version of the second law of thermodynamics. The problem of defining a practical temperature scale was approached by finding a physical quantity that depends on temperature, such as the length of a column of mercury. We assumed that this varies linearly with temperature between the normal melting point of ice (defined as 0 °C) and the normal boiling point of water (defined as 100 °C), and so constructed a practical temperature scale.

These definitions were only used as a starting point, but they allowed us to explore the properties of gases. Once this was done, an ideal gas temperature scale could be established, based on the ideal gas equation of state. In the limiting case of low pressures, all gases behave in an ideal way, so this scale is less arbitrary than one based on the expansion of any particular substance, such as mercury. Note, also, that the ideal gas temperature scale measures temperature in kelvin, starting from 0 K at absolute zero. Absolute zero is the lowest conceivable temperature and, in classical physics, corresponds to all particles coming to rest.

Statistical mechanics provides a deeper microscopic understanding of temperature. Here, temperature is recognized as a quantity that determines the distribution of energy among the particles of a system. If we compare two phase cells of different energies, the Boltzmann distribution law shows that the phase cell of lower energy is always more likely to be occupied than the phase cell of higher energy. However, it is the temperature that determines the precise odds. If the temperature is low, the low-energy phase cell is *much* more likely to be occupied than the high-energy phase cell; if the temperature is high, the chances are more evenly balanced. At absolute zero, only the lowest-energy phase cells are occupied (corresponding to all particles being at rest); in the limit of infinite temperature, all phase cells are equally likely. This behaviour is seen in the equilibrium distribution of translational energy among the molecules of an ideal gas (given by the Maxwell–Boltzmann energy distribution function). As the temperature rises, this distribution function becomes broader and flatter, and its peak moves to higher energies. The average translational energy per molecule is proportional to the absolute temperature. It is worth emphasizing the fact that the Boltzmann distribution law and the Maxwell–Boltzmann energy distribution only apply to systems that are in equilibrium. This means that temperature can only be defined for systems that are in equilibrium, or are at least close to equilibrium. Systems that are far from equilibrium simply have no temperature.

A final insight into the meaning of temperature is provided by thermodynamics, where temperature enters into the definition of entropy. In an isothermal reversible process, the change of entropy is the heat transferred, divided by the absolute temperature. Thus, a given quantity of heat transferred produces a greater entropy change at low temperatures than at high temperatures. The efficiency of heat engines

Temperature is related to:

- the direction of heat flow
- the ideal gas equation of state
- the Boltzmann factor
- the definition of entropy
- the efficiency of ideal heat engines.

also depends on the temperatures at which they operate. Kelvin proposed that the efficiency of ideal heat engines could form the basis of a fundamental, if rather impractical, definition of temperature.

In summary, there are many alternative ways of defining and interpreting temperature. The important point is that all these definitions are consistent. While some ways of measuring temperature are more convenient than others, all interpretations are important in widening the significance of temperature, and clarifying its meaning.

2.6 The physics of fluids

By definition, a fluid is a sample of matter that has the *ability* to flow (though it may not actually be flowing in a given case). Fluids include both liquids and gases. Unlike the thermodynamic systems that were of interest in Chapters 2 and 3, interesting fluid systems tend to have properties that vary with position within the system. Often, a physicist working on fluid systems will attempt to understand how certain variables vary with position, for example, how air pressure varies around the wing of an aircraft.

Two key variables that are of interest in studying fluid systems are the pressure and velocity of the flow. Our investigation of fluids began with static fluids, in which the velocity of flow is equal to zero everywhere. We balanced forces due to gravity with forces due to pressure gradients to find out how pressure varies with depth in a lake and with altitude in the atmosphere. The lake was straightforward to analyse because water can be considered to be an incompressible fluid, of constant density. This led to the conclusion that the pressure in a lake is proportional to depth. We cannot assume that the atmosphere has a constant density, but it is not a bad approximation to assume that both temperature and gravity are constant throughout a thin atmosphere. The resulting model, which is called a thin isothermal atmosphere, shows an exponential decrease of pressure with height.

Fluid flow is a complex phenomenon, but progress can be made by introducing another model — that of an ideal fluid, flowing steadily with constant density, no viscosity, and no energy dissipation or eddies. Even in this simplified case, it is not easy to apply Newton's laws directly and we relied instead on the principles of conservation of mass and conservation of energy which lead to the equation of continuity and Bernoulli's equation. The effect of these principles can be illustrated by the flow of a fluid through a pipe of varying cross-section: the equation of continuity shows that the flow is fastest in the narrowest parts of the pipe, and Bernoulli's equation then shows that the pressure is lowest there. Bernoulli's equation is also relevant to the problem of flight. A wing is designed so that the airflow over the upper surface is faster than that over the lower surface, and Bernoulli's equation then shows that the pressure above the wing is lower than below the wing, which means that the wing experiences lift.

Real fluids have viscosity, which leads to the dissipation of energy and the creation of vortices, so their flow is not ideal. Although it is difficult to analyse such flows directly, scaling arguments can be used. By studying flows around scale models, we can find out about flows around full-size objects. Each pattern of flow is characterized by a single dimensionless number, the Reynolds number. If a flow in the scale model has the same Reynolds number as the real flow, then the patterns of the flow will be the same in both cases. More precisely, the flows are identical when expressed in terms of dimensionless scaled variables (scaled by the typical lengths, flow speeds, etc. that apply in each case). This is the principle of dynamic similarity.

A closer look at non-ideal flow around solid objects reveals the existence of a thin boundary layer next to the object. Within the boundary layer, viscous effects dominate and the flow is highly non-ideal. Outside the boundary layer, the flow is almost ideal, except for the fact that vortices and eddies can drift in from the boundary layer. The shedding of vortices is something to be avoided in many applications, since it leads to increased drag.

Many real flows, such as the smoke rising from the tip of a lighted cigarette, show a transition from steady to turbulent flow. The onset of turbulence is characterized by a break-up of the steady flow into a highly irregular and unpredictable pattern. We can predict when turbulence is likely to occur by using the Reynolds number. At low values of the Reynolds number, flows are steady, and as we move to higher values (by, for example, decreasing the viscosity) flows become progressively more complex until at high values of the Reynolds number flows are highly turbulent. However, there is no general theory of turbulence, and this is an area in which our understanding of the physical world is far from complete.

3 Writing precise physical statements

One of the skills that all physicists need is the ability to make precise statements about physical systems and the laws they obey. The language used should be clear and unambiguous and, of course, the meaning should be accurate, and free from exceptions. This is a surprisingly difficult thing to achieve, partly because we all tend to talk loosely in everyday life, and partly because it is easy to forget the details needed to turn a half-truth into a whole truth. This makes it all the more important to spend some time developing this skill. You should regard it as a key communications skill, even more important for physicists than writing extended essays. We will look at some specific examples in a moment, but first we can offer some general advice.

1 Limit the task In unpacking the meaning of a term, you might be tempted to track quantities backwards through many layers of definition. For example, if an exam question asked you to state the ideal gas equation of state, you might be tempted to give subsidiary definitions for pressure, absolute temperature and mole; and the definition of pressure might even lead you to produce a definition of force. In this direction madness lies — but no extra marks! In the context of exams and assignments, you would not be expected to track backwards in this way. If the question had wanted you to define the terms pressure, absolute temperature and mole, it would have said so. In other contexts, as for all communication skills, it all depends on your intended audience. Generally speaking, tracking back should not be necessary unless you think the audience is likely to be ill-prepared for the central statement you wish to make.

2 Define your symbols If you are asked to state a law of physics, such as Newton's law of universal gravitation or Boltzmann's distribution law, it is natural to write down an equation. When you do so, remember to state what your symbols mean. There is no need to worry about standard mathematical symbols, such as π or e — they can be taken as read, but symbols representing physical quantities and constants should always be interpreted in words. This avoids ambiguities: for example, the symbol P is commonly used for momentum, pressure, probability and power; when I was at school, it was even used for force! So you cannot assume that a reader will interpret an equation correctly, unless you define your symbols.

Of course, it helps to use conventional symbols *wherever possible*, but there is a danger in becoming over-reliant on this. For example, what would you do if momentum, pressure, probability and power all appeared in the same question? You would need to adopt a more flexible approach and use symbols other than P for all but one of these quantities. It is important to have an understanding of the subject that is able to cope with such irritations. You will also find that some textbooks use different symbols to the ones chosen for this course, so it is worth recognizing that an equation in physics is more than a collection of frozen symbols: the symbols all represent physical quantities, and it is the *relationship* between the physical quantities that counts.

3 Use mathematical sense to avoid physical nonsense It is worth explicitly checking that any equations you write down are mathematically consistent. For example, if the left-hand side of the equation is a vector quantity, the right-hand side had better be a vector quantity too — and that means they should both include some wavy underlining. Sometimes it is relatively easy to check that the units on both sides of an equation agree. Remember, it makes no sense to add, subtract or equate quantities with different sets of units (energy cannot be added to momentum), and if a function like sin, cos or exp appears in an equation, its argument should be a dimensionless quantity.

4 Include any special restrictions Some statements are only valid in special circumstances, and it is essential to say what those circumstances are. For example, if you were asked to state the law of conservation of momentum, you might be tempted to reply that *the momentum of a system is conserved*, or *the momentum of a system remains constant in time*. But that would not be good enough. The momentum of a car, for example, continually changes as the driver accelerates or brakes, and so is certainly not conserved. In order to make a statement of general validity, you would need a more carefully worded version: *if a system experiences no net external force, then its momentum remains constant in time*.

As your knowledge of physics grows, you may wish to point out that certain statements, regarded as being valid in one field of physics, are not valid in general. For example, some statements made in Newtonian mechanics do not remain valid in special relativity or quantum mechanics. Of course, there would be no harm in pointing this out, but we would not normally expect you to do so. As far as this course is concerned, you can generally assume that the context (e.g. Newtonian mechanics or thermodynamics or statistical mechanics) is understood, and give a definition appropriate for that context.

5 An example is not a definition If you are asked to define something, like chaotic motion for example, it is not sufficient to give an example, or consider a specific case. What is needed is a statement that *defines* chaotic motion and distinguishes it from all other types of motion that are not chaotic. Later in the course, you will meet the idea of an electric field. When asked to define an electric field, it is surprising how many people specify the electric field produced by a single charged particle. This is not a general definition of an electric field, valid under all circumstances, and would merit few marks in an exam.

6 Avoid slogans — relate quantities to systems It is tempting to write down slogans that are easy to remember and which capture part of the truth, but which don't really define anything. For example, Newton's third law could be stated as: *action and reaction are equal and opposite*. Apart from the grammatical oddity (how can two things that are equal also be opposite?), the concepts of action and reaction are left as abstract notions, without any sense that they are forces, acting on objects. A much better statement is: *if object A exerts a force on object B (the action) then object B also exerts a force on object A (the reaction). These forces are equal in magnitude and opposite in direction.*

In general, if you use a word like 'force', it is sensible to ask yourself some questions: force on what, due to what, of what magnitude, and in what direction? If you use a word like 'energy', you should ask yourself: energy of what system? And mention of a system may lead you ask whether a special type of system will be needed — one that is free from external influences, for example.

7 Don't be shy — make clear statements In some fields of physics (especially modern physics), beginners tend to be very tentative in their statements. This is understandable, as some of the ideas encountered seem fantastic, almost incredible. When stating a definition, though, it is not a good idea to let these doubts show through. It does not help the reader if you hedge your bets by using phrases like 'it is almost as if' or 'the apparent effect of this is'. A clear, confident statement is usually preferred. Phrases like 'is a measure of' or 'is related to', convey partial knowledge, but are less clear than specifying a quantity or an equation explicitly. Avoid such vagueness if you can.

8 Read through your statements It is always worth reading through your statements, once they have been prepared, to check that there are no obvious ambiguities, loopholes or omissions.

You may feel that the best approach to making definitions would be to learn every law, definition and principle by heart. This would allow you to quote statements in a form that you know is correct, assuming you can trust your source. However, by itself, learning by rote is a pointless exercise. It is unlikely to be successful because you are almost bound to forget some details unless you have a deeper understanding, and it will not help you to construct new definitions about unfamiliar situations. In practice, you will probably proceed partly from memory, but reinforced by the advice given above.

Making definitions is a skill which different people approach in different ways, so it is worth spending a moment or two reflecting on how you actually go about the process yourself. As a concrete example, try answering the following example, and as you do so, try to think and make brief notes about the stages that you have had to go through. You might wish to review Section 2 of Chapter 3 before doing this, but don't be tempted to copy definitions out of the text.

Example 5.1

Describe what is meant by the terms 'system', 'adiabatic system' and 'closed system' as they are used in thermodynamics.

Solution

Here is my attempt, along with my thoughts on how I approached this question.

I can remember that a 'system' in thermodynamics means the region that we are interested in at the present moment. Perhaps a statement such as the following would suffice.

'A system is the region that is of immediate interest in a thermodynamic analysis.'

I thought that it might be useful to add the words 'in a thermodynamic analysis' just to cover myself against other types of system. As for closed and adiabatic systems, I can remember that a closed system is one in which no material enters or leaves the system, and so the mass of the system must be constant. An adiabatic system is one in which no heat enters or leaves the system. So the following might be appropriate:

- an adiabatic system is one in which there is no heat transfer;
- a closed system is one in which the mass is constant.

Casting a critical eye over these two sentences, I can see some problems. The statement about adiabatic systems is ambiguous because I haven't said where heat is being transferred to or from. The statement could be read as meaning there is no heat transfer within the system, but this is not what is meant by an adiabatic system — there *can* be heat transfer within an adiabatic system, but there *cannot* be any heat transfer between the system and its environment.

In the statement about closed systems, I have fallen into a trap of assuming that a constant mass means that no matter is entering or leaving the system. We *can* have a system of constant mass where matter is being transferred, provided that the rate of flow into the system is the same as the rate of flow out of the system, but this would not be a closed system. So there is an error in this statement that needs to be rectified.

I can address both of these problems by rewriting my statements as:

- an adiabatic system is one that does not exchange any heat with its environment;
- a closed system is one that does not exchange any matter with its environment.

Casting a critical eye over this, I now see that I have now introduced the term 'environment' which is clearly linked to the concept of system. It might be prudent to revise my first statement so that this relationship is made clear:

'A system is the part of the world that is of immediate interest in a thermodynamic analysis. It is distinct from, but may interact with, the environment. The environment is the rest of the world apart from the system, and is the region that is not of immediate interest.'

Your final answer should be similar to this but, of course, the process by which you got to your answer may very well be different to mine. It is important to review what you have written until you are satisfied that it passes the tests of being unambiguous, accurate and free from loopholes. You will find questions in Section 4 to help you develop your skills in framing clear definitions and statements. First, here are some more examples that illustrate good practice.

● State Newton's law of universal gravitation.

❍ Newton's law of universal gravitation states that the gravitational force on any particle, A, due to any other particle, B, points from A towards B and has magnitude:

$$F = \frac{Gm_{\mathrm{A}}m_{\mathrm{B}}}{d^2}$$

where m_{A} is the mass of particle A, m_{B} is the mass of particle B, d is the distance between the particles and G is the universal constant of gravitation. ■

Comment: *I have been careful to identify all the terms that appear in the equation and to give both the direction and magnitude of the force. I prefer to talk about the force on a given particle, rather than using the less-focused phrase 'force between particles'. It would be fine to specify the force by a vector equation, but this is not essential, so long as both the magnitude and direction of the force are defined.*

- State Boltzmann's distribution law in classical statistical mechanics.

❍ Boltzmann's distribution law states that the probability p of finding a given particle in a given phase cell is

$$p = A \exp(-E/kT)$$

where E is the energy of the particle in the given phase cell, k is Boltzmann's constant, T is the absolute temperature and A is a constant, independent of the phase cell. The system is assumed to be in equilibrium. ■

In certain cases, depending on your audience, you might decide that it would be helpful to explain what a phase cell is. If so, you could add: a phase cell is a specification of the position and velocity of a particle, to within a chosen precision.

Comment: *Boltzmann's distribution law has to be stated carefully. It is important to make it clear that the probability is that for finding a given particle in a given phase cell when the system is in a state of equilibrium. Following the advice given earlier, I decided it was inessential to give subsidiary definitions for probability, Boltzmann's constant, absolute temperature or phase cell (see also the marginal note above).*

- In what way does Bernoulli's principle explain the lift experienced by a wing in level flight?

❍ Bernoulli's principle applies to ideal flows in which gravity plays a negligible role. It states that the pressure in an ideal flow is lower where the speed of flow is higher, and *vice versa*. A wing is designed so that the air flowing over the upper surface has a greater speed (relative to the wing) than the air flowing over the lower surface. The pressure just above the wing is therefore lower than the pressure just below the wing. This pressure difference across the wing produces an upwards force known as lift. ■

Comment: *It seemed sensible to begin with a statement of Bernoulli's principle. While it would do no harm to quote Bernoulli's equation, this is not essential because Bernoulli's principle itself involves no equations (see Chapter 4). The remaining argument links lift to a pressure difference across the wings accounted for by Bernoulli's principle.*

4 Basic skills and knowledge test

The following questions should help in consolidating your understanding of the concepts and skills introduced in this book. Questions 5.1 to 5.3 are designed to give practice in interpreting and writing precise physical statements. Questions 5.4 to 5.6 are about the mathematics of probability. The remaining questions cover the physical concepts introduced in this book. Some questions are longer and more involved than others, and in particular, it is recommended that Questions 5.28 to 5.30 are solved using the problem-solving technique that was introduced in *Predicting motion*.

Question 5.1 Identify any errors and omissions in the following statements.

(a) *The first law of thermodynamics.* The sum of the heat transferred and the work done depends on the initial and final states of the system, but not on the process by which the change is brought about.

(b) *A reversible process.* A process from state A to state B is reversible if it is possible to find another process that takes the system from state B back to state A.

(c) *The second law of thermodynamics (Clausius's statement).* No cyclic process is possible which results in the transfer of a quantity of heat from a cooler body to a hotter one.

Question 5.2 Give concise descriptions of what is meant by the following terms. You may want to refer back in the book to refresh your memory, but attempt to write descriptions in your own words.

(a) The radial density function.

(b) The fractional frequency of an outcome in a random process.

(c) Equilibrium in a thermodynamic system.

(d) A quasi-static process in a thermodynamic system.

(e) The principle of dynamic similarity.

Question 5.3 Without referring back in the book, attempt to give the best definitions you can for items (a)–(d) below. Look over what you have written and reflect on whether it is clear, unambiguous and free from loopholes.

(a) Avogadro's hypothesis.

(b) Joule's law of ideal gases.

(c) The third law of thermodynamics.

(d) Archimedes' principle.

Critically compare your definitions with those given in the answer to this question. The words need not correspond exactly, but you should consider whether the meanings of your statements are identical to those given in the answer. Make a note of any significant differences and, without looking again at the definitions given in the answer, make any necessary revisions to your definitions for (a)–(d).

Question 5.4 A game of chance involves guessing which one of four light bulbs will be illuminated on pressing a switch. The game has four differently coloured bulbs: red, yellow, green and blue which light up at random. The probability of a given bulb being lit in a single attempt is different for each bulb, and the probabilities for the red, yellow and green bulbs are 0.20, 0.24 and 0.30 respectively.

(a) What is the probability of the blue bulb being lit in a single attempt?

(b) What is the probability of the same bulb (of any colour) being lit in two given plays of the game?

(c) What is the probability of the green bulb being lit up on each of four given plays of the game?

Question 5.5 A seven-sided top has equal probabilities of landing on any one of its sides. On spinning the top and letting it come to rest, a number n is obtained from the side on which the top lands. The sides are labelled −3, −2, −1, 0, 1, 2, 3. Find the value of $\langle n \rangle$, $\langle n \rangle^2$ and $\langle n^2 \rangle$.

Question 5.6 Five dice are subjected to tests to see if they are fair. Table 5.1 shows the total number of times that each die was rolled, along with the frequency of occurrence of the six possible scores. Would you conclude that any of the dice are not fair?

Table 5.1 Data for Question 5.6.

Die	Total no. of throws	No. of throws which produce the given score					
		1	2	3	4	5	6
A	150	22	30	24	25	23	26
B	1200	206	214	202	189	191	198
C	2400	415	393	419	398	380	395
D	4800	786	819	1093	612	780	710
E	9600	1629	1640	1601	1563	1597	1570

Question 5.7 The charge on the nucleus of an oxygen atom is $+1.28 \times 10^{-18}$ C. Use this information to determine the number of protons in a nucleus of oxygen. Explain why some oxygen atoms have a mass of 16 amu but others are found (rarely) with masses of 17 amu and 18 amu.

Question 5.8 A pure sample of 0.040 mol of a pure molecular substance has a mass of 0.0288 kg. Calculate:

(a) the relative molecular mass of the substance;

(b) the number of molecules that would be contained in a pure sample of the substance of mass 0.100 kg.

Question 5.9 Estimate the typical pressure exerted on the ground by a human foot. Assume the person is standing still, wearing flat shoes and has both feet flat on the ground.

Question 5.10 A cubic metre of material in an interstellar cloud contains 10^{11} molecules. Assuming that the cloud is purely molecular hydrogen, with a relative molecular mass of 2.0, calculate the density of the cloud in SI units. Assuming that the cloud is spherical and homogeneous and given that it has total mass of 2×10^{32} kg, what is the radius of the cloud?

Question 5.11 Calculate the volume of a sample of 1.00 mol of an ideal gas at the following temperatures and pressures: (a) $T = 273$ K, $P = 1.00 \times 10^5$ Pa; (b) $T = 300$ K, $P = 7.32 \times 10^4$ Pa; (c) $T = 373$ K, $P = 9.10 \times 10^4$ Pa.

Question 5.12 Calculate the mean translational energy per molecule in an ideal gas at the following temperatures: (a) 250 K; (b) 500 K; (c) 750 K.

Question 5.13 Calculate the number of molecules in a sample of ideal gas which has a volume of $0.070\,\text{m}^3$, a pressure of 2.6×10^5 Pa and a temperature of 320 K.

Question 5.14 How would you argue against the following statement? 'The use of the ideal gas law as a basis for a practical temperature scale is as arbitrary as using the length of a mercury column in a mercury-in-glass thermometer.'

Question 5.15 Several assumptions are made in developing the simple gas model; in particular it is assumed that collisions between molecules and the walls of the vessel that contains the gas are elastic. Briefly discuss why this assumption is made and whether or not it is a valid assumption to make for ideal gases.

Question 5.16 Sketch a graph which shows the variation with time of the magnitude of the force acting on one wall of a cubical container (with sides of length L) which arises from the collision of a single molecule which is initially travelling with a velocity v in a direction perpendicular to the wall. Show the variation over the time taken for the molecule to make at least two collisions. On the same graph, show the variation with time of the magnitude of the force acting on the wall due to a molecule of the same type, with an initial velocity which is again perpendicular to the wall, but which has a greater magnitude.

Question 5.17 Two phase cells in a gas in equilibrium are labelled A and B. The cells are such that the probability of finding a given molecule in cell A is twice the probability of finding the molecule in cell B. Find an expression for the difference in energy (ΔE) between the energy of cell B and that of cell A in terms of the temperature T of the gas.

Question 5.18 Without giving mathematical details, explain why the distribution function for molecular energies in an ideal gas is not given simply by Boltzmann's distribution law.

Question 5.19 A fixed amount of heat is supplied to a sample of an unknown ideal gas, initially at a temperature of 290 K. The gas is in a cylinder with a fixed piston, so that its volume is held constant, and it is found that the temperature rises to 406 K. The experiment is then repeated with the same initial conditions, but this time the piston is free to move so that the pressure remains constant throughout. In this case, the final temperature is found to be 373 K. Is this gas monatomic or diatomic? Explain your answer.

Question 5.20 A mass of 0.200 kg of an unknown metal, X, is heated to 373.0 K and is then dropped into 0.400 kg of water which is at a temperature of 295.6 K. If the final temperature of the water and metal is 299.9 K and the specific heat of water is 4200 J kg^{-1}, calculate the specific heat capacity of the unknown metal.

Question 5.21 Which of the following processes are reversible (according to strict thermodynamic usage) and which can be reversed?

(a) The transfer of energy by compression of an ideal spring, free from friction and any other dissipative effects.

(b) The transfer of energy to a real flywheel, initially at rest.

(c) The transfer of energy by the quasi-static compression of an ideal gas under ideal conditions, free from friction and other dissipative effects.

(d) The transfer of energy by the stirring of a viscous fluid.

Question 5.22 An idealized thermodynamic system consists of a fixed mass of ideal gas in a cylinder with a piston. Work may be done on the gas in two ways: either reversibly or irreversibly. The way in which work may be done reversibly is by means of a slow compression of the gas using the piston. The way in which work may be done irreversibly is by stirring the gas using a paddle. The initial state of the gas is a pressure P_A and volume V_A, and the pressure of the medium surrounding the cylinder is also P_A. There is no flow of heat to or from the gas.

(a) Sketch a schematic P–V diagram and indicate on it the other states that may be reached by a reversible change from the initial state.

(b) Indicate the region of the P–V diagram that can be reached by irreversible changes.

(c) Indicate the region of the P–V diagram which cannot be reached by either type of change.

(d) Whose formulation of which law of thermodynamics is (c) an example of?

(e) How would the system have to be modified to allow the inaccessible region of the P–V diagram to be reached?

Question 5.23 Calculate the entropy changes associated with the melting of 1 mol of lithium (Li) and 1 mol of mercury (Hg) at their normal melting points.

Table 5.2 Data for Question 5.23.

Metal	Relative atomic mass	Melting point/K	Specific latent heat of melting/J kg^{-1}
Li	6.9	452	6.667×10^5
Hg	200.6	234	1.162×10^4

Question 5.24 (a) Use Equation 3.30b to show that the change in entropy of a fixed mass of ideal gas in which the pressure is increased at constant volume is given by

$$\Delta S = n\,C_{V,\mathrm{m}} \log_e\left(\frac{P_2}{P_1}\right).$$

(b) Hence calculate the change in entropy if 1 mol of a typical diatomic ideal gas undergoes a change in which its pressure is halved at constant volume. Is the entropy of the final state higher or lower than the initial state?

Question 5.25 You see an advert for a car which says that this model has a revolutionary new petrol engine which is ‘80% efficient’. You are intrigued by this, and further find out that the maximum temperature during the operating cycle of the engine is about 700 °C. Discuss whether the advertiser’s claim is plausible.

Question 5.26 A balloon for scientific research has a volume of 400 000 m^3. The balloon floats at a very high altitude where the mean air density is 1.4×10^{-2} kg m^{-3}. The balloon is filled with the monatomic gas helium (which has a relative atomic mass of 4.0) at a pressure of 1.0×10^3 Pa (corresponding to the atmospheric pressure at this altitude) and a temperature of 250 K. If no unbalanced force acts on the balloon, what is the mass of its payload? (You may ignore the mass of the skin of the balloon.)

Question 5.27 A single-engine aircraft has a typical flying speed of $150\,\mathrm{km\,h^{-1}}$. This type of aircraft is also the basis of a radio-controlled model, which is constructed at a scale of 1 : 12. The typical flying speed of the model is $50\,\mathrm{km\,h^{-1}}$. At these speeds, would you expect the airflow around the real aircraft and the scale model to be dynamically similar?

The following questions should be attempted using the problem-solving technique that was introduced in Predicting motion.

Question 5.28 A piston contains 0.05 mol of a monatomic ideal gas at a temperature of 500 K. A process occurs in which the temperature rises to 700 K. Calculate the heat transferred to the gas and the work done by the gas if (a) the volume remains constant during this change, and (b) the pressure remains constant during this change. Attempt to solve this question using the first law of thermodynamics, *without* quoting the results for the molar specific heats of ideal gases.

Question 5.29 A small building has a rectangular flat roof of length 6 m and width 5 m. The mass of the roof is 600 kg. The building has been designed so that the roof can withstand an upward force of 8 kN. Discuss whether this roof would be safe in the event of a wind speed of $100\,\mathrm{km\,h^{-1}}$. Take the density of air to be $1.2\,\mathrm{kg\,m^{-3}}$.

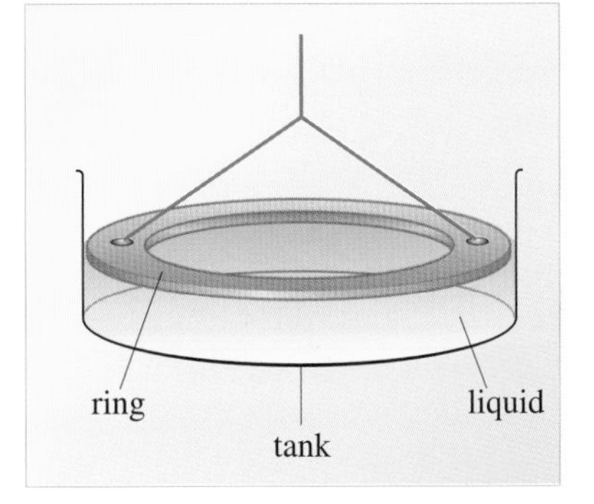

Figure 5.1 The apparatus used to measure viscosity as described in Question 5.30. (Not drawn to scale.)

Question 5.30 An experiment to measure viscosity consists of the apparatus shown in Figure 5.1. An annular ring with inside and outside radii of 98 cm and 102 cm respectively can be rotated in an annular tank. The tank itself is fixed, and contains a liquid which has a depth of 4.0 mm. The lower face of the ring is in contact with the upper surface of the liquid in the tank. It is found that in order to rotate the ring at a constant angular speed of $0.1\,\mathrm{rad\,s^{-1}}$, it is necessary to apply a torque of magnitude $0.63\,\mathrm{N\,m}$ to the ring. Calculate the viscosity of the liquid in the tank. (You may assume that the viscous forces act only between the lower face of the ring and the bottom of the tank. Neglect any effects due to surface tension.) ■

5 Interactive questions

Open University students should leave the text at this point and use the interactive questions package for *Classical physics of matter*. The interactive questions package includes a random number feature that alters the values used in many of the questions each time those questions are accessed. This means that if you try the questions again, as part of your end-of-course revision for instance, you will find that many of them will have changed, at least in their numerical content. When you have completed a selection of these questions, you should return to the text. You should not spend more than 2 hours on this package.

6 *Physica* problems

Open University students should leave the text at this point and tackle a selection of *Physica* problems that relate to *Classical physics of matter*. You should not spend more than 2 hours on this package.

Answers and comments

Q1.1 The ion has speed

$$v = 0.6\,\text{m}/3.0 \times 10^{-3}\,\text{s} = 2.0 \times 10^{2}\,\text{m}\,\text{s}^{-1}.$$

The (translational) kinetic energy of the ion is

$$E = \tfrac{1}{2}mv^2$$

so the mass of the ion is

$$m = \frac{2E}{v^2} = \frac{2 \times 2.0 \times 10^{-19}\,\text{J}}{(2.0 \times 10^{2}\,\text{m}\,\text{s}^{-1})^2} = 1.0 \times 10^{-23}\,\text{kg}.$$

The momentum of the ion has magnitude

$$mv = (1.0 \times 10^{-23}\,\text{kg}) \times (2.0 \times 10^{2}\,\text{m}\,\text{s}^{-1}) = 2.0 \times 10^{-21}\,\text{kg}\,\text{m}\,\text{s}^{-1}.$$

Q1.2 Relative to the fixed centre of mass, atom B has position

$$x = \frac{m_A}{m_A + m_B}r$$

and equilibrium position

$$x_0 = \frac{m_A}{m_A + m_B}r_0$$

so the force acting on this atom is

$$F_r = -C(r - r_0) = -C\frac{(m_A + m_B)}{m_A}(x - x_0).$$

Thus, the effective force constant for atom B is $k = (m_A + m_B)C/m_A$, and the frequency of simple harmonic vibration is

$$f = \frac{1}{2\pi}\sqrt{\frac{k}{m_B}} = \frac{1}{2\pi}\sqrt{\frac{C(m_A + m_B)}{m_A m_B}}.$$

(Note, for atom A, the effective force constant is $(m_A + m_B)C/m_B$, leading to the frequency

$$f = \frac{1}{2\pi}\sqrt{\frac{C(m_A + m_B)}{m_B m_A}}$$

which is the same as for atom B. Not surprisingly, both atoms oscillate about the centre of mass with the same frequency.)

Q1.3 By direct measurement, the gradient of the graph in Figure 1.9 is roughly $-1.6 \times 10^{2}\,\text{N}\,\text{m}^{-1}$ so the force acting on an atom can be written as

$$F_r = -C\,(r - r_0)$$

where $C = 1.6 \times 10^{2}\,\text{N}\,\text{m}^{-1}$.

Since the masses of the atoms are $m_A = 2.2 \times 10^{-25}\,\text{kg}$ and $m_B = 8.8 \times 10^{-26}\,\text{kg}$ and $m_A + m_B = 3.08 \times 10^{-25}\,\text{kg}$, the formula derived in Question 1.2 gives

$$f = \frac{1}{2\pi}\sqrt{\frac{C(m_A + m_B)}{m_B m_A}} = \frac{1}{2\pi}\sqrt{\frac{1.6 \times 10^{2}\,\text{N}\,\text{m}^{-1} \times 3.08 \times 10^{-25}\,\text{kg}}{2.2 \times 10^{-25}\,\text{kg} \times 8.8 \times 10^{-26}\,\text{kg}}} = 8.0 \times 10^{12}\,\text{Hz}.$$

Q1.4 When the distance between the atoms exceeds the equilibrium separation, r_0, they attract one another. The maximum attractive force corresponds to the lowest point in the force–separation curve shown in Figure 1.9. Taking a reading directly from this graph, the maximum attractive force has magnitude

$$|F_r| = 5.0 \times 10^{-9}\,\text{N}.$$

This is the minimum force needed to overcome the attraction between the atoms and split the molecule into separate atoms.

The energy needed to split the molecule apart is the binding energy, E_b, given by the magnitude of the lowest point in the potential energy–separation curve shown in Figure 1.11. Directly from this graph, this energy is

$$E_b = 1.4 \times 10^{-18}\,\text{J}.$$

Q1.5 The force exerted on the floor is the weight of the woman, and its magnitude is:

$$F = mg = 56\,\text{kg} \times 9.8\,\text{m}\,\text{s}^{-2} = 5.5 \times 10^{2}\,\text{N}.$$

This force acts vertically downwards on a horizontal area

$$A = 4.9 \times 10^{-5}\,\text{m}^2.$$

The pressure is therefore

$$P = \frac{F}{A} = 1.1 \times 10^{7}\,\text{N}\,\text{m}^{-2} = 1.1 \times 10^{7}\,\text{Pa}.$$

This is about 100 times greater than normal atmospheric pressure.

Q1.6 From Equation 1.5, the temperature is estimated to be

$$\theta = 0\,°\text{C} + \left(\frac{4.2 - 3.3}{6.9 - 3.3}\right) \times 100\,°\text{C} = 25\,°\text{C}.$$

Q1.7 Let V_0 be the initial volume of the bar before it is heated, and V_1 be the expanded volume of the bar after heating. Then

$$V_1 - V_0 = \alpha V \Delta\theta.$$

During the subsequent compression, the change in volume is

$$V_0 - V_1 = -\beta V \Delta P.$$

The volume V on the right-hand sides of these equations is a representative volume of the block during the expansion or compression: it can be taken to be V_0 or V_1, or an average of the two. (For a small change in volume, the precise choice makes practically no difference.)

Adding the above two equations then gives

$$0 = \alpha V \Delta\theta - \beta V \Delta P$$

so

$$\Delta P = \frac{\alpha\Delta\theta}{\beta} = \frac{3.6\times10^{-6}\,°\text{C}^{-1}\times5\,°\text{C}}{3.0\times10^{-11}\,\text{Pa}^{-1}} = 6\times10^{5}\,\text{Pa}.$$

Q1.8 Under normal conditions, the pressure is about 10^5 Pa (see Figure 1.30) and the temperature is about 295 K. The equation of state for an ideal gas (Equation 1.11) then gives

$$V = \frac{nRT}{P} = \frac{1\,\text{mol}\times8.314\,\text{J K}^{-1}\,\text{mol}^{-1}\times295\,\text{K}}{10^5\,\text{Pa}} = 2.45\times10^{-2}\,\text{m}^3.$$

This is approximately one cubic foot ($2.9\times10^{-2}\,\text{m}^3$).

Q1.9 (a) The number of moles of gas is

$$n = \frac{PV}{RT} = \frac{1.00\times10^5\,\text{Pa}\times5.00\,\text{m}^3}{8.314\,\text{J K}^{-1}\,\text{mol}^{-1}\times300\,\text{K}} = 200\,\text{mol}.$$

(b) The number of molecules in the gas is

$$200\,\text{mol}\times6.022\times10^{23}\,\text{mol}^{-1} = 1.20\times10^{26}.$$

(c) The total mass of the gas is

$$M = 200\,\text{mol}\times(2.02\times10^{-3}\,\text{kg mol}^{-1}) = 0.404\,\text{kg}$$

so the density of the gas is

$$\rho = \frac{M}{V} = \frac{0.404\,\text{kg}}{5.00\,\text{m}^3} = 0.0808\,\text{kg m}^{-3}.$$

Q1.10 1 mol of oxygen gas has a mass of 32×10^{-3} kg, so 1.6 kg of oxygen corresponds to

$$n = 1.6/(32\times10^{-3})\,\text{mol} = 50\,\text{mol}.$$

Oxygen is diatomic so Equation 1.16a gives

$$\Delta U = \tfrac{5}{2}nR\,\Delta T = \tfrac{5}{2}\times50\,\text{mol}\times8.314\,\text{J K}^{-1}\,\text{mol}^{-1}\times100\,\text{K} = 1.0\times10^5\,\text{J}.$$

Q1.11 1 mol of helium gas has mass 4×10^{-3} kg so 0.4 kg of helium corresponds to $n = 0.4/4\times10^{-3}$ mol = 100 mol. The internal energy of a monatomic gas is $U = \tfrac{3}{2}nRT$ so the increase in temperature is

$$\Delta T = \frac{2}{3}\frac{\Delta U}{nR} = \frac{2}{3}\times\frac{6.5\times10^3}{100\times8.314}\,\text{K} = 5.2\,\text{K}.$$

The initial temperature is 273 K so the final temperature of the gas is 278.2 K.

The final volume of the helium is found from the ideal gas equation of state. Let the initial volume and temperature be V_1 and T_1, and the final volume and temperature be V_2 and T_2. Then

$$\frac{V_1}{T_1} = \frac{nR}{P} = \frac{V_2}{T_2}$$

so the final volume of the gas is

$$V_2 = \frac{T_2}{T_1}V_1 = \frac{278.2}{273}\times3.2\times10^{-3}\,\text{m}^3 = 3.3\times10^{-3}\,\text{m}^3.$$

Q1.12 At point B_1, the system is in the gas phase. As the system is compressed, the volume decreases and the pressure rises. When the point B_2 is reached, solid starts to appear, and the system becomes a mixture of two phases — solid and gas. Further compression can be brought about without an increase in pressure and steadily converts the gas into solid, until at point B_3 the system is entirely in the solid phase. In the solid phase, compression is much more difficult, so the pressure increases very substantially as we reduce the volume a little more to reach point B_4. (The direct conversion of a solid to a gas is known as *sublimation*. It is less familiar than melting or solidification, but can be seen in the laboratory when solid carbon dioxide, also known as 'dry ice', is converted directly into gaseous carbon dioxide.)

Q1.13 (a) In the gas phase the radial density function is almost constant, except at very small distances, where it rises very slightly, and then drops sharply to zero, indicating a strong repulsion between molecules that are squeezed closely together. In the liquid phase there are oscillations in the radial density function extending over a few molecular diameters. These reveal transient short-range order in the liquid. In the crystalline solid phase the radial density function is a series of sharp peaks, extending over many thousands of atoms, indicating a regularly ordered structure with long-range order.

(b) In the gas phase the molecules move freely apart from occasional collisions with one another, so they have random, zigzag paths. In a liquid, a molecule spends much of its time jogging to and fro in the vicinity of its nearest neighbours, but occasionally leaves this group and migrates to another set of neighbouring molecules. In a crystalline solid, atoms oscillate in the vicinity of regularly ordered fixed equilibrium positions.

Q1.14 The ideal gas equation of state can be used in the form

$$N = \frac{PV}{kT} = \frac{10^5\,\text{Pa}\times 30\,\text{m}^3}{1.381\times 10^{-23}\,\text{J K}^{-1}\times 295\,\text{K}}$$
$$= 7.4\times 10^{26}\ \text{molecules.}$$

Only 20% of these molecules are oxygen molecules. Since oxygen is diatomic, each molecule contains two atoms, so the total number of oxygen atoms in the air in the room is

$$2\times 0.2\times 7.4\times 10^{26} = 3.0\times 10^{26}\ \text{oxygen atoms.}$$

Q1.15 Let the *total* summer pressure in the tyre be P_s and the *total* winter pressure be P_w. Then since no gas escapes and the volume of the tyre is assumed to be constant,

$$\frac{P_s}{T_w} = \frac{nR}{V} = \frac{P_w}{T_w}.$$

So, $$P_w = \frac{T_w}{T_s}\times P_s.$$

Combining the summer overpressure with atmospheric pressure gives $P_s = 2.9\times 10^5$ Pa, so

$$P_w = \frac{273-20}{273+30}\times 2.9\times 10^5\,\text{Pa} = 2.4\times 10^5\,\text{Pa}.$$

The winter overpressure in the tyre is therefore

$$2.4\times 10^5\,\text{Pa} - 1.0\times 10^5\,\text{Pa} = 1.4\times 10^5\,\text{Pa}.$$

Q1.16 Let the initial pressure, temperature and volume of the gas be P_1, T_1 and V_1, and the final pressure, temperature and volume of the gas be P_2, T_2 and V_2. Applying the ideal gas equation of state to both the initial and final states gives

$$\frac{P_1V_1}{T_1} = nR = \frac{P_2V_2}{T_2}.$$

The volume of the gas doubles, so $V_2 = 2V_1$ and the pressure is constant so $P_2 = P_1$. Thus, the final temperature of the gas is

$$T_2 = \frac{P_2}{P_1}\times\frac{V_2}{V_1}\times T_1 = \frac{P_1}{P_1}\times\frac{2V_1}{V_1}\times T_1 = 2T_1$$

and the increase in temperature is

$$\Delta T = T_2 - T_1 = 2T_1 - T_1 = T_1 = 295\,\text{K}.$$

One mole of nitrogen gas has a mass of 28×10^{-3} kg, so 1.5 kg of nitrogen gas is $1.5\,\text{kg}/(28\times 10^{-3}\,\text{kg mol}^{-1}) =$ 53.6 mol.

Nitrogen is diatomic so its internal energy increases by

$$\Delta U = \tfrac{5}{2}nR\,\Delta T$$
$$= \tfrac{5}{2}\times 53.6\,\text{mol}\times 8.314\,\text{J K}^{-1}\,\text{mol}^{-1}\times 295\,\text{K}$$
$$= 3.3\times 10^5\,\text{J}.$$

Q1.17 (a) Using the equation of state of an ideal gas, we find

$$T = \frac{PV}{nR} = \frac{1.8\times 10^5\,\text{Pa}\times 0.25\,\text{m}^3}{15\,\text{mol}\times 8.314\,\text{J K}^{-1}\,\text{mol}^{-1}} = 360\,\text{K}.$$

(b) The average molecular kinetic energy is the total internal energy of the gas, divided by the number of molecules. Since the gas is monatomic,

$$\frac{U}{N} = \tfrac{3}{2}kT = 1.5\times 1.381\times 10^{-23}\,\text{J K}^{-1}\times 360\,\text{K}$$
$$= 7.5\times 10^{-21}\,\text{J}.$$

Q1.18 (a) At point B_4, the system is in the solid phase. As the temperature is increased at constant pressure, the volume increases. The solid then starts to melt and is a mixture of solid and liquid. As the solid melts, its temperature remains fixed, but it expands and it also absorbs latent heat of melting. Eventually, all the solid has been converted to liquid, and the liquid expands as the temperature rises; this brings the system to A_4.

(b) At any given point on the constant pressure path in the solid phase, we can read off the volume, V, of the system, and note the change in volume, ΔV, that accompanies a small increase in temperature, ΔT. The constant pressure expansivity is then given by

$$\alpha = \frac{1}{V}\frac{\Delta V}{\Delta T}.$$

Q2.1 Table 2.1 showed that the probabilities of recording 1, 2, 3, 4, 5 and 6 are 21/56, 15/56, 10/56, 6/56, 3/56 and 1/56 respectively. The predicted average value is found by multiplying the possible values by their probabilities and adding the results together. This gives

$$\begin{aligned}\langle n\rangle &= \tfrac{21}{56}\times 1+\tfrac{15}{56}\times 2+\tfrac{10}{56}\times 3+\tfrac{6}{56}\times 4+\tfrac{3}{56}\times 5+\tfrac{1}{56}\times 6\\ &= \tfrac{126}{56}\\ &= 2.25.\end{aligned}$$

This result makes good sense. We are restricting the total score on the four dice to be 9, so it is not surprising that the average score on each die is 9/4 = 2.25.

Q2.2 (a) Each required keystroke has a probability of 1/27 and occurs independently of the other required keystrokes. So, applying the multiplication rule for probabilities, the probability of the sentence being typed is

$$\left(\tfrac{1}{27}\right)^{39} = 1.5\times 10^{-56}.$$

(b) A year contains $365 \times 24 \times 60 \times 60\,\text{s} \approx 3 \times 10^7\,\text{s}$. Thus, each monkey can make about 3×10^6 attempts in a year. The total number of attempts of all the billion monkeys throughout the entire age of the Universe would be

$$3 \times 10^6 \times 10^9 \times 1.5 \times 10^{10} = 4.5 \times 10^{25}.$$

This is far short of the number needed to give them a reasonable chance of success. For that, they would need something like 10^{56} attempts. Even if the number of monkeys were increased by a factor of a billion, they somehow speeded up by a factor of a billion, and they carried on for a billion times the current age of the Universe, they would still have only an outside chance of succeeding.

Comment: *This type of calculation leads to the conclusion that some things are so unlikely as to be 'effectively impossible'. This is not to say they are logically impossible, or forbidden from happening. They could happen, but it is safe to assume that no-one will ever see them take place. Many probabilities associated with large collections of atoms are of this type, and this allows us to use probability to predict results with practical certainty.*

Q2.3 The nitrogen molecule behaves as a projectile, initially moving horizontally. The horizontal displacement is

$$s_x = u_x t$$

so the time taken is

$$t = \frac{s_x}{u_x} = \frac{0.5\,\text{m}}{500\,\text{m}\,\text{s}^{-1}} = 1.0\times 10^{-3}\,\text{s}.$$

With the y-axis pointing vertically downwards, the vertical displacement is

$$\begin{aligned}s_y &= \tfrac{1}{2}gt^2 = \tfrac{1}{2}\times 9.8\,\text{m}\,\text{s}^{-2}\times(1.0\times 10^{-3}\,\text{s})^2\\ &= 5\times 10^{-6}\,\text{m}.\end{aligned}$$

This is scarcely noticeable, and the path would appear straight in this case. So gravity can be neglected.

At a pressure of one atmosphere, there are very frequent molecule–molecule collisions, giving very much shorter times between collisions. Since vertical displacement is proportional to t^2, it is even more true that we can neglect gravity in this case. Assumption 3 is well justified.

Q2.4 The simple gas model gives

$$PV = \tfrac{2}{3}N\langle E_{\text{trans}}\rangle.$$

The volume of the air in the room is $3\,\text{m} \times 4\,\text{m} \times 2.5\,\text{m} = 30\,\text{m}^3$, so the total translational energy of the molecules in the air of the room is

$$\begin{aligned}N\langle E_{\text{trans}}\rangle &= \tfrac{3}{2}PV = \tfrac{3}{2}\times 1\times 10^5\,\text{Pa}\times 30\,\text{m}^3\\ &= 4.5\times 10^6\,\text{J}.\end{aligned}$$

Q2.5 Let the drum have radius r and period of rotation T. Then the speed of a point on the rim of the drum is $2\pi r/T$. The darkest point on the film is a distance d around the drum away from the zero mark. It is made by molecules moving at the most probable speed, which have a transit time of

$$t = \frac{d\times T}{2\pi r}.$$

During this time, the molecules travel at speed v_{mp} for a distance $2r$ across the drum, so

$$\begin{aligned}v_{\text{mp}} &= \frac{2r}{t} = \frac{4\pi r^2}{d\times T} = \frac{4\pi\times(0.15\,\text{m})^2}{95.5\times 10^{-3}\,\text{m}\times(1/100)\,\text{s}}\\ &= 300\,\text{m}\,\text{s}^{-1}.\end{aligned}$$

This is the speed at which the bismuth molecules most commonly move.

Q2.6 Your answer should include most of the following points. The principal *similarities* are:

1. There is a single peak in both distributions at a most probable value.
2. There are very few molecules at the low and high ends of the range.
3. As the temperature rises, the peak shifts to a higher value on the horizontal axis, and becomes lower.
4. As the temperature rises, the range covered by the distribution spreads out to higher values.

The principal *differences* are:

1. Energy distributions do not depend on which gas is considered, only on temperature.
2. If molecular mass is reduced, the speed distribution changes as if temperature had risen.
3. The energy distribution is more asymmetric than the speed distribution.
4. Energy distributions rise steeply at zero energy, where speed distributions are much flatter.

Q2.7 (a) The area under the graph of $g(E)$ against E between E_1 and E_2 is the fraction of molecules with translational energies in the range from E_1 to E_2.

(b) The integral $\int_{E_1}^{E_2} g(E)\,\mathrm{d}E$ is equal to the area under the graph of $g(E)$ against E between E_1 and E_2, so it is also the fraction of molecules with translational energies in the range from E_1 to E_2.

Q2.8 (a) The total area under the graph of $g(E)$ against E is equal to 1.

(b) The integral $\int_0^{\infty} g(E)\,\mathrm{d}E$ is equal to the total area under the graph of $g(E)$ against E, so it is also equal to 1.

Q2.9 The integral $\int_0^{\infty} Eg(E)\,\mathrm{d}E$ is the average translational energy of a molecule in the gas.

Q2.10 The Boltzmann distribution law states that

$$\frac{p_2}{p_1} = \mathrm{e}^{-(E_2 - E_1)/kT}.$$

In phase cell 1,

$$\begin{aligned} v_1^2 &= (100\,\mathrm{m\,s^{-1}})^2 + (100\,\mathrm{m\,s^{-1}})^2 + (100\,\mathrm{m\,s^{-1}})^2 \\ &= 3.00 \times 10^4\,\mathrm{m^2\,s^{-2}} \end{aligned}$$

so the translational energy is

$$\begin{aligned} E_1 &= \tfrac{1}{2} m v_1^2 \\ &= \tfrac{1}{2} \times 4.65 \times 10^{-26}\,\mathrm{kg} \times 3.00 \times 10^4\,\mathrm{m^2\,s^{-2}} \\ &= 6.98 \times 10^{-22}\,\mathrm{J}. \end{aligned}$$

In phase cell 2,

$$\begin{aligned} v_2^2 &= (200\,\mathrm{m\,s^{-1}})^2 + (200\,\mathrm{m\,s^{-1}})^2 + (200\,\mathrm{m\,s^{-1}})^2 \\ &= 1.20 \times 10^5\,\mathrm{m^2\,s^{-2}} \end{aligned}$$

so the translational energy is

$$\begin{aligned} E_2 &= \tfrac{1}{2} m v_2^2 = \tfrac{1}{2} \times 4.65 \times 10^{-26}\,\mathrm{kg} \times 1.20 \times 10^5\,\mathrm{m^2\,s^{-2}} \\ &= 2.79 \times 10^{-21}\,\mathrm{J}. \end{aligned}$$

Thus
$$\begin{aligned} \frac{E_2 - E_1}{kT} &= \frac{2.79 \times 10^{-21}\,\mathrm{J} - 0.698 \times 10^{-21}\,\mathrm{J}}{1.38 \times 10^{-23}\,\mathrm{J\,K^{-1}} \times 300\,\mathrm{K}} \\ &= 0.505. \end{aligned}$$

So $\quad p_2 = \mathrm{e}^{-0.505} p_1 = 0.6 \times 10^{-10} = 6.0 \times 10^{-11}.$

Q2.11 The translational energy is related to the speed by

$$E = \tfrac{1}{2} m v^2.$$

Using Equation 2.44, the average of the square of the speed is

$$\langle v^2 \rangle = \frac{2}{m}\langle E \rangle = \frac{2}{m} \times \frac{3}{2} kT = \frac{3kT}{m}.$$

From Equation 2.36, the square of the average speed is

$$\langle v \rangle^2 = \frac{8kT}{\pi m}.$$

Thus,

$$\begin{aligned} \Delta v_{\mathrm{sd}} &= \sqrt{\langle v^2 \rangle - \langle v \rangle^2} = \sqrt{\frac{3kT}{m} - \frac{8kT}{\pi m}} \\ &= 0.67\sqrt{\frac{kT}{m}} \end{aligned}$$

which is proportional to $\sqrt{T/m}$.

Comment: *This shows explicitly that the typical width of the speed distribution increases as the temperature rises and the molecular mass decreases.*

Q2.12 The Maxwell–Boltzmann energy distribution function is

$$g(E) = \frac{2}{\sqrt{\pi}}\left(\frac{1}{kT}\right)^{3/2} E^{1/2} \mathrm{e}^{-E/kT}.$$

The peak of the distribution occurs at $E = kT/2$, so the peak value is

$$\frac{2}{\sqrt{\pi}}\left(\frac{1}{kT}\right)^{3/2}\left(\frac{kT}{2}\right)^{1/2} \mathrm{e}^{-1/2} = \sqrt{\frac{2}{\pi}}\,\mathrm{e}^{-1/2} \times \frac{1}{kT}$$

which is inversely proportional to the absolute temperature.

Q2.13 The peak of the energy distribution occurs at $kT/2$. From the positions of the peaks in Figure 2.32, we deduce that:

$$\tfrac{1}{2} kT_1 = 2.1 \times 10^{-21}\,\mathrm{J}$$

$$\tfrac{1}{2} kT_2 = 2.8 \times 10^{-21}\,\mathrm{J}.$$

Thus, $T_1 = \dfrac{2 \times 2.1 \times 10^{-21}\,\text{J}}{1.38 \times 10^{-23}\,\text{J K}^{-1}} = 304\,\text{K}$

$$T_2 = \frac{2 \times 2.8 \times 10^{-21}\,\text{J}}{1.38 \times 10^{-23}\,\text{J K}^{-1}} = 406\,\text{K}.$$

According to the equipartition of energy theorem, the molar internal energy of a gas with f effective degrees of freedom is

$$U_\text{m} = \frac{f}{2} RT.$$

With five effective degrees of freedom, the energy input into the nitrogen gas is

$$\begin{aligned} \Delta U_\text{m} &= \tfrac{5}{2} R(T_2 - T_1) \\ &= \tfrac{5}{2} \times 8.314\,\text{J K}^{-1}\,\text{mol}^{-1} \times (406\,\text{K} - 304\,\text{K}) \\ &= 2.1 \times 10^3\,\text{J mol}^{-1}. \end{aligned}$$

The final pressure of the gas is given by the equation of state for one mole:

$$\begin{aligned} P &= \frac{nRT_2}{V} \\ &= \frac{1.0\,\text{mol} \times 8.314\,\text{J K}^{-1}\,\text{mol}^{-1} \times 406\,\text{K}}{1.0 \times 10^{-2}\,\text{m}^3} \\ &= 3.4 \times 10^5\,\text{Pa}. \end{aligned}$$

Q2.14 The equipartition of energy theorem shows that

$$\tfrac{1}{2} k_s \langle x^2 \rangle = \tfrac{1}{2} kT$$

so $$\langle x^2 \rangle = \frac{kT}{k_s}.$$

Lindemann's criterion allows us to relate the melting temperature T_m to the equilibrium interatomic spacing:

$$\frac{kT_\text{m}}{k_s} = \frac{d^2}{100}$$

so $$T_\text{m} = \frac{k_s d^2}{100k}.$$

In the case of tungsten,

$$T_\text{m} = \frac{100\,\text{N m}^{-1} \times (2.5 \times 10^{-10}\,\text{m})^2}{100 \times 1.38 \times 10^{-23}\,\text{J K}^{-1}} = 4500\,\text{K}.$$

Lindemann's criterion only a gives a rough estimate in any given case, but it gives a reasonable account of the trends across a range of elements, as illustrated in Table 2.5:

Table 2.5 For use with Q2.14.

Element	Lindemann estimate of melting temperature	Measured melting temperature
aluminium	880 K	933 K
silver	1000 K	1234 K
copper	1200 K	1356 K
tungsten	4500 K	3650 K

Q2.15 (a) The molecule is equally likely to be in either half of the tube, so the probability is 1/2 = 0.50.

(b) For two independent molecules, the probability is $(1/2)^2 = 0.25$.

(c) For eight independent molecules, the probability is $(1/2)^8 = 3.9 \times 10^{-3}$.

(d) For 100 independent molecules, the probability is $(1/2)^{100} = 7.9 \times 10^{-31}$, which is far too small for there to be any reasonable chance of seeing it.

Q2.16 (a) The air in the stationary car obeys the equation

$$PV = \tfrac{2}{3} N \langle E_\text{trans} \rangle$$

so the total translational energy of all the molecules in this air is

$$N \langle E_\text{trans} \rangle = \tfrac{3}{2} PV.$$

Since the air is at atmospheric pressure,

$$\begin{aligned} N \langle E_\text{trans} \rangle &= \tfrac{3}{2} \times 1.0 \times 10^5\,\text{Pa} \times 6.6\,\text{m}^3 \\ &= 9.9 \times 10^5\,\text{J}. \end{aligned}$$

(b) For simplicity, suppose that air is made up of molecules of mass m. Consider the moving car from the viewpoint of a stationary pedestrian. Let the velocity of a molecule in this reference frame be $\boldsymbol{V}$, while the velocity in the frame of the moving car is $\boldsymbol{v}$. If the car has velocity $\boldsymbol{u}$,

$$\boldsymbol{V} = \boldsymbol{v} + \boldsymbol{u}.$$

The translational energy of the molecule in the frame of the pedestrian is

$$\tfrac{1}{2} mV^2 = \tfrac{1}{2} m |\boldsymbol{v} + \boldsymbol{u}|^2 = \tfrac{1}{2} m (v^2 + 2\boldsymbol{v} \cdot \boldsymbol{u} + u^2).$$

We now average over all the molecules in the air. This gives

$$\left\langle \tfrac{1}{2} mV^2 \right\rangle = \left\langle \tfrac{1}{2} mv^2 \right\rangle + m \langle \boldsymbol{v} \rangle \cdot \boldsymbol{u} + \tfrac{1}{2} mu^2.$$

In the frame of the moving car, the molecules move in random directions so the average of $\boldsymbol{v}$ is zero. Thus,

$$\left\langle \tfrac{1}{2} mV^2 \right\rangle = \left\langle \tfrac{1}{2} mv^2 \right\rangle + \tfrac{1}{2} mu^2.$$

The average translational energy of a molecule is greater in the moving car than in the stationary car by an amount $\frac{1}{2}mu^2$.

Thus, $\Delta E_{\text{trans}} = \frac{1}{2}mu^2$.

$$\frac{\Delta E_{\text{trans}}}{E_{\text{trans}}} = \frac{\frac{1}{2}mu^2}{\frac{3}{2}kT}$$

$$= \frac{4.65 \times 10^{-26}\,\text{kg} \times (30\,\text{m}\,\text{s}^{-1})^2}{3 \times 1.38 \times 10^{-23}\,\text{J}\,\text{K}^{-1} \times 300\,\text{K}}$$

$$= 0.0034.$$

This refers to the average translational energy per molecule; the total translational energy of the gas increases by the same small factor.

Q2.17 We use the simple gas model and ignore collisions between molecules. In a small time interval Δt, the only molecules that strike the surface, of area A, are those that are heading in the right direction, and are close enough to the surface to reach it in the given time interval. At the start of the time interval, the molecules destined to strike the surface all lie within a volume

$$\Delta V = A \times \langle v \rangle \Delta t.$$

If there are N molecules in volume V, the number of molecules in the volume ΔV is

$$\frac{N}{V} A \times \langle v \rangle \Delta t.$$

The Joule classification allows us to concentrate on the one-sixth fraction of molecules that are heading directly toward the surface. The number of such molecules arriving in time Δt is

$$\Delta N = \frac{1}{6}\frac{N}{V} A \times \langle v \rangle \Delta t.$$

So the rate of arrival of molecules at the surface is

$$\frac{\Delta N}{\Delta t} = \frac{1}{6}\frac{N}{V} A\langle v \rangle.$$

From the ideal gas equation of state, the number of molecules per unit volume is

$$\frac{N}{V} = \frac{P}{kT} = \frac{10^5\,\text{Pa}}{1.38 \times 10^{-23}\,\text{J}\,\text{K}^{-1} \times 300\,\text{K}}$$

$$= 2.4 \times 10^{25}\,\text{m}^{-3}.$$

Thus, $$\frac{\Delta N}{\Delta t} = \frac{1}{6} \times 2.4 \times 10^{25}\,\text{m}^{-3} \times (10^{-3}\,\text{m})^2 \times 500\,\text{m}\,\text{s}^{-1}$$

$$= 2.0 \times 10^{21}\,\text{s}^{-1}.$$

This calculation helps us picture the origin of pressure in a gas. Every second, there is a vast number of collisions with a solid surface, each transferring a tiny amount of momentum to it.

Q2.18 One mole of bismuth with mass $M = 209 \times 10^{-3}$ kg contains Avogadro's number of atoms, and thus the mass m of one atom is $209 \times 10^{-3}\,\text{kg} / 6.02 \times 10^{23} = 3.47 \times 10^{-25}$ kg. From Equation 2.35, $v_{\text{mp}} = (2kT/m)^{1/2}$ from which we find that $mv_{\text{mp}}^2 = 2kT$. So the furnace must be at a temperature:

$$T = mv_{\text{mp}}^2 / (2k)$$

$$= 3.47 \times 10^{-25}\,\text{kg} \times (296\,\text{m}\,\text{s}^{-1})^2 / (2 \times 1.381 \times 10^{-23}\,\text{J}\,\text{K}^{-1})$$

$$= 1100\,\text{K}.$$

Comment: *The treatment in this chapter is oversimplified by assuming only one type of bismuth molecule in the beam. The real results for such an experiment were analysed to show that the beam is 44% Bi, 54% Bi_2 and 2% Bi_3 for a furnace temperature of 1100 K. The three separate distributions for these masses are superposed in the deposit observed. The blackest points for Bi_2 and Bi_3 are much further round the drum, at about 135 mm and 165 mm rather than the 95.5 mm for Bi atoms.*

Q2.19 One mole contains Avogadro's number, 6.022×10^{23} molecules. Reading from Figure 2.34, at an energy of 1.00×10^{-20} J, the energy distribution function has the value $4.3 \times 10^{19}\,\text{J}^{-1}$. The fraction of molecules with energies between E_1 and $E_1 + \Delta E$, where $E_1 = 1.00 \times 10^{-20}$ J and $\Delta E = 1.00 \times 10^{-25}$ J, is

$$g(E)\,\Delta E = 4.3 \times 10^{19}\,\text{J}^{-1} \times 1.00 \times 10^{-25}\,\text{J}$$

$$= 4.3 \times 10^{-6}.$$

So the number of molecules in this energy range is predicted to be

$$N = 6.022 \times 10^{23} \times 4.3 \times 10^{-6} = 2.6 \times 10^{18}.$$

A variation in this number by one part in 10^8 means a fluctuation

$$\Delta N = 2.6 \times 10^{18} \times 10^{-8} = 2.6 \times 10^{10}.$$

This is more than 10 times $\sqrt{N} = 1.6 \times 10^9$, so such a large fluctuation would be astonishing.

Q2.20 It is convenient to use the energy distribution function, rather than the speed distribution function, so that we can make use of the integral given in the question. Since

$$E_1 = \frac{1}{2}mv_1^2$$

the fraction, F, of molecules with speeds greater than v_1 is the same as the fraction of molecules with translational energy greater than E_1. From the Maxwell–Boltzmann energy distribution function (Equations 2.41 and 2.42), this is given by

$$F = \frac{2}{\sqrt{\pi}}\left(\frac{1}{kT}\right)^{3/2}\int_{E_1}^{\infty}\sqrt{E}\mathrm{e}^{-E/kT}\,\mathrm{d}E.$$

Using the approximation given in the question, we have

$$F = \frac{2}{\sqrt{\pi}}\left(\frac{1}{kT}\right)^{3/2} \times kT\sqrt{E_1}\mathrm{e}^{-E_1/kT}$$

$$= \frac{2}{\sqrt{\pi}}\sqrt{\frac{E_1}{kT}}\mathrm{e}^{-E_1/kT}.$$

For an oxygen molecule moving at the escape speed,

$$\frac{E_1}{kT} = \frac{\frac{1}{2}mv_1^2}{kT}$$

$$= \frac{2.66\times10^{-26}\,\mathrm{kg}\times(1.12\times10^4\,\mathrm{m\,s^{-1}})^2}{2\times1.38\times10^{-23}\,\mathrm{J\,K^{-1}}\times1500\,\mathrm{K}}$$

$$= 80.6$$

so $$F = \frac{2}{\sqrt{\pi}}\sqrt{80.6}\times\mathrm{e}^{-80.6} = 1.00\times10^{-34}.$$

This probability is so small that the possibility of losing significant amounts of oxygen in this way can be safely neglected.

Q3.1 Writing $P = 1.06 \times 10^5$ Pa, and $\Delta V = 3.50 \times 10^{-3}\,\mathrm{m}^3$, Equation 3.5 tells us the work done *on* the gas is

$$W = -P\,\Delta V = -1.06\times10^5\,\mathrm{Pa}\times3.50\times10^{-3}\,\mathrm{m}^3$$

$$= -3.71\times10^2\,\mathrm{J}.$$

It follows that the work done *by* the gas ($-W$) is 3.71×10^2 J.

Q3.2 The temperature rise shows that the internal energy of the gas has increased. However, the first law of thermodynamics tells us that the increase in internal energy might have been caused by heat or work, or a combination of the two. The given information does not allow us to say how much work has been done on the gas, so it is equally impossible to say how much heat has been transferred.

Q3.3 In each case, the change in internal energy of the gas is

$$\Delta U = (3/2)nR\,\Delta T$$

$$= 1.50\times1.00\,\mathrm{mol}\times8.314\,\mathrm{J\,K^{-1}\,mol^{-1}}\times50\,\mathrm{K}$$

$$= 624\,\mathrm{J}.$$

(a) With V constant, no work is done, so the first law implies that all the additional internal energy must have been transferred as heat. Thus

$$Q = \Delta U = 624\,\mathrm{J}.$$

(b) In this case, the final volume of the gas may be obtained by using the equation of state $PV = nRT$, and is given by

$$V = nRT/P = \frac{1.00\,\mathrm{mol}\times8.31\,\mathrm{J\,K^{-1}\,mol^{-1}}\times350\,\mathrm{K}}{(1.00\times10^5\,\mathrm{Pa})}$$

i.e. $V = 2.91 \times 10^{-2}\,\mathrm{m}^3$.

As a result of the increase in volume, the work done *on* the gas is

$$W = -P\,\Delta V = -(1.00\times10^5\,\mathrm{Pa})\times(0.0291-0.0249)\,\mathrm{m}^3$$
$$= -4.2\times10^2\,\mathrm{J}.$$

(The negative value of W indicates that positive work is done *by* the gas.)

Using the first law in the form $\Delta U = Q + W$, we see that in this case

$$Q = \Delta U - W = 6.25\times10^2\,\mathrm{J} + 4.2\times10^2\,\mathrm{J} = 1.05\times10^3\,\mathrm{J}.$$

Q3.4 The combined heat capacity of the kettle and the water is

$$(0.90\times913 + 1.3\times4.19\times10^3)\,\mathrm{J\,K^{-1}} = 6.27\times10^3\,\mathrm{J\,K^{-1}}.$$

So the energy required to heat the two to 100 °C (i.e. $\Delta T = 88$ K) is

$$6.27\times10^3\,\mathrm{J\,K^{-1}}\times88\,\mathrm{K} = 5.5\times10^5\,\mathrm{J}.$$

As a check, you should note that this implies a typical (3 kW) kettle would take about three minutes to boil the water, which seems reasonable.

Q3.5 In a constant pressure expansion, the work done on the gas is $W = -P\,\Delta V$ (Equation 3.5). Since the increase in volume in this case is $V_2 - V_1$, it follows that $W = -P(V_2 - V_1)$, as required. In a constant pressure process, the heat transferred to the gas is $Q = C_P\,\Delta T$. Since the change in temperature is $\Delta T = (T_2 - T_1)$, the heat transferred is $Q = C_P(T_2 - T_1)$. Since the heat capacity at constant volume is $C_V = \Delta U/\Delta T$, the change in internal energy of the gas is $\Delta U = C_V\Delta T = C_V(T_2 - T_1)$. According to the first law of thermodynamics, $\Delta U = Q + W$, so $\Delta U - Q = W$. From the above results,

$$\Delta U - Q = C_V(T_2-T_1) - C_P(T_2-T_1) = (C_V - C_P)(T_2-T_1).$$

Using Equation 3.18 for the difference in heat capacities, and the equation of state of an ideal gas,

$$\Delta U - Q = -\,nR(T_2-T_1) = -(P_2V_2 - P_1V_1).$$

For a constant pressure process, $P_2 = P_1 = P$, so we conclude that

$$\Delta U - Q = -P(V_2 - V_1) = W$$

in agreement with the first law of thermodynamics.

Q3.6 In an isothermal process, the internal energy of the ideal gas does not change. It therefore follows that any energy transferred to or from the gas by work (due to the expansion or compression of the gas) must be compensated for by an equal but oppositely directed transfer of energy in the form of heat.

Q3.7 (a) Rearranging $PV = nRT$, we see that $V = nRT/P$. Using this to eliminate V from the adiabatic condition,

$$P\left[\frac{nRT}{P}\right]^{\gamma} = A,$$

i.e. $$T^{\gamma}P^{1-\gamma} = \frac{A}{(nR)^{\gamma}} = \text{constant}.$$

(b) Similarly, $P = nRT/V$, so the adiabatic condition may also be written

$$\left[\frac{nRT}{V}\right]V^{\gamma} = A,$$

i.e. $$TV^{\gamma-1} = \frac{A}{nR} = \text{constant}.$$

Q3.8 (a) Smashing an egg cannot be reversed and is therefore irreversible. (b) Moving a chest of drawers can be reversed, but friction will have caused irreversible changes in the environment of the drawers. (c) This (highly idealized) adiabatic process can be reversed and is reversible. (d) You can certainly reverse the turning of the pages and the scanning of your eyes, so if you taught yourself to read backwards this process could be reversed. The authors cling to the hope, however, that the process of reading will have left some irreversible record in your brain.

Q3.9 The equality will hold if the process is reversible (i.e. if the system and its environment can be returned to their original states). In fact, the change in universal entropy (i.e. the entropy of a system and its environment) is a measure of the irreversibility of a process.

Q3.10 (a) For a monatomic ideal gas, $C_{V,\mathrm{m}} = 3R/2$ and $C_{P,\mathrm{m}} = 5R/2$. It therefore follows from Equation 3.30a that the final entropy of the gas is

$$S_2 = (8.31\,\mathrm{J\,K^{-1}})\left[\tfrac{5}{2}\log_e(2) + \tfrac{3}{2}\log_e(2)\right] + 500\,\mathrm{J\,K^{-1}}$$

i.e. $$S_2 = (8.31\,\mathrm{J\,K^{-1}})\,[2.77] + 500\,\mathrm{J\,K^{-1}} = 523\,\mathrm{J\,K^{-1}}.$$

(b) None. The value of $500\,\mathrm{J\,K^{-1}}$ represents an arbitrary choice for the entropy of the reference state. Only *differences* in entropy calculated from Equation 3.30a are physically significant.

Q3.11 (a) Rearranging $PV = nRT$, we see that:

$$P = \frac{nRT}{V}$$

so $$\frac{P_2}{P_1} = \left(\frac{T_2}{T_1}\right)\Big/\left(\frac{V_2}{V_1}\right)$$

and $$\log_e\left(\frac{P_2}{P_1}\right) = \log_e\left(\frac{T_2}{T_1}\right) - \log_e\left(\frac{V_2}{V_1}\right).$$

Using this result in Equation 3.30a gives

$$S_2 - S_1 = C_V\log_e\left(\frac{P_2}{P_1}\right) + C_P\log_e\left(\frac{V_2}{V_1}\right)$$

$$= C_V\log_e\left(\frac{T_2}{T_1}\right) + (C_P - C_V)\log_e\left(\frac{V_2}{V_1}\right).$$

Equation 3.18 for the difference in heat capacities in an ideal gas finally gives

$$S_2 - S_1 = C_V\log_e\left(\frac{T_2}{T_1}\right) + nR\log_e\left(\frac{V_2}{V_1}\right).$$

(b) At constant temperature, $T_2 = T_1$, so in this special case,

$$S_2 - S_1 = nR\log_e\left(\frac{V_2}{V_1}\right).$$

Comparing with the entry for Q in row 3 of Table 3.3, we see that

$$\frac{Q}{T} = nR\log_e\left(\frac{V_2}{V_1}\right) = S_2 - S_1,$$

as expected from the general definition of entropy (Equation 3.27).

Q3.12 The argument is essentially the same as that already provided in the text except that when analysing the equation

$$\Delta S_{\mathrm{Univ}} = Q\left(\frac{1}{T_c} - \frac{1}{T_h}\right)$$

the term in brackets will now be negative. Consequently, Q must also be negative, leading to the conclusion that heat flows from the system to the environment.

Q3.13 Using Equation 3.39, we see that $T_c/T_h = 1 - \eta$, from which it follows that $T_h = T_c/(1 - \eta)$. Thus, in this case

$$T_h = 295\,\text{K}/(1 - 0.45) = 536\,\text{K}.$$

Q3.14 The idea is not a good one, not least because it won't work. No matter how I use the hot air from the back of the refrigerator, I will not be able to generate enough electrical energy to keep the refrigerator operating — even if the generator is 100% efficient (which it won't be). The reason I can be confident of this is that if I could generate the necessary external work from the rejected heat, then I could make the 'subsystem' for doing this part of the refrigerator itself. My enlarged refrigerator, including the electrical generator, would then operate in a cycle and would transfer heat from a colder to a hotter body without having any other effect. Such a device is explicitly forbidden by Clausius's statement of the second law of thermodynamics.

Q3.15 $3800\,\text{Cal} = 3.800 \times 10^6\,\text{cal} = 3.800 \times 10^6 \times 4.19\,\text{J}$. This is the energy expended in 24 hours, so the amount expended per second is

$$\frac{3.800 \times 10^6 \times 4.19\,\text{J}}{60 \times 60 \times 24\,\text{s}} = 184\,\text{W}.$$

Q3.16 To raise the weight by 1 m once requires $60 \times 9.8 \times 1$ J, so to raise it n times requires $60 \times 9.8n$ J. This must equal $15\,000\,\text{kJ} = 15 \times 10^6\,\text{J}$, so

$$n = 15 \times 10^6/60 \times 9.8 = 25\,500.$$

The moral is: if you want to lose weight, go on a diet!

Q3.17 The work done *by* the gas is $+P\,\Delta V$ at constant pressure. So: for path ABC (work is done only when V changes), $W = 10\,\text{Pa} \times (6\,\text{m}^3 - 1\,\text{m}^3) = 50\,\text{J}$. For path ADC, similarly, $W = 2\,\text{Pa} \times (6\,\text{m}^3 - 1\,\text{m}^3) = 10\,\text{J}$.

Q3.18 For a monatomic ideal gas, $\gamma = 5/3$. The adiabatic condition implies $PV^\gamma = \text{constant}$, so $P_1 V_1^\gamma = P_2 V_2^\gamma$. Hence $V_2/V_1 = (P_1/P_2)^{1/\gamma} = (1/8)^{3/5} = 0.287$ and the final volume is $V_2 = 0.287 \times 1.00\,\text{m}^3 = 2.87 \times 10^{-1}\,\text{m}^3$. It was shown in Q3.7 that the adiabatic relation between V and T is of the form $TV^{\gamma-1} = \text{constant}$, so $T_1 V_1^{\gamma-1} = T_2 V_2^{\gamma-1}$. Hence $T_2/T_1 = (V_1/V_2)^{\gamma-1} = (1/0.287)^{2/3} = 2.30$, and the final temperature is $T_2 = 2.30 \times 300\,\text{K} = 690\,\text{K}$. The process is reversible, so the change in the entropy of the Universe must be zero. In addition, this particular reversible process is adiabatic, so the change in the entropy of the gas will also be zero.

Q3.19 The four statements (associated with Carathéodory, Boltzmann, Kelvin and Clausius, in reverse historical order) can be found in Sections 3.2, 3.3, 4.1 and 4.2 respectively. The Carathéodory statement is the most economical and leads directly to the concept of entropy, though it cannot be described as intuitively appealing. The Boltzmann statement is highly memorable, but not particularly plausible without a lot of supporting argument. The Kelvin statement sounds like a reasonable generalization of engineering experience, but the fact that it implies the existence of entropy is far from obvious. Clausius's statement sounds like a generalization of everyday experience and may therefore be regarded as even more plausible than Kelvin's statement, but the fact that it implies the existence of entropy is equally obscure.

Q3.20 The sketch is shown in Figure 3.44, where the vertical lines are isotherms and the horizontal lines are adiabats. The area is just that of a rectangle, and is given by $(T_h - T_c)\Delta S$, where $\Delta S = S_2 - S_1$. Now, in the case of the Carnot cycle, $\Delta S = Q_h/T_h$, so we find

$$(T_h - T_c)\Delta S = (T_h - T_c)(Q_h/T_h) = \eta Q_h = W_{out}$$

which is the work performed by the gas during the cycle. (There is no overall change in entropy over a complete cycle since the entropy change between C and D is reversed in the step between A and B.

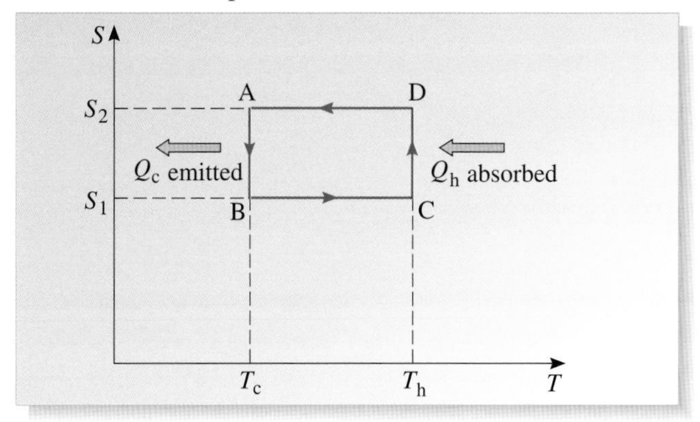

Figure 3.44 The S–T diagram of the Carnot cycle.

Q3.21 If each stage acts as a Carnot cycle, then, for stage 1, $Q_H/T_H = Q_M/T_M$, and for stage 2, $Q_M/T_M = Q_L/T_L$ so that finally $Q_H/T_H = Q_L/T_L$. Hence only the initial and final temperatures matter, and the overall efficiency is $\eta = (T_H - T_L)/T_H$.

Q4.1 The pressure due to the overlying seawater, 50 m below the surface, is

$$\Delta P = \rho g z = 1025\,\text{kg m}^{-3} \times 9.81\,\text{m s}^{-2} \times 50\,\text{m}$$
$$= 5.0 \times 10^5\,\text{Pa}.$$

The escape hatch experiences an additional pressure due to the atmosphere but this is exactly balanced by the pressure of air inside the submarine, which is maintained at 1 atmosphere. The escape hatch has area

$$A = \pi r^2 = \pi \times (0.375\,\text{m})^2 = 0.44\,\text{m}^2$$

so the force needed to open it is

$$F = A\Delta P = 0.44\,\text{m}^2 \times 5.0 \times 10^5\,\text{Pa} = 2.2 \times 10^5\,\text{N}.$$

Q4.2 From the barometric formula, the pressure at ten times the scale height is $P(0)\exp(-10)$. Taking the pressure at ground level to be 10^5 Pa gives a pressure of 4.5 Pa.

The atmosphere can be taken to obey the ideal gas equation of state:

$$PV = NkT$$

so the number of molecules per unit volume is

$$\frac{N}{V} = \frac{P}{kT} = \frac{4.5\,\mathrm{Pa}}{1.38\times10^{-23}\,\mathrm{J\,K^{-1}}\times 270\,\mathrm{K}} = 1.2\times10^{21}\,\mathrm{m^{-3}}.$$

One cubic millimetre = $(1\,\mathrm{mm})^3 = (10^{-3}\,\mathrm{m})^3 = 10^{-9}\,\mathrm{m}^3$. So in one cubic millimetre, there are 1.2×10^{12} molecules. Although the atmosphere is very thin at this height, it cannot be classified as a good vacuum.

Q4.3 The pressure increases by a factor of 3 as the spacecraft descends 25 km, from 100 km to 75 km above the surface of the planet. Since the pressure depends exponentially on height, each descent of 25 km corresponds to increasing the pressure by the same factor of 3. Thus, descending from 100 km to the surface of the planet increases the pressure by a total factor of $3\times3\times3\times3 = 81$. The pressure at the surface is therefore $2.1\times10^5\,\mathrm{Pa}\times81 = 1.7\times10^7$ Pa. Since this exceeds the design tolerance of the spacecraft, landing would not be advisable.

Q4.4 The fraction of the iceberg beneath the water is found from Equation 4.10:

$$\frac{V_0}{V} = \frac{\rho}{\rho_0} = \frac{860}{1025} = 0.84.$$

This is close to 5/6. Other rules of thumb are often quoted, but these probably use inappropriate densities for the iceberg or the seawater. In practice, the density of the iceberg is less than the density of pure ice because of trapped air, while the density of seawater is greater than that of pure water because of dissolved salts.

Q4.5 As explained in the text, the total downward force on the block is

$$F_x = \rho V g - \rho_0 V_0 g$$

where ρ and ρ_0 are densities of the block and the water, and V and V_0 are the total and submerged volumes of the block. It is convenient to write this equation in the form:

$$F_x = \rho_0 g(V_{eq} - V_0)$$

where $V_{eq} = \dfrac{\rho}{\rho_0}V$ is the submerged volume of the block in equilibrium (i.e. when F_x is zero).

When the block is over-submerged, $V_0 > V_{eq}$, and F_x is negative, indicating an upward force. When the block is under-submerged, $V_0 < V_{eq}$, and F_x is positive, indicating a downward force. In both cases, the force acts in a direction that tends to restore equilibrium. The magnitude of the force is proportional to the difference between the submerged volume and the equilibrium submerged volume. Because the block has a constant cross-section, this is proportional to the vertical displacement of the block. We therefore have a situation in which the restoring force is proportional to the displacement. So, the block oscillates in simple harmonic motion.

Q4.6 If the diameters of the input and output pipes are d_1 and d_2 respectively, then

$$A_1 = \tfrac{1}{4}\pi d_1^2$$

$$A_2 = 2\times\tfrac{1}{4}\pi d_2^2$$

so the equation of continuity gives

$$\frac{v_2}{v_1} = \frac{(\frac{1}{4}\pi d_1^2)}{(2\times\frac{1}{4}\pi d_2^2)} = \frac{d_1^2}{2d_2^2} = \frac{19^2}{2\times12^2} = 1.25.$$

Q4.7 The equation of continuity gives

$$v_2 = \frac{A_1}{A_2}v_1 = \frac{6}{12}\times2.0\,\mathrm{m\,s^{-1}} = 1.0\,\mathrm{m\,s^{-1}}.$$

Bernoulli's equation gives

$$P_1 + \tfrac{1}{2}\rho v_1^2 + \rho g h_1 = P_2 + \tfrac{1}{2}\rho v_2^2 + \rho g h_2$$

so

$$P_2 - P_1 = \tfrac{1}{2}\rho(v_1^2 - v_2^2) + \rho g(h_1 - h_2)$$

$$= \tfrac{1}{2}\times1000\,\mathrm{kg\,m^{-3}}\,[(2.0\,\mathrm{m\,s^{-1}})^2 - (1.0\,\mathrm{m\,s^{-1}})^2] + (1000\,\mathrm{kg\,m^{-3}})\times(9.8\,\mathrm{m\,s^{-2}})\times(0.5\,\mathrm{m})$$

$$= 6400\,\mathrm{Pa}.$$

Q4.8 Using Bernoulli's equation, the pressures inside and outside the bin are related by

$$P_{in} = P_{out} + \tfrac{1}{2}\rho v^2.$$

The lid lifts off when the force provided by this pressure difference becomes greater than the weight of the lid. Taking the lid to have radius r and mass M gives the critical condition

$$(P_{in} - P_{out})\,\pi r^2 = Mg.$$

The lid will therefore be lifted off if the flow speed exceeds the value

$$v = \sqrt{\left(\frac{2}{\rho}\right)\left(\frac{Mg}{\pi r^2}\right)}$$

$$= \sqrt{\left(\frac{2}{1.2\,\text{kg}\,\text{m}^{-3}}\right)\left(\frac{1.0\,\text{kg}\times 9.8\,\text{m}\,\text{s}^{-2}}{\pi(0.25\,\text{m})^2}\right)}$$

$= 9.1\,\text{m}\,\text{s}^{-1}$.

Q4.9 This question is very like the previous one, but set in a different context. Using Bernoulli's equation, the pressures on the upper and lower wing surfaces are related by

$$P_\text{U} + \tfrac{1}{2}\rho v_\text{U}^2 = P_\text{L} + \tfrac{1}{2}\rho v_\text{L}^2\,.$$

Since the plane is in steady level flight, the force provided by the pressure difference across the wings must balance the plane's weight. Thus,

$$(P_\text{L} - P_\text{U})2A = Mg$$

where A is the area of a single wing and M is the total mass of the plane (the factor of 2 appears because both wings support the weight). Combining these equations gives

$$\tfrac{1}{2}\rho v_\text{U}^2 = \tfrac{1}{2}\rho v_\text{L}^2 + \frac{Mg}{2A}$$

so

$$v_\text{U} = \sqrt{v_\text{L}^2 + \frac{Mg}{\rho A}}$$

$$= \sqrt{(300\,\text{m}\,\text{s}^{-1})^2 + \frac{60\times 10^3\,\text{kg}\times 9.81\,\text{m}\,\text{s}^{-2}}{1.2\,\text{kg}\,\text{m}^{-3}\times 20\,\text{m}\times 2\,\text{m}}}$$

$= 320\,\text{m}\,\text{s}^{-1}$.

Surprisingly, a relatively small difference in flow speed (around 6%) creates forces that are large enough to support a heavy plane.

Q4.10 Rearranging Equation 4.16:

$$\Delta P = \frac{4\eta L v}{R^2 - r^2}.$$

In the current case, $L = 12$ m, $R = 0.015$ m, $r = 0.010$ m, $v = 0.15\,\text{m}\,\text{s}^{-1}$ and η is $1.0\times 10^{-3}\,\text{kg}\,\text{m}^{-1}\,\text{s}^{-1}$, so the pressure drop is

$$\Delta P = \frac{(4\times 1.0\times 10^{-3}\,\text{kg}\,\text{m}^{-1}\,\text{s}^{-1}\times 12\,\text{m}\times 0.15\,\text{m}\,\text{s}^{-1})}{((0.015\,\text{m})^2 - (0.010\,\text{m})^2)}$$

$= 58$ Pa.

Q4.11 Using the definition of the Reynolds number

$$\frac{F_\text{D}}{\rho L_0^2 v_0^2} = \frac{6\pi}{(\rho L_0 v_0/\eta)}$$

so $F_\text{D} = 6\pi\eta L_0 v_0$,

which shows that the drag force is proportional to the radius, L_0, and the speed, v_0, of the sphere. This result is known as Stokes's law and only applies for $Re < 1$. The upper limit ($Re = 1$) is reached for

$$v_0 = \frac{\eta}{\rho L_0} = \frac{1.0\times 10^{-3}\,\text{kg}\,\text{m}^{-1}\,\text{s}^{-1}}{1.0\times 10^{3}\,\text{kg}\,\text{m}^{-3}\times 10^{-4}\,\text{m}}$$

$= 10^{-2}\,\text{m}\,\text{s}^{-1}$,

so the sphere must move slower than $1\,\text{cm}\,\text{s}^{-1}$ for the above result to be valid. (At larger Reynolds numbers, $F_\text{D}/(\rho L_0^2 v_0^2)$ is roughly constant.)

Q4.12 (a) The pressure exerted by the water at depth z is $P_0 + \rho g z$, where P_0 is the atmospheric pressure and ρ is the density of water. Taking $\rho = 10^3\,\text{kg}\,\text{m}^{-3}$ and $g = 10\,\text{m}\,\text{s}^{-2}$, the pressure at the bottom of the dam is $P_0 + 4\times 10^5\,\text{Pa} = 5P_0$. Figure 4.41 shows a graph of pressure against depth in the dam.

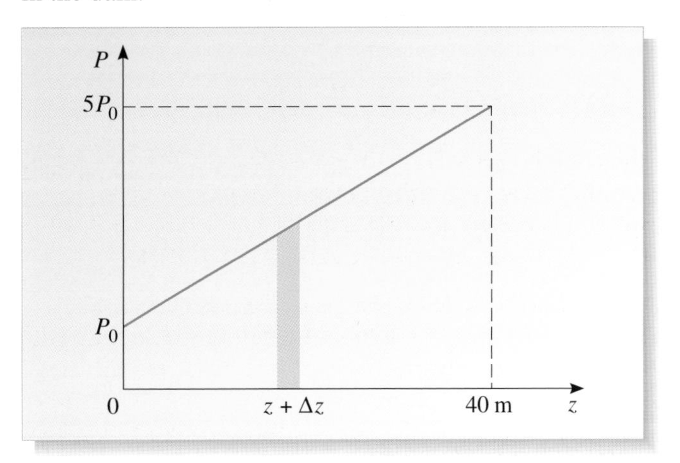

Figure 4.41 Pressure experienced by the vertical dam wall of Q4.12.

(b) To calculate the total force on the wall, consider the element of water between depths z and $z + \Delta z$, where Δz is small. The magnitude of the force exerted by this element on the wall of the dam is $P(z)L\,\Delta z$, where $L = 500$ m. This is equal to L times the area of the shaded strip in Figure 4.41. The total magnitude of the force on the wall of the dam is therefore L times the total area under the graph in Figure 4.41. Letting $h = 40$ m be the height of the dam, and calculating the area under the graph as the sum of the area of a rectangle and a triangle, the total magnitude of the force is

$$F = L\times(P_0 h + \tfrac{1}{2}4P_0 h) = 3P_0 Lh$$

$$= 3\times 10^5\,\text{Pa}\times 500\,\text{m}\times 40\,\text{m} = 6\times 10^9\,\text{N}.$$

Q4.13 The barometric formula can be written as

$$\frac{P(z)}{P(0)} = \exp(-z/\lambda).$$

Taking logs on both sides,

$$\log_e\left(\frac{P(z)}{P(0)}\right) = -\frac{z}{\lambda}$$

so $$z = -\lambda \log_e\left(\frac{P(z)}{P(0)}\right).$$

The height at which the pressure reaches a value that is 1% of the pressure at ground level is

$$z = -7.6\,\text{km} \times \log_e(0.01) = 35\,\text{km}.$$

Q4.14 If the depth of the bottom of the barge below the water surface (the draught) is d, then the volume of water displaced is $20\,\text{m} \times 3\,\text{m} \times d$, and the mass of water displaced is $60\,\text{m}^2 \times d \times 10^3\,\text{kg}\,\text{m}^{-3} = 6 \times 10^4 d\,\text{kg}\,\text{m}^{-1}$. For the unladen barge, this must be equal to 10^4 kg. Thus

$$d = \frac{10^4\,\text{kg}}{6 \times 10^4\,\text{kg}\,\text{m}^{-1}} = 0.17\,\text{m}.$$

If the draught of the laden barge is 1 m, then the volume of water displaced is $60\,\text{m}^3$, and the total mass of water displaced is 6×10^4 kg. The unladen barge has a mass of 10^4 kg so the load that can be carried is 5×10^4 kg.

Q4.15 Let the speed of the water be v and the total cross-sectional area of the tubes be A. Then the volume of water conducted upwards in time Δt is $vA\,\Delta t$ and the mass of water conducted upwards in time Δt is $\rho vA\,\Delta t$, where ρ is the density of water. The rate of flow of mass upwards is $dm/dt = \rho vA$. Hence,

$$v = \frac{dm/dt}{\rho A} = \frac{10^{-2}\,\text{kg}\,\text{s}^{-1}}{10^3\,\text{kg}\,\text{m}^{-3} \times 10^{-3}\,\text{m}^2} = 10^{-2}\,\text{m}\,\text{s}^{-1}.$$

Q4.16 Let A_1, v_1 and P_1 be the cross-sectional area, speed of flow and pressure in the widest part of the pipe and A_2, v_2 and P_2 be the corresponding quantities in the narrowest part of the pipe. Then the equation of continuity gives

$$A_1 v_1 = A_2 v_2$$

and Bernoulli's equation gives

$$P_1 + \tfrac{1}{2}\rho v_1^2 = P_2 + \tfrac{1}{2}\rho v_2^2.$$

Combining these equations gives

$$P_2 = P_1 + \tfrac{1}{2}\rho v_1^2\left(1 - \frac{A_1^2}{A_2^2}\right).$$

For the cross-sectional areas given in the question, $A_1 = 5A_2$, so

$$P_2 = P_1 - 12\rho v_1^2.$$

The pressure in the narrowest part of the tube is predicted to reach zero when the speed in the widest part of the tube is

$$v_1 = \sqrt{\frac{P_1}{12\rho}} = \sqrt{\frac{3 \times 10^5\,\text{Pa}}{12 \times 10^3\,\text{kg}\,\text{m}^{-3}}} = 5\,\text{m}\,\text{s}^{-1}.$$

In practice, cavitation sets in just before zero pressure is reached. When this happens, the fluid is no longer incompressible, so the ideal fluid model breaks down and the above equations are invalid. This is just as well, as they predict a negative value for P_2 when $v_1 > 5\,\text{m}\,\text{s}^{-1}$, which is not possible — pressures must always be positive.

Q4.17 The Reynolds number for motion of the bug through water is

$$\begin{aligned} Re &= \frac{\rho L_0 v_0}{\eta} \\ &= \frac{10^3\,\text{kg}\,\text{m}^{-3} \times 10^{-6}\,\text{m} \times 10^{-4}\,\text{m}\,\text{s}^{-1}}{10^{-3}\,\text{kg}\,\text{m}^{-1}\,\text{s}^{-1}} \\ &= 10^{-4}. \end{aligned}$$

Viscous behaviour is dominant when Re is much smaller than 1, so it is fair to describe the bug as perceiving its surroundings as being very viscous. Dynamical similarity can be obtained with the larger sphere provided

$$\begin{aligned} \eta &= \frac{\rho L_0 v_0}{Re} \\ &= \frac{10^3\,\text{kg}\,\text{m}^{-3} \times 10^{-2}\,\text{m} \times 10^{-3}\,\text{m}\,\text{s}^{-1}}{10^{-4}} \\ &= 10^2\,\text{kg}\,\text{m}^{-1}\,\text{s}^{-1}, \end{aligned}$$

which corresponds to an extremely viscous fluid.

Q5.1 (a) A major omission from this statement of the first law of thermodynamics is that no reference is made to the fact that the change must be between *equilibrium* states of the system. This qualification is necessary and a statement of the first law of thermodynamics needs to state clearly that the law applies to changes between equilibrium states of a system.

A second problem is that the statement is unclear about the direction of heat flow and about whether work is done on the system or by the system. We need to say that it is the sum of the heat transferred *to* the system and the work done *on* the system which is dependent only on the initial and final equilibrium states of the system.

(b) The problem with this statement is that it does not mention possible changes in the environment of the system, and therefore suggests that reversible processes are much more common than is the case. The correct definition requires both the system *and the environment of the system* to be returned to their initial states.

(c) This statement is clearly wrong, since it would imply that a domestic refrigerator could never be built. A cyclic process *can* result in the transfer of heat from a cooler body to a hotter one. However, it is impossible to find a cyclic process whose *sole* result is the transfer of heat from a cooler body to a hotter one. (In addition to the transfer of heat, some other process such as the performance of work must also occur.)

Q5.2 (a) The radial density function is the average number of molecules per unit volume, given as a function of the radial distance from a given molecule.

(b) In a finite set of experiments which measure the outcome of a random process, the fractional frequency is the number of times that a given outcome is observed divided by the number of times that the experiment is run.

(c) A thermodynamic system is said to be in equilibrium if it is in a stable and unchanging state.

(d) A quasi-static process in a thermodynamic system is one which proceeds sufficiently slowly that at every stage of the process the system is, to a good approximation, in an equilibrium state.

(e) The principle of dynamic similarity states that two flows that are geometrically similar will show similar patterns of flow when expressed in terms of their scaled variables, provided that the value of the Reynolds number is the same for both flows.

Q5.3 (a) Avogadro's hypothesis: Equal volumes of different gases, at a given temperature and pressure, contain the same number of molecules.

(b) Joule's law of ideal gases: In equilibrium, the internal energy of an ideal gas is independent of the pressure and volume and can be written as

$$U = nF(T)$$

where n is the number of moles of gas and $F(T)$ is some function of temperature.

(c) The third law of thermodynamics: It is impossible to reduce the temperature of any system to absolute zero by a finite number of operations.

(d) Archimedes' principle: If a body is immersed, in part or wholly in a fluid, the magnitude of its apparent weight is decreased by the magnitude of the weight of the fluid displaced by the body.

Q5.4 (a) The sum of probabilities for all four bulbs must equal 1, so

$$p_{\text{red}} + p_{\text{yellow}} + p_{\text{green}} + p_{\text{blue}} = 1$$

thus $\quad p_{\text{blue}} = 1 - (0.20 + 0.24 + 0.30) = 1 - 0.74 = 0.26.$

The probability that the blue bulb will be lit is 0.26.

(b) The probability of the same bulb (of any colour) being lit in two plays of the game is the sum of the probabilities of the different ways that this can be achieved. So it is the sum of the probabilities of getting the same colour bulb twice:

$$\begin{aligned} p_{\text{two}} &= p_{\text{red}}^2 + p_{\text{yellow}}^2 + p_{\text{green}}^2 + p_{\text{blue}}^2 \\ &= (0.20)^2 + (0.24)^2 + (0.30)^2 + (0.26)^2 \\ &= 0.2552. \end{aligned}$$

To two significant figures, the probability of the same bulbs being illuminated on two given plays is 0.26.

(c) The probability of the green lamp being illuminated on four plays of the game is:

$$p = p_{\text{green}} \times p_{\text{green}} \times p_{\text{green}} \times p_{\text{green}} = p_{\text{green}}^4 = 8.1 \times 10^{-3}.$$

Q5.5 The probability of the top landing on any one side is 1/7, so

$$\begin{aligned} \langle n \rangle &= \tfrac{1}{7} \times (-3) + \tfrac{1}{7} \times (-2) + \tfrac{1}{7} \times (-1) + \tfrac{1}{7} \times (0) \\ &\quad + \tfrac{1}{7} \times (1) + \tfrac{1}{7} \times (2) + \tfrac{1}{7} \times (3) \\ &= 0. \end{aligned}$$

The mean value of n is 0, and the value of $\langle n \rangle^2$ is simply $0 \times 0 = 0$.

To find the mean value of n^2, we adopt a similar approach as for $\langle n \rangle$, but using the squares of the scores instead:

$$\begin{aligned} \langle n^2 \rangle &= \tfrac{1}{7} \times (9) + \tfrac{1}{7} \times (4) + \tfrac{1}{7} \times (1) + \tfrac{1}{7} \times (0) \\ &\quad + \tfrac{1}{7} \times (1) + \tfrac{1}{7} \times (4) + \tfrac{1}{7} \times (9) \\ &= 4. \end{aligned}$$

So, the value of $\langle n^2 \rangle$ is 4.

Q5.6 In order to analyse the data for the dice, we should determine the expected range of variation in the numbers of times (n) that we obtain a given score based on the assumption that the dice are fair. The values of n are obtained by dividing the total numbers of throws by six. We know that a variation of $\pm\sqrt{n}$ is not unusual, whereas a variation of $\pm 10\sqrt{n}$ would be highly unusual. We can construct a table of expected variation as shown in Table 5.3. By comparing the expected variation with the observed results, we can see that dice A, B, C and E all show variation which lies in the interval $(n - \sqrt{n})$ to $(n + \sqrt{n})$ and so we cannot claim that any of these dice are unfair. Die D requires more careful consideration. In this case the largest variation is 293 (=1093 – 800), which is more than $10\sqrt{n}$ (= 280). This is a highly unusual result and we would be justified in assuming that die D is not equally weighted.

Table 5.3 The expected variation in the numbers of times a given score should be obtained with fair dice.

Die	Total no. of throws	n	$\sqrt{n}$
A	150	25	5
B	1200	200	14
C	2400	400	20
D	4800	800	28
E	9600	1600	40

Q5.7 The number of protons in the nucleus is found by dividing the charge on the nucleus by the charge of the proton ($+e$):

$$\text{number of protons} = \frac{1.28 \times 10^{-18}\,\text{C}}{1.602 \times 10^{-19}\,\text{C}} = 7.99.$$

Since the number of protons in a nucleus must be a whole number, the element has 8 protons.

While all oxygen atoms contain 8 protons, the number of neutrons present in the nucleus can vary, giving different isotopes of oxygen. The most common isotope has a mass of 16 amu (8 protons + 8 neutrons), but rarer isotopes which contain a greater number of neutrons also exist and these correspond to the atoms with masses of 17 amu (8 protons + 9 neutrons) and 18 amu (8 protons + 10 neutrons).

Q5.8 (a) The relative molecular mass M_r of a substance is defined by:

$$\text{mass of 1 mol} = M_r \times 10^{-3}\,\text{kg}.$$

In this case,

$$0.040 \times M_r \times 10^{-3}\,\text{kg} = 0.0288\,\text{kg}$$

$$M_r = 720.$$

So the molecule has a relative molecular mass of 720.

(b) The number of molecules N contained in a sample of mass m is

$$N = N_m \frac{m}{M_r \times 10^{-3}\,\text{kg}}$$

where N_m is Avogadro's constant. So, in this case,

$$N = 6.022 \times 10^{23}\,\text{mol}^{-1} \times \frac{0.100\,\text{kg}}{720 \times 10^{-3}\,\text{kg mol}^{-1}}$$

$$= 8.36 \times 10^{22}.$$

Q5.9 In order to calculate the pressure P, we use

$$P = \frac{F}{A}$$

where the force magnitude is given by $F = mg$. We need to make estimates of m and A. A typical human has a mass of about 70 kg, and we can approximate the human foot as a rectangle with a length of 25 cm and width of 10 cm. The pressure is then,

$$P = \frac{70\,\text{kg} \times 9.8\,\text{m s}^{-2}}{2 \times 0.25\,\text{m} \times 0.1\,\text{m}} = 1.37 \times 10^4\,\text{N m}^{-2}.$$

So the pressure on the sole of a shoe of a stationary person is about 1.4×10^4 Pa. (Your own estimate may differ slightly from this, but should be of the same order of magnitude.)

Q5.10 The number of molecules per cubic metre is 10^{11}. Each molecule has a mass of

$$M_r \times 1\,\text{amu} = 2.0 \times 1.6603 \times 10^{-27}\,\text{kg} = 3.32 \times 10^{-27}\,\text{kg}.$$

The mass per cubic metre is then:

$$\rho = 3.32 \times 10^{-27} \times 10^{11}\,\text{kg m}^{-3} = 3.32 \times 10^{-16}\,\text{kg m}^{-3}.$$

So the density of the cloud is $3 \times 10^{-16}\,\text{kg m}^{-3}$. While this may seem very low by terrestrial standards, this would actually be a dense cloud in the interstellar medium. (Note that the assumption that the cloud is purely molecular hydrogen is a simplification. The composition (by mass) of a real interstellar cloud would be about 73% hydrogen, 25% helium and 2% heavier elements.)

To find the radius R of a spherical cloud of mass M_c and density ρ, we use

$$M_c = \tfrac{4}{3}\pi R^3 \rho$$

$$R = \left(\frac{3M_c}{4\pi\rho}\right)^{1/3}$$

$$R = \left(\frac{3 \times 2 \times 10^{32}\,\text{kg}}{4 \times \pi \times 3.32 \times 10^{-16}\,\text{kg m}^{-3}}\right)^{1/3} = 5.2 \times 10^{15}\,\text{m}.$$

So the radius of this cloud would be 5×10^{15} m.

Q5.11 Rearranging the ideal gas equation to give an expression for volume

$$V = \frac{nRT}{P}.$$

In this case, $n = 1.00$ mol.

(a)
$$V = \frac{1.00\,\text{mol} \times 8.314\,\text{J K}^{-1}\,\text{mol}^{-1} \times 273\,\text{K}}{1.00 \times 10^5\,\text{Pa}}$$

$$= 2.27 \times 10^{-2}\,\text{m}^3.$$

(b) $$V = \frac{1.00\,\text{mol} \times 8.314\,\text{J K}^{-1}\,\text{mol}^{-1} \times 300\,\text{K}}{7.32 \times 10^4\,\text{Pa}}$$
$$= 3.41 \times 10^{-2}\,\text{m}^3.$$

(c) $$V = \frac{1.00\,\text{mol} \times 8.314\,\text{J K}^{-1}\,\text{mol}^{-1} \times 373\,\text{K}}{9.10 \times 10^4\,\text{Pa}}$$
$$= 3.41 \times 10^{-2}\,\text{m}^3.$$

Q5.12 The mean translational energy per molecule of a gas is given by

$$\langle E_{\text{trans}} \rangle = \tfrac{3}{2} kT.$$

For case (a) then,

$$\langle E_{\text{trans}} \rangle = \tfrac{3}{2} \times 1.381 \times 10^{-23}\,\text{J K}^{-1} \times 250\,\text{K} = 5.18 \times 10^{-21}\,\text{J}.$$

For case (b), it can be seen that the mean translational energy is directly proportional to the absolute temperature, so the mean translational energy per molecule at 500 K will be twice this value, giving $\langle E_{\text{trans}} \rangle = 1.04 \times 10^{-20}$ J. Similarly, for case (c), by multiplying the result (a) by 3, we find that the mean translational energy per molecule at 750 K is $\langle E_{\text{trans}} \rangle = 1.55 \times 10^{-20}$ J.

Q5.13 We can use the ideal gas law to find the number of moles of gas in this sample, and then multiply by Avogadro's constant to find the total number of molecules. The number of molecules is then

$$N_{\text{sample}} = N_{\text{m}} n = N_{\text{m}} \times \frac{PV}{RT}$$

$$N_{\text{sample}} = 6.022 \times 10^{23}\,\text{mol}^{-1} \times \frac{2.6 \times 10^5\,\text{Pa} \times 0.070\,\text{m}^3}{8.314\,\text{J K}^{-1}\,\text{mol}^{-1} \times 320\,\text{K}}$$
$$= 4.12 \times 10^{24}.$$

The number of molecules in this sample is then 4.1×10^{24}.

Q5.14 Most practical thermometers are based on the principle that a physical property of the thermometer varies linearly with temperature. It is necessary to calibrate the thermometer at two fixed points, and once this has been done, the chosen physical property can be used to measure the temperature. The problem with this approach is that a variety of different physical properties could be used to define temperature, and these definitions could be inconsistent with one another. Furthermore, in many cases, the exact physical significance of the temperature is unclear. For this reason, a standard technique, that of the ideal gas temperature scale, has been adopted against which all other temperature scales are defined. The reason for choosing this method is, however, not arbitrary, but has a sound theoretical basis. For example, if temperature is interpreted as the parameter, T, that appears in Boltzmann's distribution law, the ideal gas equation of state can be derived using the methods of statistical mechanics. So the ideal gas temperature scale has a much clearer interpretation than a scale based on an arbitrarily chosen physical property, such as the length of the column of mercury in a mercury-in-glass thermometer.

Q5.15 The assumption that collisions between molecules and the walls of the vessel containing the gas are elastic is made because this is the simplest assumption that can be made about collisions. The model gives a good description of the behaviour of an ideal gas, so it would seem the assumption is justified. However, detailed study of processes near surfaces reveals that, in fact, individual molecules do not collide elastically with the wall of the container, but can adhere to the surface for long periods of time in a layer which is a few molecules thick. Molecules arrive at the surface from random directions with a range of velocities. However, molecules also leave this surface layer at random, in such a way that the processes of arrival and departure have the same average effect on the surface as if individual molecules were undergoing elastic collisions.

Q5.16 The variation of the magnitude of the force acting on one wall of a cubic container which arises from the collision of a single molecule is shown in blue in Figure 5.2. The variation is a series of equally spaced spikes of equal height. Since the molecule is travelling perpendicularly to the wall, the time interval between the spikes is $\Delta t_1 = 2L/v$. In the case of a molecule which is again travelling perpendicularly to the wall, but with a greater speed, the time interval between the spikes will be shorter. Since the molecule is travelling faster, the magnitude of the force acting on the wall will be greater, giving larger spikes.

Q5.17 We can apply Boltzmann's law to both phase cells. If the energies of the cells A and B are E_{A} and E_{B} respectively, the probabilities of finding a molecule in these phase cells are

$$p_{\text{A}} = D\text{e}^{-E_{\text{A}}/kT}$$

$$p_{\text{B}} = D\text{e}^{-E_{\text{B}}/kT}$$

where D is a constant of proportionality (the normalization factor).

So $$\frac{p_{\text{A}}}{p_{\text{B}}} = \frac{\text{e}^{-E_{\text{A}}/kT}}{\text{e}^{-E_{\text{B}}/kT}}$$

$$\frac{p_{\text{A}}}{p_{\text{B}}} = \text{e}^{-E_{\text{A}}/kT} \times \text{e}^{+E_{\text{B}}/kT}$$

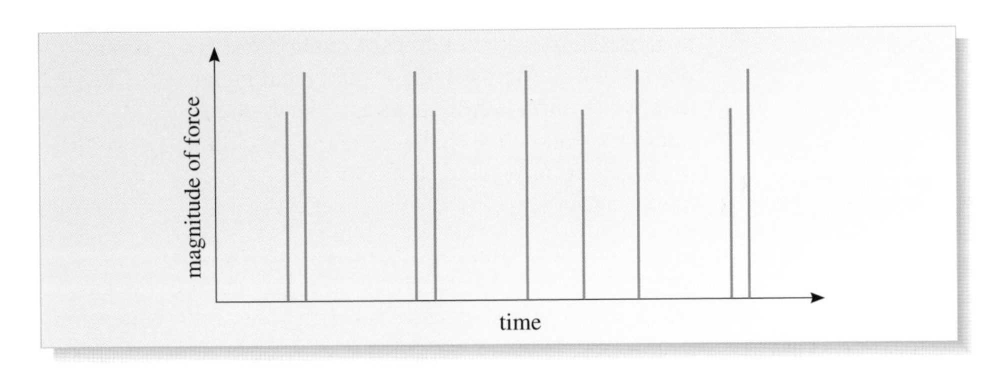

Figure 5.2 The variation of the force acting on the wall of a container due to collisions by two molecules.

$$\frac{p_A}{p_B} = e^{(E_B - E_A)/kT}.$$

But we know that $p_A/p_B = 2$, and that $\Delta E = E_B - E_A$, so

$$2 = e^{\Delta E/kT}.$$

We want to rearrange to obtain an expression for ΔE. This can be done by taking logarithms of both sides of this equation

$$\log_e 2 = \log_e (e^{\Delta E/kT})$$

$$\log_e 2 = \frac{\Delta E}{kT}$$

$$\Delta E = (\log_e 2) \times kT \cdot$$

So the difference in energy between these two phase cells is $(\log_e 2) \times kT = 0.69\,kT$.

Q5.18 Boltzmann's distribution law describes the probability that a molecule in a gas will occupy a given *phase cell*. The distribution function of molecular energies however, represents the probability that a molecule will have a given *energy*. The two distributions are different because the number of phase cells changes with energy. In order to calculate the molecular energy distribution function (the Maxwell–Boltzmann distribution of energies), it is necessary to take account of both the Boltzmann distribution *and* the way in which the number of phase cells increases with energy.

Q5.19 We want to evaluate a quantity that will allow us to discriminate between monatomic and diatomic gases. One such quantity is γ, the ratio of the heat capacities at constant pressure and constant volume. We are given information about temperature changes which occur under these conditions for a fixed amount of heat input, so we should be able to find an expression for γ.

Let the temperature changes at constant volume and constant pressure be ΔT_V and ΔT_P respectively. If the sample consists of n moles of gas, then the quantity of heat Q required to cause the observed changes in temperature is given by:

$$Q = C_V\,\Delta T_V$$

and $$Q = C_P\,\Delta T_P.$$

However, the quantity of heat is the same in both cases, so we can equate these two expressions giving

$$C_V\,\Delta T_V = C_P\,\Delta T_P$$

$$\frac{C_P}{C_V} = \frac{\Delta T_V}{\Delta T_P}.$$

Numerically,

$$\gamma = \frac{C_P}{C_V} = \frac{406\,\text{K} - 290\,\text{K}}{373\,\text{K} - 290\,\text{K}} = \frac{116\,\text{K}}{83\,\text{K}} = 1.398.$$

This is close to the value of 1.40 expected for an ideal diatomic gas, so we conclude that the gas is diatomic.

Q5.20 In this process, heat flows from the metal to the water. The quantity of heat transferred from the metal X is

$$Q = m_X C_X \Delta T_X$$

where m_X, C_X and ΔT_X are respectively the mass, specific heat and temperature change of metal X.

The quantity of heat transferred to the water is

$$Q = m_W C_W \Delta T_W$$

where m_W, C_W and ΔT_W are respectively the mass, specific heat and temperature change of the water. Equating these two expressions for the heat transferred,

$$m_X C_X \Delta T_X = m_W C_W \Delta T_W$$

and so $$C_X = \frac{m_W\, C_W\, \Delta T_W}{m_X\, \Delta T_X}.$$

Numerically,

$$C_X = \frac{0.40\,\text{kg} \times 4.20 \times 10^3\,\text{J K}^{-1}\,\text{kg}^{-1} \times (299.9 - 295.6)\,\text{K}}{0.20\,\text{kg} \times (373.0 - 299.9)\,\text{K}}$$

$C_X = 494.1\,\text{J K}^{-1}\,\text{kg}^{-1}$.

So the specific heat of metal X is $4.9 \times 10^2\,\text{J K}^{-1}\,\text{kg}^{-1}$.

Q5.21 (a) The transfer of energy by compression of an ideal spring can be reversed and is a reversible process, since it can be reversed without any change in the environment.

(b) The transfer of energy to a real flywheel can be reversed but this is not a reversible process. It is possible to return the flywheel to its original state but, in a real flywheel, some energy will be dissipated as heat due to friction at the bearings, so it is not possible to return the system to its original state without a change in the environment.

(c) The transfer of energy by quasi-static compression of an ideal gas under ideal conditions, free from friction and other dissipative effects, can be reversed and is a reversible process, since it is possible to return to the original state without any change in the environment.

(d) The transfer of energy by stirring of a viscous fluid cannot be reversed — it is impossible to 'unstir' a liquid. The process is therefore also irreversible.

Q5.22 (a) Since no heat is transferred to or from the gas, the processes under consideration are adiabatic. Because the system contains an ideal gas, any reversible adiabatic change will be subject to the adiabatic condition:

$$PV^{\gamma} = \text{constant}.$$

We know that the initial state is at pressure P_A and volume V_A, so the pressure P and volume V of the gas must satisfy

$$PV^{\gamma} = P_A V_A^{\gamma}.$$

This curve is sketched in Figure 5.3, with the point (V_A, P_A) indicated.

(b) Irreversible changes, such as doing work on the gas using the paddle, can only cause the system to reach a state which is above the line $PV^{\gamma} = P_A V_A^{\gamma}$. This is shown by the green shaded area in Figure 5.3.

(c) The region below the line $PV^{\gamma} = P_A V_A^{\gamma}$ cannot be reached by any combination of reversible and irreversible changes in an adiabatic system. The pink shaded region on Figure 5.3 thus represents states which are not accessible through any adiabatic change.

(d) The fact that around the point (V_A, P_A) there exist states which are adiabatically inaccessible is an example of Carathéodory's statement of the second law of thermodynamics (namely that 'in the neighbourhood of any state of a thermodynamic system there are states which are adiabatically inaccessible'.)

(e) In order to allow the inaccessible region of the P–V diagram to be reached, the system would have to be modified to allow non-adiabatic changes to occur. In particular, the system would have to allow changes in which heat could be transferred *from* the gas.

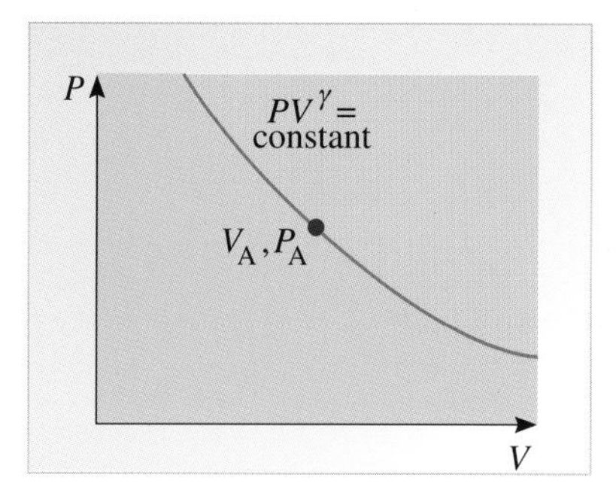

Figure 5.3 The P–V diagram for the changes described in Q5.22.

Q5.23 The change in entropy ΔS associated with an isothermal reversible process is given by

$$\Delta S = \frac{Q_{\text{rev}}}{T}$$

where Q_{rev} is the heat flow into the system and T is the temperature at which the process occurs. In both cases, the quantity of heat transferred is the molar latent heat of melting $l_{\text{melt,m}}$, which is related to the specific latent heat of melting l_{melt} by

$$l_{\text{melt,m}} = M_r \times 10^{-3}\,\text{kg mol}^{-1} \times l_{\text{melt}}$$

where M_r is the relative atomic mass. So, the change in entropy associated with the melting of 1 mol of a substance at a temperature T_{melt} is

$$\Delta S = \frac{1\,\text{mol} \times M_r \times 10^{-3}\,\text{kg mol}^{-1} \times l_{\text{melt}}}{T_{\text{melt}}}.$$

(a) For lithium, we have $M_r = 6.9$, $l_{\text{melt}} = 6.667 \times 10^5\,\text{J kg}^{-1}$ and $T_{\text{melt}} = 452\,\text{K}$, so

$$\Delta S = \frac{1\,\text{mol} \times 6.9 \times 10^{-3}\,\text{kg mol}^{-1} \times 6.667 \times 10^5\,\text{J kg}^{-1}}{452\,\text{K}}$$

$\Delta S = 10.2\,\text{J K}^{-1}$.

(b) For mercury, we have $M_r = 200.6$, $l_{\text{melt}} = 1.162 \times 10^4\,\text{J kg}^{-1}$ and $T_{\text{melt}} = 234\,\text{K}$, so

$$\Delta S = \frac{1\,\text{mol} \times 200.6 \times 10^{-3}\,\text{kg mol}^{-1} \times 1.162 \times 10^4\,\text{J kg}^{-1}}{234\,\text{K}}$$

$\Delta S = 9.96\,\text{J K}^{-1}$.

Thus, despite these metals having quite different properties, we see that on melting the entropy change per mole is actually very similar.

Q5.24 (a) The system consists of a fixed mass of an ideal gas. Let the initial and final states be indicated by the subscripts 1 and 2 respectively. The entropy of the second state is given by Equation 3.25:

$$S_2 = C_V \log_e\left(\frac{P_2}{P_1}\right) + C_P \log_e\left(\frac{V_2}{V_1}\right) + S_1.$$

So, the change in entropy ΔS is

$$\begin{aligned}\Delta S &= S_2 - S_1\\ &= C_V \log_e\left(\frac{P_2}{P_1}\right) + C_P \log_e\left(\frac{V_2}{V_1}\right).\end{aligned}$$

If the gas pressure is increased at constant volume then $V_2 = V_1$. So the entropy change is

$$\begin{aligned}\Delta S &= C_V \log_e\left(\frac{P_2}{P_1}\right) + C_P \log_e\left(\frac{V_1}{V_1}\right)\\ &= C_V \log_e\left(\frac{P_2}{P_1}\right) + C_P \log_e(1)\end{aligned}$$

Since $\log_e(1) = 0$, this simplifies to

$$\Delta S = C_V \log_e\left(\frac{P_2}{P_1}\right) = n C_{V,\mathrm{m}} \log_e\left(\frac{P_2}{P_1}\right)$$

as required.

(b) If 1 mol of a diatomic ideal gas undergoes a change in which its pressure is halved at constant volume, then $P_2 = P_1/2$. The change in entropy is then

$$\begin{aligned}\Delta S &= n C_{V,\mathrm{m}} \log_e\left(\frac{P_1/2}{P_1}\right)\\ &= n C_{V,\mathrm{m}} \log_e(\tfrac{1}{2}).\end{aligned}$$

This is a typical diatomic gas so $C_{V,\mathrm{m}} = \frac{5}{2}R$, and hence

$$\Delta S = \tfrac{5}{2} Rn \log_e(\tfrac{1}{2}).$$

This is a molar sample of gas, so $n = 1$, and numerically we find

$$\begin{aligned}\Delta S &= \tfrac{5}{2} \times 8.314\,\mathrm{J\,K^{-1}\,mol^{-1}} \times 1\,\mathrm{mol} \times (-0.693)\\ &= -14.4\,\mathrm{J\,K^{-1}}.\end{aligned}$$

So the final state has an entropy which is $14.4\,\mathrm{J\,K^{-1}}$ lower than the initial state. Of course, this must be at least compensated by entropy increases in the environment.

Q5.25 The most efficient heat engine that can operate between two temperatures is the Carnot engine. Even though we don't know what sort of processes go on within the petrol engine advertised, we know it cannot be any more efficient than a Carnot engine. The efficiency of a Carnot engine is given by

$$\eta = 1 - \frac{T_\mathrm{c}}{T_\mathrm{h}}.$$

We have very few details about the petrol engine, but we know that the maximum temperature that is reached in its operating cycle is about 700 °C, so we can adopt a T_h value of about 1000 K.

The lower operating temperature of any cycle that the engine performs must be at least the typical air temperature, and would probably be somewhat higher than this, but we can estimate an upper limit to the efficiency based on $T_\mathrm{c} = 300$ K:

$$\eta = 1 - \frac{300\,\mathrm{K}}{1000\,\mathrm{K}} = 0.7.$$

So, the maximum theoretical efficiency of a Carnot engine operating between these two temperatures is 70%. Since a real engine must be less efficient (and in practice is considerably less efficient) than a Carnot engine, we can conclude that the advertiser's claim that this engine is 80% efficient cannot be true.

Q5.26 Suppose that the mass of the payload is M, the volume of the balloon is V_b and that the densities of the helium in the balloon and the air at this altitude are ρ_b and ρ_0 respectively. If the x-axis is chosen to point vertically upwards, then the total force acting on the balloon plus payload is

$$F_x = (\rho_0 - \rho_\mathrm{b}) V_\mathrm{b} g - Mg.$$

No unbalanced force acts on the balloon, so $F_x = 0$, giving

$$M = (\rho_0 - \rho_\mathrm{b}) V_\mathrm{b}.$$

In order to find the density of the helium within the balloon, we use the form of the equation of state for an ideal gas given in Equation 1.13:

$$\rho_\mathrm{b} = \frac{mP}{kT}$$

where m is the mass of a single helium atom. The mass of the payload is then

$$M = \left(\rho_0 - \frac{mP}{kT}\right) V_\mathrm{b}.$$

Numerically, for this case

$$\frac{mP}{kT} = \frac{4.0 \times 1.66 \times 10^{-27}\,\mathrm{kg} \times 1.0 \times 10^{3}\,\mathrm{Pa}}{1.38 \times 10^{-23}\,\mathrm{J\,K^{-1}} \times 250\,\mathrm{K}}$$
$$= 1.92 \times 10^{-3}\,\mathrm{kg\,m^{-3}}$$
$$M = (1.4 \times 10^{-2}\,\mathrm{kg\,m^{-3}} - 1.92 \times 10^{-3}\,\mathrm{kg\,m^{-3}}) \times 4.0 \times 10^{5}\,\mathrm{m^{3}}$$
$$= 4.83 \times 10^{3}\,\mathrm{kg}.$$

So, the mass of the payload is 4.8×10^3 kg.

Q5.27 Two fluid flows would be dynamically similar only if the Reynolds number for the two flows were identical. The Reynolds number is given by

$$Re = \frac{\rho L_0 v_0}{\eta}.$$

The ratio of the Reynolds numbers for the two flows is:

$$\frac{Re_{\mathrm{r}}}{Re_{\mathrm{m}}} = \frac{\rho L_{\mathrm{r}} v_{\mathrm{r}}}{\eta} \times \frac{\eta}{\rho L_{\mathrm{m}} v_{\mathrm{m}}} = \frac{L_{\mathrm{r}} v_{\mathrm{r}}}{L_{\mathrm{m}} v_{\mathrm{m}}}$$

where the subscripts r and m denote parameters of the flow around the real aircraft and the scale model respectively.

In this case, we have a 1 : 12 scale model, so $L_{\mathrm{r}}/L_{\mathrm{m}} = 12$. The ratio of flying speeds is $v_{\mathrm{r}}/v_{\mathrm{m}} = 150\,\mathrm{km\,h^{-1}}/50\,\mathrm{km\,h^{-1}} = 3$. So, the ratio of the Reynolds numbers for the real and model flows is $Re_{\mathrm{r}}/Re_{\mathrm{m}} = 36$. Hence, we would *not* expect the airflow around the model to be dynamically similar to that around the real aircraft.

Q5.28 **Preparation** The first law of thermodynamics:

$$Q = \Delta U - W$$
$$W = -P\,\Delta V$$
$$Q = \Delta U + P\,\Delta V.$$

For a monatomic ideal gas, we have

$$U = \tfrac{3}{2}nRT.$$

Known values: $n = 0.05$ mol, $\Delta T = 700\,\mathrm{K} - 500\,\mathrm{K} = 200\,\mathrm{K}$.

Working If we consider changes in which the quantity of gas is fixed, a change ΔU in internal energy results from a change in temperature

$$\Delta U = \tfrac{3}{2}nR\,\Delta T.$$

(a) Change at constant volume. In this case, the change in volume is zero, so

$$Q = \Delta U$$

and using the equation for ΔU

$$Q = \tfrac{3}{2}nR\,\Delta T.$$

Numerically,

$$Q = \tfrac{3}{2} \times 0.05\,\mathrm{mol} \times 200\,\mathrm{K} \times 8.314\,\mathrm{J\,K^{-1}\,mol^{-1}}$$
$$= 124.7\,\mathrm{J}.$$

So, the heat required to make this change is 125 J.

The work done by the gas is zero (since the volume of the gas is constant).

(b) Change at constant pressure. This is a monatomic ideal gas, and so using the ideal gas law

$$V = \frac{nR}{P}T.$$

Since P is constant, any change in volume is related solely to the change in temperature, so

$$\Delta V = \frac{nR}{P}\Delta T.$$

So, the first law of thermodynamics may be written in this case as

$$Q = \Delta U + P\left(\frac{nR}{P}\right)\Delta T$$
$$Q = \Delta U + nR\,\Delta T$$

but we also have an expression for ΔU, so

$$Q = \tfrac{3}{2}nR\,\Delta T + nR\,\Delta T = \tfrac{5}{2}nR\,\Delta T.$$

Numerically,

$$Q = \tfrac{5}{2} \times 0.05\,\mathrm{mol} \times 200\,\mathrm{K} \times 8.314\,\mathrm{J\,K^{-1}\,mol^{-1}}$$
$$= 207.9\,\mathrm{J}.$$

The heat required in this case is 208 J.

The work done *on* the gas is:

$$W = -P\Delta V = -nR\Delta T$$

so

$$W = -0.05\,\mathrm{mol} \times 8.314\,\mathrm{J\,K^{-1}\,mol^{-1}} \times 200\,\mathrm{K}$$
$$= -83.14\,\mathrm{J}.$$

The work done *by* the gas is therefore 83 J.

Checking The heat required to cause a change in temperature at constant pressure is greater than the heat required to cause the same change in temperature at constant volume. We would expect this from what we know about the heat capacities of ideal gases. (In fact, the problem is more easily solved using the results for the specific heat capacities of ideal gases.) Moreover, the difference in these heat transfers is equal to the work done by the gas in part (b), as expected from the first law.

Q5.29 Preparation Useful equations and principles:

$$W = Mg$$

$$P = F/A$$

$$P + \tfrac{1}{2}\rho v^2 = \text{constant}.$$

Data:

Area of roof: $A = 5\,\text{m} \times 6\,\text{m} = 30\,\text{m}^2$.

Magnitude of force required: $F_{\text{max}} = 8\,\text{kN}$.

Maximum wind speed = $100\,\text{km}\,\text{h}^{-1} = 27.8\,\text{m}\,\text{s}^{-1}$.

Plan of attack: Use Bernoulli's equation to determine the pressure difference between the inside and outside of the roof, if the wind outside has a speed of $100\,\text{km}\,\text{h}^{-1}$. Then calculate the forces acting on the roof, and determine whether the resultant force exceeds the force required for the roof to fail.

Working Let conditions inside and outside the building be represented by subscripts of 'in' and 'out' respectively. Assuming that the air inside the building is static (i.e. that $v_{\text{in}} = 0$), we can write Bernoulli's equation as:

$$P_{\text{in}} = P_{\text{out}} + \tfrac{1}{2}\rho v_{\text{out}}^2.$$

The magnitude of the resultant force F acting on the roof is given by

$$F = (P_{\text{in}} - P_{\text{out}})A - Mg.$$

Using Bernoulli's equation to substitute for $(P_{\text{in}} - P_{\text{out}})$, we get

$$F = \tfrac{1}{2}A\rho v_{\text{out}}^2 - Mg.$$

Using known values, we obtain

$$F = \tfrac{1}{2} \times 30\,\text{m}^2 \times 1.2\,\text{kg}\,\text{m}^{-3} \times (27.8\,\text{m}\,\text{s}^{-1})^2$$
$$= -600\,\text{kg} \times 9.8\,\text{m}\,\text{s}^{-2}$$

$$F = 8.03 \times 10^3\,\text{N}.$$

So in a wind with a speed of $100\,\text{km}\,\text{h}^{-1}$, the resultant force acting upwards on the roof will be 8.0 kN. This is just sufficient to blow the roof off, and so this building would not be safe in these conditions.

Checking If the speed of the air flowing above the roof were greater, our equations show that the magnitude of the force would be greater, which makes good sense. The units of the force turn out to be $\text{kg}\,\text{m}\,\text{s}^{-2} = \text{N}$.

Q5.30 Preparation Useful equations

$$\boldsymbol{\Gamma} = \boldsymbol{r} \times \boldsymbol{F}$$
$$\boldsymbol{v} = \boldsymbol{\omega} \times \boldsymbol{r}$$
$$F = \eta A\left|\frac{\Delta v}{\Delta x}\right|.$$

Working The angular velocity of the ring is constant, and so the applied torque must equal the force due to viscous forces acting on the ring:

$$\boldsymbol{\Gamma}_{\text{applied}} + \boldsymbol{\Gamma}_{\text{viscous}} = \boldsymbol{0}. \qquad (5.1)$$

$$\left|\boldsymbol{\Gamma}_{\text{applied}}\right| = \left|\boldsymbol{\Gamma}_{\text{viscous}}\right|.$$

Let the bottom of the ring and the bottom of the tank be separated by a distance Δx.

Because the difference between the inner and outer radii of the ring is much smaller than the mean radius, it is reasonable to assume that the speed at which the ring is moving at its inner and outer radii is approximately the speed at the mean radius (r_{ring}).

The difference in speed between the ring and the tank is Δv. Since the tank is stationary, this is just the speed of the ring (at the mean radius)

$$\Delta v = v_{\text{ring}} = \omega\, r_{\text{ring}}.$$

The force F acting on the ring due to viscosity is then

$$F = \eta A\left|\frac{\Delta v}{\Delta x}\right|$$

$$F = \eta A\left|\frac{\omega\, r_{\text{ring}}}{\Delta x}\right|.$$

The magnitude of the force acts so as to slow the rotation of the ring. The direction of the viscous force is perpendicular to the radius vector. Hence the magnitude of the torque experienced by the ring is

$$\Gamma_{\text{viscous}} = \eta A\frac{\omega\, r_{\text{ring}}^2}{|\Delta x|}. \qquad (5.2)$$

We can ignore the modulus sign, since ω, r_{ring} and Δx are all positive quantities.

If the difference between inner and outer radii of the ring is Δr, then the area of the ring is

$$A = 2\pi r_{\text{ring}}\,\Delta r.$$

Substituting this into Equation 5.2, and using Equation 5.1:

$$\Gamma_{\text{applied}} = \frac{2\pi r_{\text{ring}}^3 \Delta r \eta \omega}{\Delta x}$$

and rearranging for η

$$\eta = \frac{\Gamma_{\text{applied}}\Delta x}{2\pi r_{\text{ring}}^3\,\Delta r \omega}.$$

Numerically, for this experiment:

$$\eta = \frac{0.63\,\mathrm{N\,m} \times 4 \times 10^{-3}\,\mathrm{m}}{2\pi \times (1.0\,\mathrm{m})^3 \times (1.02\,\mathrm{m} - 0.98\,\mathrm{m}) \times 0.10\,\mathrm{s}^{-1}}$$

$$= \frac{2.52 \times 10^{-3}\,\mathrm{N\,m}^2}{2.51 \times 10^{-2}\,\mathrm{m}^4\,\mathrm{s}^{-1}}$$

$$= 0.10\,\mathrm{N\,m}^{-2}\,\mathrm{s}$$

$$= 0.1\,\mathrm{kg\,m}^{-1}\,\mathrm{s}^{-1}.$$

So, the liquid has a viscosity of $0.10\,\mathrm{kg\,m}^{-1}\,\mathrm{s}^{-1}$.

Checking The units of viscosity as given by the final equation are $\mathrm{N\,m}^{-2}\,\mathrm{s}$. This is equivalent to the usual units of viscosity ($\mathrm{kg\,m}^{-1}\,\mathrm{s}^{-1}$) since $1\ \mathrm{N} = 1\ \mathrm{kg\,m\,s}^{-2}$. The equation for the viscosity shows that if the applied torque required increases, then the viscosity of the liquid also increases as expected. The magnitude of the viscosity is rather high, but still within the range of viscosities that could be exhibited by liquids.

Acknowledgements

Grateful thanks to Mr Neil Higgins of National Power plc at Didcot Power Station, UK for information about the Didcot B station.

Grateful acknowledgement is also made to the following sources for permission to reproduce material in this block:

Figures 1.1a, 2.6, 3.3a–c, 3.14: Science Museum/Science & Society Picture Library; *Figure 1.1b:* from *Atoms and Powers* by A. Thackery, Harvard University Press, 1970 (London facsimile reprint 1953); *Figure 1.2e:* K. Seddon and D. T. Evans, Queen's University Belfast/Science Photo Library; *Figure 1.5:* Max-Planck-Institut, Stuttgart; *Figure 1.6:* courtesy IBM Corp., Research Division, Almaden Research Center; *Figure 1.7:* courtesy Kore Technology Ltd; *Figure 1.24:* Mehau Kulyk/ Science Photo Library; *Figure 1.25:* Alfred Pasteka/Science Photo Library; *Figure 1.26:* Dr Mike McNamee/Science Photo Library; *Figure 1.28:* Ann Ronan Picture Library; *Figure 1.35:* Ashmolean Museum, Oxford; *Figure 1.37:* courtesy Great Escape Ballooning Ltd; *Figure 2.1:* by permission of Prof. Dr. Dieter Flamm, University of Vienna/Oxford University Press; *Figure 2.21a:* NASA/Science Photo Library; *Figure 2.21b,c:* Photo Library International; *Figure 2.30:* from *History of Physics* by S. R. Weart and M. Phillips, American Institute of Physics, 1985; *Figures 3.1, 4.20a:* Ford Motor Co. Ltd; *Figure 3.32:* courtesy Pam Lambourne and by permission of Prof. Dr. Dieter Flamm; *Figure 3.37:* © National Power Picture Unit/Jupiter dpp; *Figure 3.42:* © 1999 The Nobel Foundation; *Figure 4.1a:* NASA; *Figure 4.1b,c:* © Derek Budd, Camera Ways Picture Library; *Figure 4.4:* courtesy Stephen Hall/Horus Editions, Ward Publications Ltd.; *Figure 4.8:* © Pat Morrow/ Mountain Camera; *Figure 4.13:* Mary Evans Picture Library/Louis Figuier 'Merveilles de la Science' vol. 3 p. 416; *Figure 4.14:* courtesy Dr Stuart Bennett; *Figure 4.17a: J. M.* Labat/Ardea London; *Figure 4.17b:* courtesy Royal Navy/ Steven Johnson, Cyberheritage; *Figure 4.18:* © Konrad Wothe/Oxford Scientific Films; *Figure 4.23:* Mary Evans Picture Library; *Figures 4.33, 4.34:* kindly supplied by Prof. J. D. Jackson, School of Engineering at the University of Manchester; *Figure 4.36:* United Engineering Trustees Inc/Cambridge University Press; *Figure 4.39b:* © Terry Heathcote/Oxford Scientific Films; *Figure 4.40a,b:* L. Prandtl (1930) from *An Introduction to Fluid Dynamics*, G. K. Batchelor, Cambridge University Press, 1967.

Index

Entries and page numbers in **bold type** refer to key words which are printed in **bold** in the text and which are defined in the Glossary.